Protein Folding, Misfolding and Aggregation
Classical Themes and Novel Approaches

RSC Biomolecular Sciences

Editorial Board:
Professor Stephen Neidle (Chairman), *The School of Pharmacy, University of London, UK*
Dr Simon F Campbell CBE, FRS
Dr Marius Clore, *National Institutes of Health, USA*
Professor David M J Lilley FRS, *University of Dundee, UK*

This Series is devoted to coverage of the interface between the chemical and biological sciences, especially structural biology, chemical biology, bio- and chemo-informatics, drug discovery and development, chemical enzymology and biophysical chemistry. Ideal as reference and state-of-the-art guides at the graduate and post-graduate level.

Titles in the Series:
Biophysical and Structural Aspects of Bioenergetics
Edited by Mårten Wikström, *University of Helsinki, Finland*
Computational and Structural Approaches to Drug Discovery: Ligand–Protein Interactions
Edited by Robert M. Stroud and Janet Finer-Moore, *University of California in San Francisco, San Francisco, CA, USA*
Exploiting Chemical Diversity for Drug Discovery
Edited by Paul A. Bartlett, *Department of Chemistry, University of California, Berkeley, USA* and Michael Entzeroth, *S*Bio Pte Ltd, Singapore*
Metabolomics, Metabonomics and Metabolite Profiling
Edited by William J. Griffiths, *The School of Pharmacy, University of London, London, UK*
Protein–Carbohydrate Interactions in Infectious Disease
Edited by Carole A. Bewley, *National Institutes of Health, Bethesda, Maryland, USA*
Protein Folding, Misfolding and Aggregation: Classical Themes and Novel Approaches
Edited by Victor Muñoz, *Department of Chemistry and Biochemistry, University of Maryland, MD, USA*
Protein-Nucleic Acid Interactions: Structural Biology
Edited by Phoebe A. Rice, *Department of Biochemistry & Molecular Biology, The University of Chicago, Chicago IL, USA* and Carl C. Correll, *Dept of Biochemistry and Molecular Biology, Rosalind Franklin University, North Chicago, IL, USA*
Quadruplex Nucleic Acids
Edited by Stephen Neidle, *The School of Pharmacy, University of London, London, UK*, and Shankar Balasubramanian, *Department of Chemistry, University of Cambridge, Cambridge, UK*
Ribozymes and RNA Catalysis
Edited by David M. J. Lilley FRS, *University of Dundee, Dundee, UK*, and Fritz Eckstein, *Max Planck-Institut for Experimental Medicine, Goettingen, Germany*
Sequence-specific DNA Binding Agents
Edited by Michael Waring, *Department of Pharmacology, University of Cambridge, Cambridge, UK*
Structural Biology of Membrane Proteins
Edited by Reinhard Grisshammer and Susan K. Buchanan, *Laboratory of Molecular Biology, National Institutes of Health, Bethesda, Maryland, USA*
Structure-based Drug Discovery: An Overview
Edited by Roderick E. Hubbard, *University of York, UK, and Vernalis (R&D) Ltd, Cambridge, UK*
Therapeutic Oligonucleotides
Edited by Jens Kurreck, *Institute for Chemistry and Biochemistry, Free University Berlin, Berlin, Germany*

Visit our website on www.rsc.org/biomolecularsciences

For further information please contact:
Sales and Customer Care, Royal Society of Chemistry, Thomas Graham House, Science Park, Milton Road, Cambridge, CB4 0WF, UK
Telephone: +44 (0)1223 432360, Fax: +44 (0)1223 426017, Email: sales@rsc.org

Protein Folding, Misfolding and Aggregation
Classical Themes and Novel Approaches

Edited by

Victor Muñoz
*Department of Chemistry and Biochemistry, University of Maryland, MD, USA
and Center for Biological Investigations, CSIC, Madrid, Spain*

RSCPublishing

ISBN: 978-0-85404-257-9

1005863072

A catalogue record for this book is available from the British Library

© Royal Society of Chemistry 2008

All rights reserved

Apart from fair dealing for the purposes of research for non-commercial purposes or for private study, criticism or review, as permitted under the Copyright, Designs and Patents Act 1988 and the Copyright and Related Rights Regulations 2003, this publication may not be reproduced, stored or transmitted, in any form or by any means, without the prior permission in writing of The Royal Society of Chemistry or the copyright owner, or in the case of reproduction in accordance with the terms of licences issued by the Copyright Licensing Agency in the UK, or in accordance with the terms of the licences issued by the appropriate Reproduction Rights Organization outside the UK. Enquiries concerning reproduction outside the terms stated here should be sent to The Royal Society of Chemistry at the address printed on this page.

Published by The Royal Society of Chemistry,
Thomas Graham House, Science Park, Milton Road,
Cambridge CB4 0WF, UK

Registered Charity Number 207890

For further information see our website at www.rsc.org

Preface

Protein folding, the process by which newly synthesized proteins fold into the specific three-dimensional structures defining their biologically active states, is an old scientific problem that can be dated back at least seven decades, namely to the experiments of Anson and Mirsky in the 1930s. It is also multifaceted, changing its definition according to the background and emphasis of the particular researcher. For the cell biologist-biochemist the *in vivo* protein folding problem consists of identifying, isolating and characterizing all components of the cellular machinery in charge of facilitating and catalyzing protein folding inside the cell. From a bioinformatics viewpoint, the folding problem could be seen as devising methods to predict with high accuracy the native three-dimensional structure of proteins from the amino acid sequence alone. Physics-inclined scientists phrase the folding problem as understanding the processes and mechanisms that control the self-organization of disordered protein molecules to form unique, exquisitely detailed structures, while avoiding their irreversible assembly into high-order aggregates. The scope of this book belongs to the last of these viewpoints.

To introduce the book it is useful to take a historical perspective, which illustrates how the prevailing views about the mechanisms of protein folding have closely followed the idiosyncrasies in the catalog of available proteins and experimental approaches. In the early days and for a long time after, folding was circumscribed to equilibrium denaturation experiments on a small group of complex proteins, such as hemoglobin, because they were readily available. A theoretical framework to interpret experiments was not available, and there was significant discussion as to whether simple models based on elementary chemical reactions could be applied to protein folding (incidentally a similar discussion has regained center stage in recent years). The development of techniques that exploited thiol chemistry to trap intermediary folding species in proteins containing disulfide bonds, together with folding coupled to prolyl-bond isomerization, opened the era of kinetic experiments. This led to the characterization of folding as a convoluted process involving multiple pathways, misfolded intermediates, and heterogeneous unfolded states. However,

---

RSC Biomolecular Sciences
Protein Folding, Misfolding and Aggregation: Classical Themes and Novel Approaches
Edited by Victor Muñoz
© Royal Society of Chemistry 2008

the question was whether such heterogeneity was intrinsic to the folding reaction or induced by the trapping reactions, which involved the formation or breakage of covalent bonds. Later on, the combination of molecular biology and stopped-flow kinetic methods with millisecond resolution changed the landscape dramatically. Many single domain proteins with neither disulfide bonds nor *cis* prolines were studied, showing what appeared to be very simple behavior. In the absence of chemical trapping, folding of small proteins looked like a two-state process. Two-state implied the existence of a high free energy barrier separating the folded and unfolded states, which seemed to agree with the still slow (seconds to milliseconds) folding kinetics observed in these proteins. In a two-state regime the only mechanistic information accessible to experiment relies on mapping out the properties of the top of the folding barrier (*i.e.* the folding "transition-state") from the effects that small free energy perturbations have on folding and unfolding rates. Combining this idea with structure-oriented site directed mutagenesis resulted in the protein engineering approach to protein folding, which was independently initiated in the labs of Goldenberg, Fersht, and Matthews, and then fully developed by the Fersht group. The approach quickly caught on among protein biochemists, who applied it to many two-state proteins. Theoreticians immediately saw this avalanche of new experimental results as an opportunity to test results from theory and computer simulations, leading to the first *de facto* connection between the worlds of experiment and theory in protein folding. At the same time research in protein misfolding and aggregation was starting to reach the status of quantitative science. This state of affairs has been portrayed in detail by several books that appeared in the 1990s and 2000, including "Protein Folding" edited by Creighton, "The Mechanisms of Protein Folding" edited by Pain, and "Protein Folding Mechanisms" edited by Richards, Eisenberg, and Kim.

However, in the last 10 years there have been important developments in the area of protein folding and aggregation that have not yet been discussed in a book of these characteristics. These advances have emerged from a close partnership between statistical theory, novel approaches that dramatically increase the temporal, structural and ensemble resolution of folding experiments, and the maturity of computer simulations, which are now capable of producing results directly comparable to experiments. Once again, these advances are changing our general perception of protein folding to one that emphasizes the stochastic nature of the process and the subtle energetic balance that eventually determines whether a protein folds, the mechanisms by which it does, and its propensity to aggregate. The aim of this book is to provide an account of these major advances as seen by some of the main contributors. The book is intended for graduate students and postdoctoral researchers actively involved in protein folding research, other scientists interested in the recent progress of the field, and instructors revamping the protein folding section of their biochemistry and biophysics courses. Chapters 1 and 2 focus on the α-helix, one of the basic structural elements found in proteins. The main attraction in investigating α-helix formation is that one encounters many of the features of protein folding but in their simplest version. These two chapters will introduce the reader to conformational ensembles, partially cooperative unfolding processes, the connection between protein energetics and stereochemistry, detailed kinetic modeling, and simple examples of the application of statistical approaches to the analysis of experimental data. Chapter 3 explains the statistical theory that, even if just judged by the number of times it is cited throughout the book, provides the conceptual backbone for most of the subsequent experimental and computational developments. Chapters 4 through 7 discuss experimental approaches for the investigation of folding mechanisms. This selection is not intended to be comprehensive, but to include techniques that either probe or exploit the stochastic nature of protein folding: classical hydrogen-exchange techniques (Chapter 4), novel ensemble-based methods to estimate folding free energy surfaces from differential scanning calorimetry (Chapter 5), fast-folding kinetic experiments and their most important findings (Chapter 6), and the application of single molecule spectroscopy to protein folding (Chapter 7). Chapters 8 and 9 deal with the impressive recent developments in computational approaches; starting from atomistic simulations of complete folding (Chapter 8) and continuing with applications to *de novo* protein design (Chapter 9). Finally, Chapters 10 and 11 are devoted to the experimental and computational investigation of the other side of the problem, that of protein misfolding and aggregation.

Contents

Preface   v

Chapter 1   **The α-Helix as the Simplest Protein Model: Helix–Coil Theory, Stability, and Design**
Andrew James Doig

1.1	Introduction		1
1.2	Structure of the α-Helix		1
	1.2.1	Capping Motifs	2
	1.2.2	Metal Binding	3
	1.2.3	The 3_{10}-Helix	3
	1.2.4	The π-Helix	3
1.3	Design of Peptide Helices		4
	1.3.1	Host–Guest Studies	4
	1.3.2	Helix Lengths	5
	1.3.3	The Helix Dipole	5
	1.3.4	Acetylation and Amidation	5
	1.3.5	Solubility	6
	1.3.6	Concentration Determination	6
	1.3.7	Helix Templates	7
1.4	Helix–Coil Theory		7
	1.4.1	Zimm–Bragg Model	8
	1.4.2	Lifson–Roig Model	8
	1.4.3	AGADIR	12
	1.4.4	Lomize–Mosberg Model	13
1.5	Forces Affecting α-Helix Stability		13
	1.5.1	Helix Interior	13
	1.5.2	Caps	14
	1.5.3	Phosphorylation	15

RSC Biomolecular Sciences
Protein Folding, Misfolding and Aggregation: Classical Themes and Novel Approaches
Edited by Victor Muñoz
© Royal Society of Chemistry 2008

	1.5.4	Non-covalent Side-chain Interactions	18
	1.5.5	Covalent Side-chain Interactions	20
	1.5.6	Capping Motifs	20
	1.5.7	Ionic Strength	20
	1.5.8	Temperature	21
	References		21

Chapter 2 Kinetics and Mechanisms of α-Helix Formation
Urmi Doshi

2.1	Introduction	28
2.2	Experimental Techniques Employed to Study Helix–Coil Kinetics	31
2.3	Theoretical Approaches to Explore Helix–Coil Kinetics	34
2.4	General Observations in Helix–Coil Kinetics	35
2.5	Kinetic Theory of the Helix–Coil Transition	37
2.6	Free-Energy Landscape for α-Helix Formation	39
2.7	Mechanisms of α-Helix Formation	41
2.8	Reaction Coordinates for α-Helix Formation	43
2.9	The Nature of the Diffusion Coefficient for α-Helix Formation	44
2.10	Implications for Protein Folding	45
References		46

Chapter 3 The Protein Folding Energy Landscape: A Primer
Peter G. Wolynes

3.1	Energy Landscape: Metaphor and Math	49
3.2	Random Sequences – Prehistoric Proteins (Possibly), but Not Most Modern Proteins	50
3.3	The Statistical Energy Landscape	50
3.4	The Energy Landscape of Long Evolved Proteins	55
3.5	Minimal Frustration, Capillarity, and Protein Topology	61
3.6	Delightful Prediction of Many of the Devilish Details of Folding	64
References		66

Chapter 4 Hydrogen Exchange Experiments: Detection and Characterization of Protein Folding Intermediates
Yawen Bai

4.1	Introduction	70
4.2	Intrinsic Exchange Rates for Unfolded Polypeptides	71

	4.3	Linderstrøm–Lang Model for Amide Hydrogen Exchange in Folded Proteins	72
	4.4	Characterization of Acid Denatured States by Hydrogen Exchange	73
		4.4.1 Apomyoglobin (AMb)	74
		4.4.2 Cytochrome c (cyt c)	75
		4.4.3 Ribonuclease H (RNase H)	75
	4.5	Pulsed-Amide H/D Exchange Method	75
		4.5.1 Cytochrome c	78
		4.5.2 Apomyoglobin	78
		4.5.3 RNase H	79
		4.5.4 Hen Egg White Lysozyme (HEWL)	79
	4.6	Native-State Hydrogen Exchange Method	80
		4.6.1 Cytochrome c	82
		4.6.2 RNase H	83
		4.6.3 Rd-apocytochrome b_{562}	83
	References		83

Chapter 5 Statistical Differential Scanning Calorimetry: Probing Protein Folding–Unfolding Ensembles
Beatriz Ibarra-Molero and Jose Manuel Sanchez-Ruiz

5.1	Differential Scanning Calorimetry (DSC) as a Tool for the Complete Energetic Description of Protein Folding/Unfolding Thermal Equilibria	85
5.2	Partition Functions of Folding/Unfolding Processes	87
5.3	The Two-state Equilibrium Model: A Historical Perspective	93
5.4	Folding Free-energy Barriers from Equilibrium DSC Experiments	96
5.5	The van 't Hoff to Calorimetric Enthalpy Ratio Revisited	98
5.6	Protein Kinetic Stability: Free-energy Barriers for Irreversible Denaturation from Scan-rate Dependent DSC	100
References		103

Chapter 6 Fast Protein Folding
Martin Gruebele

6.1	Introduction		106
6.2	Fast Folding: Why and How?		108
6.3	Fast Dynamics of Polypeptide Chains		110
	6.3.1	Loop Formation	111
	6.3.2	Protein Collapse	114

		6.3.3 Secondary Structure Formation	115
		6.3.4 Timescales	115
	6.4	Microsecond Protein Folding	115
		6.4.1 History	115
		6.4.2 Sub-millisecond Instrumentation	117
		6.4.3 Spectroscopic Signatures Used in Fast Folding	119
		6.4.4 Case Studies	121
	6.5	Downhill Folding	127
	6.6	Outlook	130
	Acknowledgement		131
	References		131

Chapter 7 Single Molecule Spectroscopy in Protein Folding: From Ensembles to Single Molecules
Benjamin Schuler

	7.1 Introduction	139
	7.2 History and Principles of Single Molecule Detection	140
	7.3 Kinetics: From Ensembles to Single Molecules	141
	7.3.1 Rate Constants and Probabilities	143
	7.4 Correlation Analysis	146
	7.5 FRET Efficiency Distributions and Distance Dynamics	149
	7.5.1 Single Molecule FRET Experiments	149
	7.5.2 Timescales and Distance Distributions	150
	7.5.3 Dynamics from Transfer Efficiency Fluctuations	153
	7.6 Pleasure, Pain, and Promise of Single Molecule Experiments	154
	Acknowledgements	156
	References	156

Chapter 8 Computer Simulations of Protein Folding
Vijay S. Pande, Eric J. Sorin, Christopher D. Snow and Young Min Rhee

	8.1 Introduction: Goals and Challenges of Simulating Protein Folding	161
	8.1.1 Simulating Protein Folding	161
	8.1.2 What are the Challenges for Atomistic Simulation?	163
	8.2 Protein Folding Models: from Atomistic to Simplified Representations	164
	8.2.1 Atomic Force Fields	164
	8.2.2 Implicit Solvation Models	166

	8.2.3	Minimalist Models	167
	8.2.4	How Accurate are the Models?	168
8.3	Sampling: Methods to Tackle the Long Timescales Involved in Folding		169
	8.3.1	Tightly Coupled Molecular Dynamics (TCMD)	169
	8.3.2	Replica Exchange Molecular Dynamics (REMD)	169
	8.3.3	High-temperature Unfolding	170
	8.3.4	Low-viscosity Simulation Coupled with Implicit Solvation Models	170
	8.3.5	Coarse-grained and Minimalist Models	170
	8.3.6	Path Sampling	171
	8.3.7	Graph-based Methods	171
	8.3.8	Markovian State Model Methods	171
8.4	Validation of Simulation Methodology: Protein Folding Kinetics		172
	8.4.1	Low-viscosity Simulations	172
	8.4.2	Estimating Rates with a Two-state Approximation	173
	8.4.3	Markovian State Models (MSMs)	176
	8.4.4	Other Approaches	177
8.5	Predicting Protein Folding Pathways		179
	8.5.1	Kinetics Simulations	179
	8.5.2	Thermodynamics Simulations	181
8.6	Conclusions		182
References			184

Chapter 9 Protein Design: Tailoring Sequence, Structure, and Folding Properties
Andreas Lehmann, Christopher J. Lanci, Thomas J. Petty II, Seung-gu Kang and Jeffery G. Saven

9.1	Introduction		188
9.2	Empirical Approaches to Protein Design		190
	9.2.1	Hierarchical Protein Design	190
	9.2.2	Combinatorial Methods	191
	9.2.3	Directed Evolution	192
	9.2.4	Intrinsic Limitations	192
9.3	Computational Approaches to Structured-based Design		193
	9.3.1	Backbone Structure and Sequence Constraints	194
	9.3.2	Residue Degrees of Freedom	194
	9.3.3	Energy Function	195
	9.3.4	Solvation	195
	9.3.5	Foldability Criteria and Negative Design	196

		9.3.6	Search and Characterization of Sequence Ensembles	197
	9.4	Recent Successes in Protein Design		198
		9.4.1	Tailored Mutations for Ultrafast Folding	198
		9.4.2	Designing Structure and Sequence	199
		9.4.3	Facilitating the Study of Membrane Proteins	201
		9.4.4	Proteins with Non-biological Components	202
		9.4.5	Symmetric Structures	202
		9.4.6	Computational Methods for Directed Evolution	203
	9.5	Outlook		204
	Acknowledgements			204
	References			205

Chapter 10 Protein Misfolding and β-Amyloid Formation
Alexandra Esteras-Chopo, Maria Teresa Pastor and Luis Serrano

	10.1	Introduction		214
	10.2	General Principles of Amyloid Formation		216
		10.2.1	Historical Perspective	216
		10.2.2	Molecular Basis of Amyloidosis: Protein Misfolding	217
		10.2.3	The Structural Architecture of Amyloid Fibrils	221
		10.2.4	Amyloid Induced Toxicity	224
		10.2.5	Experimental Techniques to Study Amyloid Formation	226
		10.2.6	Cytotoxicity Studies	228
	10.3	Experimental Studies on Amyloid Model Systems		229
		10.3.1	Diversity and Commonalities in the Amyloid Protein Family	229
		10.3.2	Protein Amyloidogenic Regions	230
	Acknowledgements			235
	References			235

Chapter 11 Scenarios for Protein Aggregation: Molecular Dynamics Simulations and Bioinformatics Analysis
Ruxandra Dima, Bogdan Tarus, G. Reddy, John E. Straub and D. Thirumalai

	11.1	Introduction		241
	11.2	Scenarios for Peptide Association		243
		11.2.1	General Ideas	243

	11.2.2	The Assembly of $A\beta_{16-22}$ Oligomers	245
	11.2.3	Dimerization of $A\beta_{10-35}$ Peptides	247
	11.2.4	Initial Stages in the PrP^C Conformational Transition	251
11.3	Conclusions		262
References			263

Subject Index   266

CHAPTER 1

The α-Helix as the Simplest Protein Model: Helix–Coil Theory, Stability, and Design

ANDREW JAMES DOIG

Faculty of Life Sciences, The University of Manchester, Jackson's Mill, PO Box 88, Sackville Street, Manchester M60 1QD, UK

1.1 Introduction

Proteins are built of regular local folds of the polypeptide chain called secondary structure. α-Helices are present in nearly all globular proteins, with ≈30% of residues found in α-helices.[1] It is such ubiquity and its structural simplicity that makes the α-helix an ideal candidate for detailed quantitative studies of the complex energetic factors involved in protein folding and stability. Here, we discuss structural features of the helix and their contributions to helix stability from studies in peptides. Some earlier reviews in this field are references 2–10.

1.2 Structure of the α-Helix

A helix combines a linear translation with an orthogonal circular rotation. In the α-helix the linear translation is a rise of 5.4 Å per turn of the helix and a circular rotation is 3.6 residues per turn. Side chains spaced $i, i+3$, $i, i+4$, and $i, i+7$ are therefore close in space and interactions between them can affect helix stability. Spacings of $i, i+2$, $i, i+5$, and $i, i+6$ place the side chain pairs on opposite faces of the helix avoiding any interaction. The helix is primarily stabilized by $i, i+4$ hydrogen bonds between backbone amide groups.

The conformation of a polypeptide can be described by the backbone dihedral angles ϕ and ψ. Most ϕ,ψ combinations are sterically excluded, leaving only the broad β region and narrower α region. The residues at the N-terminus of the α-helix are called N′-N-cap-N1-N2-N3-N4 *etc.*, where the N-cap is the residue with non-helical ϕ, ψ angles immediately preceding the N-terminus of an α-helix and N1 is the first residue with helical ϕ, ψ angles.[11] The C-terminal residues are similarly called C4-C3-C2-C1-C-cap-C′ *etc.* The N1, N2, N3, C1, C2, and C3 residues are unique because their amide groups participate in $i, i+4$ backbone–backbone hydrogen bonds using either only their CO (at the N-terminus) or NH (at the C-terminus) groups. The need for these groups to form hydrogen bonds has powerful effects on helix structure and stability.[12]

1.2.1 Capping Motifs

The amide NH groups at the helix N-terminus are satisfied predominantly by side-chain H-bond acceptors. In contrast, carbonyl CO groups at the C-terminus are satisfied primarily by backbone NH groups from the sequence following the helix.[12] The presence of such interactions would therefore stabilize helices. These interactions can be identified as specific patterns found at or near the ends of helices and are generally termed capping motifs.[11,13–17]

A common pattern of capping at the helix N-terminus is the capping box. Here, the side chain of the N-cap forms a hydrogen bond with the backbone of N3 and, reciprocally, the side chain of N3 forms a hydrogen bond with the backbone of the N-cap.[18] The definition of the capping box was expanded by Seale *et al.*[19] to include an associated hydrophobic interaction between residues N′ and N4 and is also known as a "hydrophobic staple".[20] A variant of the capping box motif is termed the "big" box with an observed hydrophobic interaction between non-polar side-chain groups in residues N4 and N″ (not N′).[19] The Pro-box motif involves three hydrophobic residues and a Pro residue at the N-cap.[21]

The two primary capping motifs found at helix C-termini are the Schellman and the $α_L$ motifs.[22–24] The Schellman motif is defined by a doubly hydrogen-bonded pattern between backbone partners, consisting of hydrogen bonds between the amide NH at C″ and the carbonyl CO at C3 and between the amide NH at C′ and the carbonyl CO at C2, respectively. The associated hydrophobic interaction is between C3 and C″. In a Schellman motif, polar residues are highly favoured at the C1 position and the C′ residue is typically glycine. If C″ is polar, the alternative $α_L$ motif is observed, defined by a hydrogen bond between the amide NH at C′ and the carbonyl CO at C3. As in the Schellman motif, the C′ residue is typically glycine, which adopts a positive value of ϕ. However, the hydrophobic interaction in an $α_L$ is heterogeneous, occurring between C3 and any of several residues external to the helix ($C^{3\prime}$, $C^{4\prime}$, or $C^{5\prime}$).[23]

A notable difference between the N- and C-terminal motifs is that at the N-terminus, helix geometry favors side-chain-to-backbone hydrogen bonding and selects for compatible polar residues.[25,26] Accordingly, the N-terminus promotes selectivity in all polar positions, especially N-cap and N3 in the capping box. In contrast, at the C-terminus, side-chain-to-backbone hydrogen bonding is disfavored. Backbone hydrogen bonds are satisfied instead by post-helical backbone groups. The C-terminus need only select for C' residues that can adopt positive values of the backbone dihedral angle ϕ, most notably Gly.[23]

1.2.2 Metal Binding

One way to stabilize helix conformations, especially in short peptides, is to introduce an artificial nucleation site composed of a few residues fixed in a helical conformation. For example, the calcium-binding loop from EF-hand proteins saturated with a lanthanide ion promotes a rigid short helical conformation at its C-terminus region.[27] This system has been used to measure enthalpic terms contributing to helical preferences of the amino acids.[28–30] In the presence of Cd ions, a synthetic peptide containing Cys-His ligands $i,i+4$ apart at the C-terminal region increased helicity (that is the average probability of finding dihedral angle pairs in values typical of α-helix) from 54% to 90%. The helicity of a similar peptide containing His-His ligands increased by up to 90% as a result of Cu and Zn binding.[31] The addition of a *cis*-Ru(III) ion to a 6-mer peptide, Ac-AHAAAHA-NH$_2$, changed the peptide conformation from random coil to 37% helix.[32] An 11-residue peptide was converted from random coil to 80% helix content by the addition of Cd ions, although the ligands used were not natural amino acids but aminodiacetic acids.[33] As(III) stabilizes helices when bound to Cys side chains spaced $i,i+4$ by -0.7 to -1.0 kcal mol^{-1}.[34] 19-Membered metallocyclic rings induce helix formation by covalently linking helical turns.[35]

1.2.3 The 3$_{10}$-Helix

3$_{10}$-Helices are stabilized by $i,i+3$ hydrogen bonds, instead of the $i,i+4$ found in α-helices, making the cylinder of the 3$_{10}$-helix narrower than α and their hydrogen bonds non-linear. 3–4% of residues in crystal structures are in 3$_{10}$-helices.[1,36] Most 3$_{10}$-helices are short, only 3 or 4 residues long, compared to a mean of 10 residues in α-helices,[1] and are commonly found as N- or C-terminal extensions to an α-helix:[1,37,38] strong amino acid preferences have been observed for different locations within the interior[36] and N- and C-caps[25] of 3$_{10}$-helices in crystal structures. The 3$_{10}$-helix is being recognized as of increasing importance in isolated peptides and even as a possible intermediate in α-helix formation.[39,40]

1.2.4 The π-Helix

In contrast to the widely occurring α- and 3$_{10}$-helices, the π-helix is extremely rare. The π-helix is unfavorable for three reasons: its dihedral angles are energetically unfavorable relative to the α-helix,[41,42] its three-dimensional structure has a 1 Å hole down the center that is too narrow for access by a water molecule resulting in the loss of van der Waals interactions, and a higher number of residues (four) must be correctly oriented before the first $i,i+5$ hydrogen bond is formed, making helix initiation more entropically unfavorable than for α- or 3_{10}-helices.[43] π-Helices are known in both peptide and proteins, however.[44–48]

1.3 Design of Peptide Helices

The earliest work on peptide helices was on long homopolymers of Glu or Lys which show coil-to-helix transitions on changing the pH from charged to neutral.[49] The neutral polypeptides are metastable and prone to aggregation, ultimately to β-sheet amyloid.[50] In 1971 Brown and Klee[51] reported that the C-peptide of ribonuclease A, which contains the first 13 residues of the protein and which forms a helix in the protein, had high helical content at 0 °C. Work on the C-peptide showed that the replacement of interior helical residues with Ala was stabilizing, indicating that a major reason why this helix was folded in isolation was the presence of three successive alanines from positions 4–6. This led to the successful design of isolated, monomeric helical peptides in aqueous solution, first containing several salt bridges and a high alanine content, based on $(EAAAK)_n$,[52,53] and then a simple sequence with a high alanine content solubilized by several lysines.[54] These "AK peptides" are based on the sequence $(AAKAA)_n$, where n is typically 2–5. The Lys side chains are spaced $i,i+5$ so they are on opposite faces of the helix, giving no charge repulsion. Hundreds of AK peptides have been studied, giving most of the available results on helix stability in peptides. The alanines in the $(EAAAK)_n$-type peptides may be removed entirely; E_4K_4 peptides, with sequences based on $(EEEEKKKK)_n$ or EAK patterns, are also helical, stabilized by large numbers of salt bridges.[5,55–57]

1.3.1 Host–Guest Studies

Extensive work from the Scheraga group has obtained helix–coil parameters using a host–guest method. Long random co-polymers were synthesized of a water soluble, non-ionic guest (poly[N^5-(3-hydroxypropyl)-L-glutamine] (PHPG) or poly[N^5-(4-hydroxybutyl)-L-glutamine] (PHBG)), together with a low (10–50%) content of the guest residue. Using the s and σ Zimm–Bragg helix–coil parameters (see below) for the host homopolymer, it was possible to calculate those for the guest using helix–coil theory as a function of temperature. The results obtained from the host–guest work are in disagreement with most of the results from short peptides of fixed sequence.

1.3.2 Helix Lengths

Helix formation in peptides is cooperative, with a nucleation penalty. Helix stability therefore tends to increase with length, in homopolymers at least. As the length of a homopolymer increases, the mean fraction helix will level off below 100%, as long helices tend to break in two. In heteropolymers, observed lengths are highly sequence dependent. As helices are at best marginally stable in monomeric peptides in aqueous solution, they are readily terminated by the introduction of a strong capping residue or a residue with a low intrinsic helical preference.

The length distribution of helices in proteins is very different from homo- and heteropolymers.[1] Most protein helices are short, with 5 to 14 residues most abundant. There is a general trend for a decrease in frequency as the length increases beyond 13 residues. Helix lengths longer than 25 are rare. There is also a preference to have close to an integral number of turns so that their N- and C-caps are on the same side of the helix.[58]

1.3.3 The Helix Dipole

The secondary amide group in a protein backbone is polarized with the oxygen negatively charged and hydrogen positively charged. In a helix, the amides are all oriented in the same direction with the positive hydrogens pointing to the N-terminus and negative oxygens pointing to the C-terminus. This can be regarded as giving a partial positive charge at the helix N-terminus and a partial negative charge at the helix C-terminus.[59-61] In general, therefore, negatively charged groups are stabilizing at the N-terminus and positively charged at the C-terminus. An alternative interpretation of these results is that favored side chains are those that can make hydrogen bonds to the free amide NH groups at N1, N2, and N3 or free CO groups at C1, C2, and C3.[62] Charged groups can form stronger hydrogen bonds than neutral groups, thus providing an alternative rationalization of the pH titration results. These hypotheses are not mutually exclusive, as a charged side chain can also function as a hydrogen bond acceptor or donor. Measurements of the amino acid preferences for the N-cap, N1, N2, and N3 positions in the helix allow a comparison to be made of the relative importance of helix dipole and hydrogen bonding interactions,[63-66] suggesting that both charge and hydrogen-bonding interactions are important.

1.3.4 Acetylation and Amidation

A simple, yet effective, way to increase the helicity of a peptide is to acetylate its N-terminus.[15,67] Acetylation removes the positive charge that is present at the helix terminus at low or neutral pH; this charge would interact unfavorably with the positive helix dipole and free N-terminal NH groups. The extra CO group from the acetyl group can form an additional hydrogen bond to the NH group, putting the acetyl at the N-cap position. This has a strong stabilizing effect by approximately 1.0 kcal mol^{-1} compared to alanine.[63,68,69]

Amidation of the peptide C-terminus is structurally analogous to N-terminal acetylation: the helix is extended by one hydrogen bond and an unfavorable charge–charge repulsion with the helix dipole is removed. The energetic benefit of amidation is rather smaller, however, with the amide group being no better than Ala and in the middle if the C-cap residues are ranked in order of stabilization effect.[63] As most helical peptides studied to date are both acetylated and amidated, and acetylation is more stabilizing than amidation, the distribution of helicity along the peptide is generally skewed so that residues near the N-terminus are more helical than those near the C-terminus.

1.3.5 Solubility

Peptide aggregation can be assayed rigorously by sedimentation equilibrium, which determines the oligomeric state of a molecule in solution. This is difficult, however, with the short peptides often used as their molecular weights are at the lower limit for this technique. A simpler method is to check a spectroscopic signal that depends on peptide structure, most obviously circular dichroism (CD), as a function of concentration. If the signal depends linearly on peptide concentration across a large range, including that used to study the peptide structure, it is safe to assume that the peptide is monomeric. An oligomer that does not change state, such as a coiled-coil, across the concentration range cannot be excluded, however. Light scattering can detect aggregation. A monomeric peptide should have a flat baseline in a UV spectrum outside the range of any chromophores in the peptide. Stock solutions of a peptide with a single tyrosine isolated from the helix region by Gly should have $A_{300}/A_{275} < 0.02$ and $A_{250} < A_{275} < 0.2$.[70]

Consideration of solubility is essential when designing helical peptides. Solubility can be achieved most easily by including polar side chains spaced $i, i+5$ in the sequence where they cannot interact. Lys, Arg and Gln are used most often for this purpose. Gln may be preferred if unwanted interactions with charged Lys or Arg may be a problem, but some AQ peptides lack sufficient solubility and AQ peptides are less helical.

The spacing of side chains in the helix is best visualized with a helical wheel, to ensure that the designed helix does not have a non-polar face that may lead to dimerization. The following webpage provides a useful resource for this: http://cit.itc.virginia.edu/~cmg/Demo/wheel/wheelApp.html

1.3.6 Concentration Determination

An accurate measurement of helix content depends on an accurate spectroscopic measurement and, equally importantly, peptide concentration. This is usually achieved by including a Tyr side chain at one end of the peptide. The extinction coefficient of Tyr at 275 nm is 1450 M^{-1} cm^{-1}.[71] If Trp is present, measurements at 281 nm can be used where the extinction coefficient of Trp is 5690 M^{-1} cm^{-1} and Tyr 1250 M^{-1} cm^{-1}.[72] Though the inclusion of aromatic residues is required for concentration determination, this can have the unwanted side effect of perturbing a CD spectrum, leading to an inaccurate determination of helix content. A simple solution to this problem is to separate the terminal Tyr from the rest of the sequence by one or more Gly residues.[73] If the aromatic residues must be included within the helical region, the CD spectrum should be corrected to remove this perturbation.[74]

1.3.7 Helix Templates

A major penalty to helix formation is the loss of entropy arising from the requirement to fix three consecutive residues to form the first hydrogen bond of the helix. Following this nucleation, propagation is much more favored as only a single residue need be restricted to form each additional hydrogen bond. A way to avoid this barrier is to synthesize a template molecule that facilitates helix initiation, by fixing hydrogen bond acceptors or donors in the correct orientation for a peptide to bond in a helical geometry. The ideal template nucleates a helix with an identical geometry to a real helix. Kemp's group applied this strategy and synthesized a proline-like template that nucleated helices when a peptide chain was covalently attached to a carboxyl group.[75–79] Bartlett *et al.* reported on a hexahydroindol-4-one template[80] that induces helicity in an appended hexameric peptide. Several other templates were less successful and could only induce helicity in organic solvents.[81–83] Their syntheses are often lengthy and difficult, partly due to the challenging requirement of orienting several dipoles to act as hydrogen bond acceptors or donors.

1.4 Helix–Coil Theory

Peptides that form helices in solution do not show a simple two-state equilibrium between a fully formed helix and a fully unfolded structure. Instead they form a complex mixture of all helix, all coil, or, most frequently, a distribution of helices of different lengths with increased probability at the center of the peptide (helix fraying). In order to interpret experiments on helical peptides and make theoretical predictions on helices it is therefore essential to use a helix–coil theory that deals with this distribution of helices. Recent reviews of helix–coil theory are references 84–86.

The simplest way to analyse the helix–coil equilibrium, still occasionally seen, is the two-state model where the equilibrium is assumed to be between a 100% helix conformation and 100% coil. This is incorrect and its use gives serious errors.

1.4.1 Zimm–Bragg Model

The two major types of helix–coil model are i) those which count hydrogen bonds, principally Zimm–Bragg (ZB),[87] and ii) those that consider residue conformations, principally Lifson–Roig (LR).[88] In the ZB theory the units being considered are peptide groups and they are classified on the basis of whether their NH groups participate in hydrogen bonds within the helix. The ZB coding is shown in Figure 1.1. A unit is given a code of 1 (*e.g.* peptide unit 5 in Figure 1.1) if its NH group forms a hydrogen bond and 0 otherwise. The first hydrogen-bonded unit proceeding from the N-terminus has a statistical weight of σs, successive hydrogen-bonded units have weights of s and non-hydrogen bonded units have weights of 1. The s-value is a propagation parameter and σ is an initiation parameter. The difficulty of nucleating a helix is captured in the ZB model by having σ smaller than s. The statistical weight of a homopolymeric helix of N hydrogen bonds is σs^{N-1}. The cost of initiation, σ, is thus paid only once for each helix, while extending the helix simply multiplies its weight by one additional s-value for each extra hydrogen bond.

1.4.2 Lifson–Roig Model

In the LR model each residue is assigned a conformation of helix (h) or coil (c), depending on whether it has helical ϕ, ψ angles. Every conformation of a peptide of N residues can be written as a string of N c's or h's, giving 2^N conformations in total. Residues are assigned statistical weights depending on their conformations and the conformations of surrounding residues. A residue in an h conformation with an h on either side has a weight of w. This can be thought of as an equilibrium constant between the helix interior and the coil. Coil residues are used as a reference and have a weight of 1. In order to form an $i, i + 4$ hydrogen bond in a helix, three successive residues need to be fixed in a helical conformation. M consecutive helical residues will therefore have $M-2$ hydrogen bonds. The two residues at the helix termini (*i.e.* those in the centre of chh or hhc conformations) are assigned weights of v (Figure 1.1). The ratio of w to v gives approximately the effect of hydrogen bonding (1.7 : 0.036 for Ala[69] or $-RT \ln (1.7/0.036) = -2.1 \, \text{kcal mol}^{-1}$). A helical homopolymer segment of M residues has a weight of $v^2 w^{M-2}$ and a population in

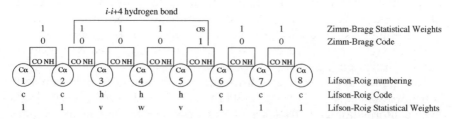

Figure 1.1 Zimm–Bragg and Lifson–Roig codes and weights for the α-helix.

the equilibrium of v^2w^{M-2} divided by the sum of the weights of every conformation (*i.e.* the partition function). In this way the population of every conformation is calculated and all properties of the helix–coil equilibrium evaluated. The LR model is easier to handle conceptually for heteropolymers since the parameters are assigned to individual residues. The substitution of one amino acid at a certain position thus changes the w- and v-values at that position. In the ZB model the initiation parameter σ is associated with several residues and s with a peptide group, rather than a residue. It is therefore easier to use the LR model when making substitutions. Indeed, most recent work has been based on this model. A further difference is that the ZB model assigns weights of zero to all conformations that contain a chc or chhc sequence. This excludes a very large number of conformations that contain a residue with helical ϕ, ψ angles but with no hydrogen bond. In LR theory, these are all considered. The ZB and LR weights are related by the following formulae:[85] $s = w/(1 + v)$; $\sigma = v^2/(1 + v)^4$.

The complete helix–coil equilibrium is handled by determining the statistical weight for every possible conformation that contains a helix plus a reference weight of 1 for the coil conformation. Each conformation considered in the helix–coil equilibrium is given a statistical weight. This indicates the stability of that conformation, with the higher the weight, the more probable the conformation. Weights are defined relative to the all-coil conformation, which is given a weight of 1. The statistical weight of a conformation can thus be regarded as an equilibrium constant relative to the coil; a weight >1 indicates the conformation is more stable than coil, <1 means less stable and $=1$ means equally stable. The population of each conformation is given by the statistical weight of that conformation divided by the sum of the statistical weights for every conformation (the partition function). Thus the greater the statistical weight, the more stable the conformation. The key to using helix–coil theory is the partition function. All the properties of a system at equilibrium are contained within the partition function, which makes it very valuable. Partition functions are extremely powerful concepts in statistical thermodynamics since they allow calculation of all properties of an equilibrium ensemble. Any property of the equilibrium can be extracted from the partition function by applying the appropriate mathematical function. In this case the properties could be the mean number of hydrogen bonds, the mean helix length, the probability that each residue is within a helix, *etc*. In particular, the mean number of residues with a weight x is given by $\frac{\partial \ln Z}{\partial \ln x}$. Circular dichroism is commonly used to give the mean helix content of a helical peptide, namely the fraction of residues that have a weight of w. LR-based models can thus be related to experimental data by equating the measured mean helix content to $\frac{\partial \ln Z}{\partial \ln w}/N$, where N is the number of residues in the peptide. Statistical weights can be regarded as equilibrium constants for the equilibrium between coil and the structure (as the reference coil weight is defined as 1). They can therefore be converted to free energies as $-RT$ ln (weight). The Lifson–Roig formalism has also been adapted to describe 3_{10}- and π-helices.[43,89,90]

1.4.2.1 The Unfolded State and Polyproline II Helix

The treatment of peptide conformations is based on Flory's isolated-pair hypothesis.[91] This states that while ϕ and ψ for a residue are strongly interdependent, giving preferred areas in a Ramachandran plot, ϕ, ψ pair is independent of the ϕ, ψ angles of its neighbors. Pappu *et al.* found that non-helical poly(Ala) chains mostly populated extended or fully helical conformations as many partly helical conformations are sterically disallowed. Such effects are not included in helix–coil theories. Helix–coil theories assign the same weight (1) to every coil residue; steric exclusion means that these should vary and be lower than 1 in many cases.

The polyproline II helix may well be an important conformation for unfolded proteins. Many recent papers have addressed this issue. Examples are references.[92–94] In particular, denatured alanine rich peptides may form a polyproline II helix.[95–101] It may therefore be valid to consider residues in helical peptides to be in three possible states (helix, coil, or polyproline II), rather than two (helix or coil). No current helix–coil model takes this into account. A scale of amino acid preferences for the polyproline II helix has been published.[102]

1.4.2.2 Single Sequence Approximation

Since helix nucleation is difficult, conformations with multiple helical segments are expected to be rare in short peptides. In the one-, or single-, helical sequence approximation, peptide conformations containing more than one helical segment are assumed not to be populated and are excluded from the partition function (*i.e.* assigned statistical weights of zero). As peptide length increases, the approximation is no longer valid since multiple helical segments can be long enough to overcome the initiation penalty. The single sequence approximation will also break down when a sequence with a high preference for a helix terminus is within the middle of the chain. Conformations with two or more helices may also often include helix–helix tertiary interactions that are ignored in all helix–coil models.

1.4.2.3 N- and C-caps

N-Capping has been added to LR theory be assigning a weight of n to the central residue in a cch triplet, as the N-cap is the non-helical residue preceding the start of a helical segment. Similarly, the C-cap is the first residue in a non-helical conformation (c) at the C-terminus of a helix. C-Cap weights (c-values) are assigned to central residues in hcc triplets.[6,68,103]

1.4.2.4 Capping Boxes

The N-terminal capping [18] includes a side-chain–backbone hydrogen bond from N3 to the N-cap ($i, i - 3$). This is included in the LR model by assigning a weight of w^*r to the chhh conformation, where r is the weight for the Ser backbone to Glu side-chain bond.[6]

1.4.2.5 Side-chain Interactions

As helices have 3.6 residues per turn, side chains spaced $i,i+3$ or $i,i+4$ are close in space. Side-chain interactions are thus possible when four or five consecutive residues are in a helix. They are included in the LR-based model by giving a weight of w^*q to hhhh quartets and w^*p to hhhhh quintets. The side-chain interaction is between the first and last side chains in these groups; the w weight is maintained to preserve the equivalence between the number of residues with a w weighting and the number of backbone helix hydrogen bonds.[104]

1.4.2.6 N1, N2, and N3 Preferences

The helix N-terminus shows significantly different residue frequencies for the N-cap, N1, N2, N3, and helix interior positions.[11,26,105,106] A complete theory for the helix should therefore include distinct preferences for the N1, N2, and N3 positions. In the original LR model, the N1 and C1 residues are both assigned the same weight, v. Shalongo and Stellwagen[107] separated these as v_N and v_C. Andersen and Tong[103] did the same and derived complete scales for these parameters from fitting experimental data, though some values were tentative. The helix initiation penalty is $v_N^*v_C$ and so v_N- and v_C-values are all small (≈ 0.04).

We added weights for the N1, N2, and N3 ($n1$, $n2$, and $n3$) positions as follows:[108] The $n1$-value is assigned to a helical residue immediately following a coil residue. The penalty for helix initiation is now $n1.v$, instead of v^2, as v remains the C1 weight. An N2 helical residue is assigned a weight of $n2.w$, instead of w. The weight w is maintained in order to keep the useful definition of the number of residues with a w weighting being equal to the number of residues with an $i,i+4$ main chain–main chain hydrogen bond. The $n2$-value is an adjustment to the weight of an N2 residue that takes into account the structures that can be adopted by side chains uniquely at this position. Similarly, an N3 residue is now assigned the weight $n3.w$, instead of w.

1.4.2.7 Helix Dipole

Helix dipole effects were added to the LR model by Scholtz et al.,[109] though they used the one sequence approximation so that only one or no dipoles in total are present. In LR models helix dipole effects are subsumed within other energies. For example, N-cap, N1, N2, and N3 energies will include a contribution from the helix dipole interaction so the energy of interaction of charged groups at this position with the dipole should not be counted in addition.

1.4.3 AGADIR

AGADIR is an LR-based helix–coil model developed by Serrano, Muñoz, and co-workers. The original model[110] included parameters for helix propensities excluding backbone hydrogen bonds (attributed to conformational entropy), backbone hydrogen bond enthalpy, side-chain interactions and a term for coil weights at the end of helical sequences (*i.e.* caps). The single sequence approximation was used. The original partition function assumed that many helical conformations did not exist, as all conformations in which the residue of interest is not part of a helix were excluded.[104,110] These were corrected in a later version, AGADIRms, which considers all possible conformations.[111] If AGADIR and LR models are both applied to the same data, to determine a side-chain interaction energy, for example, the results are similar, showing that the models are now not significantly different.[111,112] The treatment of the helix–coil equilibrium differs in a number of respects from the ZB and LR models and these have been discussed in detail by Muñoz and Serrano.[111] The minimal helix length in AGADIR is four residues in an h conformation, rather than three. The effect of this assumption is to exclude all helices which contain a single hydrogen bond; only helices with two or more hydrogen bonds are allowed. In practice, this probably makes little difference as chhhc conformations are usually unfavorable and hence have low populations. Early versions of AGADIR considered that residues following an acetyl at the N-terminus or preceding an amide at the C-terminus were always in a c conformation; this was changed to allow these to be helical.[113]

The latest version of AGADIR, AGADIR1s-2,[113] includes terms for electrostatics,[113] the helix dipole,[113,114] pH dependence,[114] temperature,[114] ionic strength,[113] N1, N2, and N3 preferences,[115] and capping motifs such as the capping box, hydrophobic staple, Schellman motif, and Pro-capping motif.[113] The free energy of a helical segment, $\Delta G_{\text{helical-segment}}$, is given by $\Delta G_{\text{helical-segment}} = \Delta G_{\text{Int}} + \Delta G_{\text{Hbond}} + \Delta G_{\text{SD}} + \Delta G_{\text{dipole}} + \Delta G_{\text{nonH}} + \Delta G_{\text{electrost}}$, which are terms for the energy required to fix a residue in helical angles (with separate terms for N1, N2, N3, and N4), backbone hydrogen bonding, side-chain interactions excluding those between charged groups, capping and helix dipole interactions, respectively. Electrostatic interactions are calculated with Coulomb's equation. Helix dipole interactions were all electrostatic interactions between the helix dipole or free N- and C-termini and groups in the helix. Interactions of the helix dipole with charged groups located outside the helical segment were also included. pH dependence calculations considered a different parameter set for charged and uncharged side chains and their pK_a values. The single sequence approximation (see above) is used again, unlike in AGADIRms. AGADIR is at present the only model that can give a prediction of helix content for any peptide sequence, thus making it very useful. It can also predict NMR chemical shifts and coupling constants.

1.4.4 Lomize–Mosberg Model

Lomize and Mosberg developed a model for calculating the stability of helices in solution.[116] Interestingly, they extended it to consider helices in micelles or a uniform non-polar droplet to model a protein core environment. Helix stability in water is calculated as the sum of main chain interactions, which is the free energy change for transferring Ala from coil to helix, the difference in energy when replacing an Ala with another residue, hydrogen bonding and electrostatic interactions between polar side chains and hydrophobic side-chain interactions. An entropic nucleation penalty of two residues per helix is included. Different energies are included for N-cap, N1-N3, C1-C3, C-cap, hydrophobic staples, Schellman motifs, and polar side-chain interactions. Hydrophobic interactions were calculated from decreases in non-polar surface area when they are brought in contact. Helix stability in micelles or non-polar droplets is found by calculating the stability in water then adding a transfer energy to the non-polar environment.

1.5 Forces Affecting α-Helix Stability

1.5.1 Helix Interior

Different approaches have been used in order to determine the helical propensity or preference of individual amino acids. Scheraga and co-workers used a host–guest strategy (see above) to derive values for the helical preference of various amino acid residues. The host–guest system uses long random copolymers of a water soluble, non-ionic guest (poly[N^5-(3-hydroxypropyl)-L-glutamine] (PHPG) or poly[N^5-(4-hydroxybutyl)-L-glutamine] (PHBG)), together with a low (10–50%) content of the guest residue. The Zimm–Bragg model s and σ values of the host homopolymer are used to compute those for the guest.[117] This work has been criticized as the host side chains can interact with each other.[118] The introduction of a guest residue thus removes host–host interactions and replaces them with PHBG–guest or PHPG–guest side-chain interactions that may obscure the intrinsic helix propensities.

Rohl et al.[69] used many alanine-based peptides with the general sequences Ac-(AAKAA)$_m$Y-NH$_2$ (or with Q instead of K) to measure interior helix propensities. Substitutions in the helix interior and subsequent measures of helicity using CD spectroscopy in both water and 40% (v/v) trifluoroethanol (TFE) allowed the calculation of both the Lifson–Roig w parameter and the stabilization energy for all 20 amino acids. Kallenbach and co-workers used synthetic peptides of the form succinyl-YSEEEEKAKKAXAEEAEKKKK-NH$_2$, where substitutions at X allowed determination of helix stabilizing energies for common amino acids.[56] Stellwagen and co-workers made substitutions in position 9 of Ac-Y(EAAAK)$_3$A-NH$_2$.[53]

In 1998 Pace and Scholtz[119] gathered information from many different sources and derived a scale for the propensity of each amino acid in the helix interior. This is summarized in Table 1.1. The values are in $\Delta(\Delta G)$ relative to alanine

Table 1.1 Summary of other experimental helix propensities (relative to alanine).

Amino acid	Helix propensity $\Delta(\Delta G)$ (kcal mol^{-1}) (taken from reference 119)
Ala	0.00
Arg$^+$	0.21
Leu	0.21
Met	0.24
Lys$^+$	0.26
Gln	0.39
Glu	0.40
Ile	0.41
Trp	0.49
Ser	0.50
Tyr	0.53
Phe	0.54
Val	0.61
His	0.61
Asn	0.65
Thr	0.66
Cys	0.68
Asp	0.69
Gly	1.00
Pro	3.16

because it is generally (though not universally) agreed that this amino acid has the highest helical propensity. (The s values from the Zimm–Bragg model, as derived by the Scheraga group,[117] do not agree with other scales, alanine having the highest helix propensity and all other residues having lower values (a positive $\Delta(\Delta G)$ value relative to Ala)). Proline and glycine have the lowest helical propensity. The most controversial of these differences over the years has been that of alanine. Host guest analysis showing alanine to be effectively helix-neutral has been supported by data from some other groups, notably the templated helices of Kemp and co-workers.[75] The use of template-nucleated helices has been criticized by Rohl et al.,[120] who argued that the low apparent helix propensity of alanine is a consequence of properties of the template–helix junction. Kemp and co-workers[121] used templates to investigate the helix-forming tendency of polyalanine. Below six residues Ala had a low helix propensity, but when the limit of six was exceeded an increase was found. This suggested that there is a length-dependent term in the helicity of polyalanine. Alternatively, any destabilizing effect of the template is less significant in longer helices.

1.5.2 Caps

Some capping preferences were measured in proteins using barnase[13] and T4 lysozyme,[14] giving slightly varying results. The Kallenbach group[122] substituted several amino acids at the N-cap position in peptide models in the presence of a capping box. They found that Ser and Arg are the most stabilizing residues, whilst Gly and Ala are less stabilizing. The results are in agreement with the results of Forood et al.,[16] who found that the trend in α-helix inducing ability at the N-cap is Asp > Asn > Ser > Glu > Gln > Ala. A more comprehensive work to determine the preferences for all 20 amino acids at the N-cap position used peptides with a sequence of NH_2-XAKAAAAKAAAAKAAGY-$CONH_2$.[15,63,68] N-capping free energies ranged from Asn (best) to Gln (worst) (Table 1.2).

We have used a similar approach using peptide models to probe the preferences at N1[64], N2,[65] and N3[66] using peptides with sequences of CH_3CO-XAAAAQAAAAQAAGY-$CONH_2$, CH_3CO-AXAAAAKAAAAKAAGY-$CONH_2$, and CH_3CO-AAXAAAAKAAAAKAGY-$CONH_2$, respectively. The results have given N1, N2, and N3 preferences for most amino acids for these positions (Table 1.2) and these agree well with preferences seen in protein structures, with the exception of Pro at N1. Petukhov et al. similarly obtained N1, N2, and N3 preferences for non-polar and uncharged polar residues by applying AGADIR to experimental helical peptide data, and found similar results.[115,123] The complete sequences of peptides used can be seen in the table footnote. At N1, N2, and N3, Asp and Glu as well as Ala are preferred, presumably because negative side chains interact favorably with the helix dipole or NH groups, while Ala has the strongest interior helix preference.

Although it is also unique in terms of the presence of unsatisfied backbone hydrogen bonds, the C-terminal region is less explored experimentally. The C-terminus of the α-helix tends to fray more than the N-terminus, making C-terminal measurements less accurate. Zhou et al.[124] found that Asn is the most favored residue at the C-cap followed by Gln > Ser ~ Ala > Gly ~ Thr. Forood et al.[16] tested a limited number of amino acids at the C-terminus (C1) finding a rank order of Arg > Lys > Ala. Doig and Baldwin[63] determined the C-capping preferences for all 20 amino acids in α-helical peptides. The thermodynamic propensities of some amino acids at C′, C-cap, C1, C2, and C3 are also included in Table 1.2.[125,126]

1.5.3 Phosphorylation

Phosphoserine is destabilizing compared to serine at interior helix positions.[127,128] We investigated the effect of placing phosphoserine at the N-cap, N1, N2, N3, and interior position in alanine-based α-helical peptides, studying both the −1 and −2 phosphoserine charge states.[129] Phosphoserine stabilizes at the N-terminal positions by as much as 2.3 kcal mol^{-1}, while it destabilizes in the helix interior by 1.2 kcal mol^{-1}, relative to serine. The rank order of free energies relative to serine at each position is N2 > N3 > N1 > N-cap > interior. Moreover, −2 phosphoserine is the most preferred residue known at each of these N-terminal positions. Experimental pK_a values for the −1 to −2

Table 1.2 Amino acid propensities at N- and C-terminal positions of the helix.

	$\Delta\Delta G$ relative to Ala for transition from coil to the position ($kcal\,mol^{-1}$)												
Residue	N-cap 63[a]	N1 64[b]	115 123[c]	N2 65[d]	115 123[e]	N3 65[f]	115 123[g]	C3 125[h]	C2 126[i]	125[j]	C1 125[k]	C-cap 63[l]	C' 172[m]
A	0	0	0	—	0	0	0	0	0	0	0	0	0
C°	-1.4											0.2	
C⁻													
D°	-1.6	1.0		0.9		—						0.2	0.3
D⁻		0.5		0.7		—							
E°	-0.7	0		-0.2		1.1							
E⁻	-0.7	1.0		-0.2		—						-0.4	0.3
F	-1.2	0.1		-0.4		0.6						-0.5	
G	-0.7	1.4	0.7	0.9	0.4	1.3	0.8	2.1	0.6	0.6	0.4	0.1	0.1
H°		1.0		—		—			1.0				-1.1
H⁺		0.7		1.8		2.6						-0.2	
I	-0.5	—	0.4	1.6	0.5	0.7	0.5	0.2	0.2	0.4	0.5	-0.1	-0.9
K⁺	0.1	0.5		0.9		0.9						-0.1	1.5
L	-0.7	0.7	0.2	0.5	0.5	0.8	0.4		0.1				-0.1
M	-0.3	0.4	0.1	0.7	0.3	0.7	0.4		0.1				
N	-1.7	0.5	0.6	1.7	0.7	—	0.7	0.5	0.7	0.4	0.3	-0.3	0.1
P	-0.4	—	0.5	—		—						0.1	-0.4
Q	2.5	0.6	0.3	0.5	0.3	1.2	0.2	0.2	-0.02	0.2	0.05	-0.5	1.2
R⁺	-0.1	0.5		0.8		—						-0.4	-0.1
S	-1.2	0.7	0.4	0.7	0.5	1.1	0.6	0.6	0.5	0.7	0.5	0.8	-0.2
		0.4											0.3

T	−0.7	0.5	0.5	0.5	1.2	0.6	0.8	0.6	0.5	0.8		1.1
V	−0.1	0.6	0.5	0.4	—	0.4	0.3	0.4	0.7	0.6	0.9	1.6
W	−1.3	0.4	0.8		4.0							0.7
Y	−0.9	—	—		1.2						−2.2	

[a] NH$_2$-XAKAAAAKAAAAKAAGY-CONH$_2$.
[b] CH$_3$CO-XAAAAQAAAAQAAGY$_2$.
[c] CH$_3$CO -XAAAAAARAAARGGY-CONH$_2$.
[d] CH$_3$CO-AXAAAAKAAAAKAAGY-CONH$_2$
[e] CH$_3$CO -AYAAAAARAAARGGY-NH$_2$.
[f] CH$_3$CO-AAYAAAAKAAAAKAGY-CONH$_2$.
[g] CH$_3$CO -AAYAAAAARAAARGGY-NH$_2$.
[h] NH$_2$-YGGSAKEAAARAAAAXAA-CONH$_2$.
[i] Substitution of residue 32 (C2 position) of α-helix of ubiquitin.
[j] NH$_2$-YGGSAKEAAARAAAAAXA-CONH$_2$.
[k] NH$_2$-YGGSAKEAAARAAAAAAX-CONH$_2$.
[l] CH$_3$CO-YGAAKAAAAKAAAAKAX-COOH.
[m] Substitution of residue 35 (C' position) of α-helix of ubiquitin.

phosphoserine transition are in the order N2 < N-Cap < N1 < N3 < interior. Phosphoserine can form highly stabilizing salt bridges to Arg [128] or Lys.[130]

1.5.4 Non-covalent Side-chain Interactions

Many studies have been performed on the stabilizing effects of interactions between amino acid side chains in α-helices. These studies have identified a number of types of interaction that stabilize the helix, including salt bridges,[52,55,57,109,131–135] hydrogen bonds,[109,135–137] hydrophobic interactions,[104,138–140] basic–aromatic interactions,[74,141] and polar/non-polar interactions.[142] The stabilizing energies of many pairs in these categories have been measured, though some have only been analysed qualitatively. As described earlier, residue side chains spaced $i,i + 3$ and $i,i + 4$ are on the same face of the α-helix, though it is the $i,i+4$ spacing that receives most attention in the literature, as this is stronger. A summary of stabilizing energies for side-chain interactions is given in Table 1.3. We give only those that have been measured in helical peptides with the side-chain interaction energies determined by applying helix–coil theory. Almost all are attractive, with the sole exception of the Lys–Lys repulsion.

1.5.4.1 Cooperativity

After individual side-chain interactions, the next most complex step is to study triplets, with residues A, B, and C, where B forms bonds to both residues A and C. The free energy of the triplet is often not the sum of the AB and BC bond energies. The first evaluation of the strength of an engineered complex salt-bridge in a peptide was reported by Mayne et al.[143] after studying a stabilizing multiple salt-bridge involving Glu3, Asp4, and Arg7 in an 11-mer α-helix. A triplet of charged Arg-Glu-Arg residues spaced $i,i + 4$, $i + 8$ or $i,i + 3$, $i + 6$ also stabilizes α-helical peptides by $-1.5\,\text{kcal mol}^{-1}$ and $-1.0\,\text{kcal mol}^{-1}$, respectively, which is more than the additive contribution of two single salt-bridges.[144] A similar stabilizing effect in an Arg-Phe-Met triplet in $i,i + 4$, $i + 8$ spacing was observed.[145] Here, the triplet energy was $0.75\,\text{kcal mol}^{-1}$ greater than the sum of the Arg-Phe and Phe-Met energies. This was attributed to both interactions favoring the same conformation of the shared central Phe. Other non-salt-bridge triplets in isolated helical peptides have also been reported, for example Glu-Phe-Arg[146] and Glu-Phe-Glu,[147] although they do not show significant effects on peptide stability. In a Glu-Lys-Glu triplet, the second potential salt-bridge provide no additional stabilization over a single interaction, as the central Lys is only able to form one bond at a time. These simple examples show that side-chain interactions can be highly non-additive when residues have the potential to form more than one bond simultaneously, and show the difficulty of predicting helix stability for typical protein sequences with multiple interactions.

The α-Helix as the Simplest Protein Model

Table 1.3 Summary of side-chain interaction energies from literature.

Interaction	$\Delta\Delta G$ (kcal mol^{-1})	Reference
Ile – Lys $(i,i+4)$	−0.22	142
Val – Lys $(i,i+4)$	−0.25	142
Ile – Arg $(i,i+4)$	−0.22	142
Phe – Met $(i,i+4)$	−0.8	104
Met – Phe $(i,i+4)$	−0.5	104
Gln – Asn $(i,i+4)$	−0.5	137
Asn – Gln $(i,i+4)$	−0.1	137
Phe – Lys $(i,i+4)$	−0.14	74
Lys – Phe $(i,i+4)$	−0.10	74
Phe – Arg $(i,i+4)$	−0.18	74
Phe – Orn $(i,i+4)$	−0.4	141
Arg – Phe $(i,i+4)$	−0.1	74
Tyr – Lys $(i,i+4)$	−0.22	74
Glu – Phe $(i,i+4)$	−0.5	147
Asp – Lys $(i,i+3)$	−0.12	167
Asp – Lys $(i,i+4)$	−0.24	167
Asp – His $(i,i+3)$	> −0.63	173
Asp – His $(i,i+4)$	> −0.63	173
Asp – Arg $(i,i+3)$	−0.8	174
Glu – His $(i,i+3)$	−0.23	167
Glu – His $(i,i+4)$	−0.10	167
Glu – Lys $(i,i+3)$	−0.38	109
Glu – Lys $(i,i+4)$	−0.44	109
Phe – His $(i,i+4)$	−1.27	135
Phe – Met $(i,i+4)$	−0.7	140
His – Asp $(i,i+3)$	−0.53	135
His – Asp $(i,i+4)$	−2.38	175
His – Glu $(i,i+3)$	−0.45	167
His – Glu $(i,i+4)$	−0.54	167
Lys – Asp $(i,i+3)$	−0.4	167
Lys – Asp $(i,i+4)$	−0.58	167
Lys – Glu $(i,i+3)$	−0.38	167
Lys – Glu $(i,i+4)$	−0.46	167
Lys – Lys $(i,i+4)$	+0.17	133
Leu – Tyr $(i,i+3)$	−0.44	107
Leu – Tyr $(i,i+4)$	−0.65	107
Met – Phe $(i,i+4)$	−0.37	140
Gln – Asp $(i,i+4)$	−0.97	136
Gln – Glu $(i,i+4)$	−0.31	109
Trp – Arg $(i,i+4)$	−0.4	147
Trp – His $(i,i+4)$	−0.8	112
Tyr – Leu $(i,i+3)$	−0.02	107
Tyr – Leu $(i,i+4)$	−0.44	107
Tyr – Val $(i,i+3)$	−0.13	107
Tyr – Val $(i,i+4)$	−0.31	107
Arg $(i,i+4)$ Glu $(i,i+4)$ Arg	−1.5	176
Arg $(i,i+3)$ Glu $(i,i+3)$ Arg	−1.0	176
Arg $(i,i+3)$ Glu $(i,i+4)$ Arg	−0.3	176
Arg $(i,i+4)$ Glu $(i,i+3)$ Arg	−0.1	176
Phosphoserine – Arg $(i,i+4)$	−0.45	128

1.5.5 Covalent Side-chain Interactions

Lactam (amide) bonds formed between NH_3^+ and CO_2^- side chains can stabilize a helix, acting in a similar way to disulfide bridges in a protein by constraining the side chains to be close, reducing the entropy of non-helical states.[148] Lactam bridges between Lys-Asp, Lys-Glu and Glu-Orn spaced $i, i + 4$ have been introduced into analogues of human growth hormone releasing factor[149] and proved to be stabilizing. The same Lys-Asp $i, i + 4$ lactam was stabilizing in other helical peptide systems,[150–153] while Lys-Glu $i, i + 4$ lactam bridges were less effective.[151] Two overlapping Lys-Asp lactams were even more effective.[154] The effect of the ring size formed by the lactam was investigated by replacing Lys with ornithine or (S)-diaminopropionic acid. A ring size of 21 or 22 atoms was most stabilizing (a Lys-Asp $i, i + 4$ lactam is 20 atoms).[149] Lactams between side chains spaced $i, i + 7$[155,156] or $i, i + 3$,[156,157] spanning two or one turns of the helix, have also been reported. Disulfide bonds spaced $i, i + 7$ have been introduced into alanine-based peptides, using (D)- and (L)-2-amino-6-mercaptohexanoic acid derivatives.[158]

Helix formation can be reversibly photoregulated. Two cysteine residues are cross-linked by an azobenzene derivative which can be photoisomerized from *trans* to *cis*, causing a large increase or decrease in the helix content of the peptide, depending on its spacing.[159–161]

1.5.6 Capping Motifs

Although the N-terminal capping box sequence stabilizes helices by inhibiting N-terminal fraying, it does not necessarily promote elongation unless accompanied by favorable hydrophobic interactions as in a "hydrophobic staple" motif.[162,163] The nature of the capping box stabilizing effect thus not only arises from reciprocal hydrogen bonds between compatible residues, but also from local interactions between side chains, helix macrodipole-charged residue interactions, and solvation.[164] Despite statistical analyses revealing that Schellman motifs are observed more frequently than expected at the helix C-terminus, this motif populates only transiently in aqueous solution, but it is formed in 30% TFE.[165] This might be due to the C-terminus being very frayed and so the increase of helical content contributed from this motif is small. The α_L motif seems to be more stable than the alternative Schellman motif.[163]

1.5.7 Ionic Strength

Electrostatic interactions between charged side chains and the helix macrodipole can stabilize the helix.[61,70,166] The interactions are alleviated by the screening effects of water, ions, and nearby protein atoms. The energetics of the interaction between fully charged ion pairs can be diminished by added salt and completely screened at 2.5 M NaCl.[134,167] Interactions of charged residues with the helix macrodipole are less affected by salt than those between charged side chains.[167,168]

1.5.8 Temperature

Thermal unfolding experiments show that the helix unfolds with increasing temperature.[169–171] There is no sign of cold denaturation, as seen with proteins. Enthalpy and entropy changes for the helix–coil transition are difficult to determine as the helix–coil transition is very broad, precluding accurate determination of high- and low-temperature baselines by calorimetry.[169] Nevertheless, isothermal titration calorimetric studies of a series of peptides that form helices when binding a nucleating La^{3+} find ΔH for helix formation to be $-1.0\,\text{kcal}\,\text{mol}^{-1}$,[27,29] in good agreement with the earlier work. This system has been used to measure the enthalpic preferences of all the amino acids.[30]

References

1. D. J. Barlow and J. M. Thornton, *J. Mol. Biol.*, 1988, **201**, 601.
2. J. M. Scholtz and R. L. Baldwin, *Ann. Rev. Biophys. Biomol. Struct.*, 1992, **21**, 95.
3. R. L. Baldwin, *Biophys. Chem.*, 1995, **55**, 127.
4. A. Chakrabartty and R. L. Baldwin, *Adv. Protein Chem.*, 1995, **46**, 141.
5. N. R. Kallenbach, P. Lyu and H. Zhou, in *Circular Dichroism and the Conformational Analysis of Biomolecules*, ed. G. D. Fasman, Plenum Press, New York, 1996, 201.
6. C. A. Rohl and R. L. Baldwin, *Meth. Enzymol.*, 1998, **295**, 1.
7. M. J. I. Andrews and A. B. Tabor, *Tetrahedron*, 1999, **55**, 11711.
8. L. Serrano, *Adv. Prot. Chem.*, 2000, **53**, 49.
9. A. J. Doig, N. Errington and T. Iqbalsyah, in *Stability and Design of α-Helices*, ed. T. Kiefhaber and J. Buchner, Weinheim, 2005.
10. J. J. Osterhout, *Prot. Pept. Lett.*, 1995, **12**, 159.
11. J. S. Richardson and D. C. Richardson, *Science*, 1988, **240**, 1648.
12. L. G. Presta and G. D. Rose, *Science*, 1988, **240**, 1632.
13. L. Serrano and A. R. Fersht, *Nature*, 1989, **342**, 296.
14. J. A. Bell, W. J. Becktel, U. Sauer, W. A. Baase and B. W. Matthews, *Biochemistry*, 1992, **31**, 3590.
15. A. Chakrabartty, A. J. Doig and R. L. Baldwin, *Proc. Natl. Acad. Sci. USA*, 1993, **90**, 11332.
16. B. Forood, E. J. Feliciano and K. P. Nambiar, *Proc. Natl. Acad. Sci. USA*, 1993, **90**, 838.
17. S. Dasgupta and J. A. Bell, *Int. J. Pept. Prot. Res.*, 1993, **41**, 499.
18. E. T. Harper and G. D. Rose, *Biochemistry*, 1993, **32**, 7605.
19. J. W. Seale, R. Srinivasan and G. D. Rose, *Prot. Sci.*, 1994, **3**, 1741.

20. V. Muñoz and L. Serrano, *Nat. Struct. Biol.*, 1995, **2**, 380.
21. A. R. Viguera and L. Serrano, *Prot. Sci.*, 1999, **8**, 1733.
22. C. Schellman, *Protein Folding*, 1980, 53.
23. R. Aurora, R. Srinivasan and G. D. Rose, *Science*, 1994, **264**, 1126.
24. R. Aurora and G. D. Rose, *Prot. Sci.*, 1998, **7**, 21.
25. A. J. Doig, M. W. MacArthur, B. J. Stapley and J. M. Thornton, *Prot. Sci.*, 1997, **6**, 147.
26. S. Penel, E. Hughes and A. J. Doig, *J. Mol. Biol.*, 1999, **287**, 127.
27. M. Siedlecka, G. Goch, A. Ejchart, H. Sticht and A. Bierzynski, *Proc. Natl. Acad. Sci. USA*, 1999, **96**, 903.
28. M. M. Lopez, D. H. Chin, R. L. Baldwin and G. I. Makhatadze, *Proc. Natl. Acad. Sci. USA*, 2002, **99**, 1298.
29. G. Goch, M. Maciejczyk, M. Oleszczuk, D. Stachowiak, J. Malicka and A. Bierzynski, *Biochemistry*, 2003, **42**, 6840.
30. J. M. Richardson, M. M. Lopez and G. I. Makhatadze, *Proc. Natl. Acad. Sci. USA*, 2005, **102**, 1413.
31. M. R. Ghadiri and C. Choi, *J. Am. Chem. Soc.*, 1990, **112**, 1630.
32. K. J. Kise and B. E. Bowler, *Biochemistry*, 2002, **41**, 15826.
33. F. Ruan, Y. Chen and P. B. Hopkins, *J. Am. Chem. Soc.*, 1990, **112**, 9403.
34. D. J. Cline, C. Thorpe and J. P. Schneider, *J. Am. Chem. Soc.*, 2003, **125**, 2923.
35. R. L. Beyer, H. N. Hoang, T. G. Appleton and D. P. Fairlie, *J. Am. Chem. Soc.*, 2004, **126**, 15096.
36. M. E. Karpen, P. L. De Haset and K. E. Neet, *Prot. Sci.*, 1992, **1**, 1333.
37. E. N. Baker and R. E. Hubbard, *Prog. Biophys. Mol. Biol.*, 1984, **44**, 97.
38. G. Némethy, D. C. Phillips, S. J. Leach and H. A. Scheraga, *Nature*, 1967, **214**, 363.
39. G. L. Millhauser, *Biochemistry*, 1995, **34**, 3872.
40. K. A. Bolin and G. L. Millhauser, *Acc. Chem. Res.*, 1999, **32**, 1027.
41. G. N. Ramachandran and V. Sasisekharan, *Adv. Prot. Chem.*, 1968, **23**, 283.
42. B. W. Low and H. J. Grenville-Wells, *Proc. Natl. Acad. Sci. USA*, 1953, **39**, 785.
43. C. A. Rohl and A. J. Doig, *Prot. Sci.*, 1996, **5**, 1687.
44. W. A. Shirley and C. L. Brooks, *Proteins: Struct. Funct. Genet.*, 1997, **28**, 59.
45. K. H. Lee, D. R. Benson and K. Kuczera, *Biochemistry*, 2000, **39**, 13737.
46. T. M. Weaver, *Prot. Sci.*, 2000, **9**, 201.
47. D. M. Morgan, D. G. Lynn, H. Miller-Auer and S. C. Meredith, *Biochemistry*, 2001, **40**, 14020.
48. M. N. Fodje and S. Al-Karadaghi, *Prot. Eng.*, 2002, **15**, 353.
49. M. L. Tiffany and S. Krimm, *Biopolymers*, 1968, **6**, 1379.
50. E. J. Spek, Y. Gong and N. R. Kallenbach, *J. Am. Chem. Soc.*, 1995, **117**, 10773.
51. J. E. Brown and W. A. Klee, *Biochemistry*, 1971, **10**, 470.

52. S. Marqusee and R. L. Baldwin, *Proc. Natl. Acad. Sci. USA*, 1987, **84**, 8898.
53. S. H. Park, W. Shalongo and E. Stellwagen, *Biochemistry*, 1993, **32**, 7048.
54. S. Marqusee, V. H. Robbins and R. L. Baldwin, *Proc. Natl. Acad. Sci. USA*, 1989, **86**, 5286.
55. P. C. Lyu, L. A. Marky and N. R. Kallenbach, *J. Am. Chem. Soc.*, 1989, **111**, 2733.
56. P. C. Lyu, M. I. Liff, L. A. Marky and N. R. Kallenbach, *Science*, 1990, **250**, 669.
57. P. J. Gans, P. C. Lyu, P. C. Manning, R. W. Woody and N. R. Kallenbach, *Biopolymers*, 1991, **31**, 1605.
58. S. Penel, R. G. Morrison, R. J. Mortishire-Smith and A. J. Doig, *J. Mol. Biol.*, 1999, **293**, 1211.
59. A. Wada, *Adv. Biophys.*, 1976, **9**, 1.
60. W. G. J. Hol, P. T. van Duijnen and H. J. C. Berendsen, *Nature*, 1978, **273**, 443.
61. J. Aqvist, H. Luecke, F. A. Quiocho and A. Warshel, *Proc. Natl. Acad. Sci. USA*, 1991, **88**, 2026.
62. E. A. Zhukovsky, M. G. Mulkerrin and L. G. Presta, *Biochemistry*, 1994, **33**, 9856.
63. A. J. Doig and R. L. Baldwin, *Prot. Sci.*, 1995, **4**, 1325.
64. D. A. E. Cochran, S. Penel and A. J. Doig, *Prot. Sci.*, 2001, **10**, 463.
65. D. A. E. Cochran and A. J. Doig, *Prot. Sci.*, 2001, **10**, 1305.
66. T. M. Iqbalsyah and A. J. Doig, *Prot. Sci.*, 2004, **13**, 32.
67. S. M. Decatur, *Biopolymers*, 2000, 180.
68. A. J. Doig, A. Chakrabartty, T. M. Klingler and R. L. Baldwin, *Biochemistry*, 1994, **33**, 3396.
69. C. A. Rohl, A. Chakrabartty and R. L. Baldwin, *Prot. Sci.*, 1996, **5**, 2623.
70. B. M. Huyghues-Despointes, J. M. Scholtz and R. L. Baldwin, *Prot. Sci.*, 1993, **2**, 1604.
71. J. R. Brandts and K. J. Kaplan, *Biochemistry*, 1973, **10**, 470.
72. H. Edelhoch, *Biochemistry*, 1967, **6**, 1948.
73. A. Chakrabartty, T. Kortemme, S. Padmanabhan and R. L. Baldwin, *Biochemistry*, 1993, **32**, 5560.
74. C. D. Andrew, S. Bhattacharjee, N. Kokkoni, J. D. Hirst, G. R. Jones and A. J. Doig, *J. Am. Chem. Soc.*, 2002, **124**, 12706.
75. D. S. Kemp, J. G. Boyd and C. C. Muendel, *Nature*, 1991, **352**, 451.
76. D. S. Kemp, T. J. Allen and S. L. Oslick, *J. Am. Chem. Soc.*, 1995, **117**, 6641.
77. K. Groebke, P. Renold, K. Y. Tsang, T. J. Allen, K. F. McClure and D. S. Kemp, *Proc. Natl. Acad. Sci. USA*, 1996, **93**, 4025.
78. D. S. Kemp, S. L. Oslick and T. J. Allen, *J. Am. Chem. Soc.*, 1996, **118**, 4249.
79. D. S. Kemp, T. J. Allen, S. L. Oslick and J. G. Boyd, *J. Am. Chem. Soc.*, 1996, **118**, 4240.

80. R. E. Austin, R. A. Maplestone, A. M. Sefler, K. Liu, W. N. Hruzewicz, C. W. Liu, H. S. Cho, D. E. Wemmer and P. A. Bartlett, *J. Am. Chem. Soc.*, 1997, **119**, 6461.
81. T. Arrhenius and A. C. Sattherthwait, *Peptides: Chemistry, Structure and Biology: Proceedings of the 11th American Peptide Symposium*, 1989, p. 870.
82. K. Muller, D. Obrecht, A. Knierzinger, C. Stankovic, C. Spiegler, W. Bannwarth, A. Trzeciak, G. Englert, A. M. Labhardt and P. Schonholzer, in *Perspectives in Medicinal Chemistry*, John Wiley & Sons, 1993.
83. D. Gani, A. Lewis, T. Rutherford, J. Wilkie, I. Stirling, T. Jenn and M. D. Ryan, *Tetrahedron*, 1998, **54**, 15793.
84. D. Poland and H. A. Scheraga, *Theory of Helix-Coil Transitions in Biopolymers*, Academic Press, New York and London, 1970.
85. H. Qian and J. A. Schellman, *J. Phys. Chem.*, 1992, **96**, 3987.
86. A. J. Doig, *Biophys. Chem.*, 2002, **101–102**, 281.
87. B. H. Zimm and J. K. Bragg, *J. Chem. Phys.*, 1959, **31**, 526.
88. S. Lifson and A. Roig, *J. Chem. Phys.*, 1961, **34**, 1963.
89. J. K. Sun and A. J. Doig, *Prot. Sci.*, 1998, **7**, 2374.
90. F. B. Sheinerman and C. L. Brooks, *J. Am. Chem. Soc.*, 1995, **117**, 10098.
91. P. J. Flory, *Statistical Mechanics of Chain Molecules*, Wiley, New York, 1969.
92. K. Chen, Z. G. Liu, C. H. Zhou, Z. S. Shi and N. R. Kallenbach, *J. Am. Chem. Soc.*, 2005, **127**, 10146.
93. A. K. Jha, A. Colubri, M. H. Zaman, S. Koide, T. R. Sosnick and K. F. Freed, *Biochemistry*, 2005, **44**, 9691.
94. A. Rath, A. R. Davidson and C. M. Deber, *Biopolymers*, 2005, **80**, 179.
95. E. W. Blanch, L. A. Morozova-Roche, D. A. E. Cochran, A. J. Doig, L. Hecht and L. D. Barron, *J. Mol. Biol.*, 2000, **301**, 553.
96. R. V. Pappu and G. D. Rose, *Prot. Sci.*, 2002, **11**, 2437.
97. Z. Shi, C. A. Olson, G. D. Rose, R. L. Baldwin and N. R. Kallenbach, *Proc. Natl. Acad. Sci. USA*, 2002, **99**, 9190.
98. K. Chen, Z. G. Liu and N. R. Kallenbach, *Proc. Natl. Acad. Sci USA*, 2004, **101**, 15352.
99. S. A. Asher, A. V. Mikhonin and S. Bykov, *J. Am. Chem. Soc.*, 2004, **126**, 8433.
100. A. Kentsis, M. Mezei, T. Gindin and R. Osman, *Proteins: Struc. Funct. Bioinf.*, 2004, **55**, 493.
101. M. Mezei, P. J. Fleming, R. Srinivasan and G. D. Rose, *Proteins: Struc. Funct. Bioinf.*, 2004, **55**, 502.
102. A. L. Rucker, C. T. Pager, M. N. Campbell, J. E. Qualis and T. P. Creamer, *Proteins: Struct. Funct. Genet.*, 2003, **53**, 68.
103. N. H. Andersen and H. Tong, *Prot. Sci.*, 1997, **6**, 1920.
104. B. J. Stapley, C. A. Rohl and A. J. Doig, *Prot. Sci.*, 1995, **4**, 2383.
105. P. Argos and J. Palau, *Int. J. Pept. Prot. Res.*, 1982, **19**, 380.
106. S. Kumar and M. Bansal, *Proteins*, 1998, **31**, 460.
107. W. Shalongo and E. Stellwagen, *Prot. Sci.*, 1995, **4**, 1161.

108. J. K. Sun, S. Penel and A. J. Doig, *Prot. Sci.*, 2000, **9**, 750.
109. J. M. Scholtz, H. Qian, V. H. Robbins and R. L. Baldwin, *Biochemistry*, 1993, **32**, 9668.
110. V. Muñoz and L. Serrano, *Nat. Struct. Biol.*, 1994, **1**, 399.
111. V. Muñoz and L. Serrano, *Biopolymers*, 1997, **41**, 495.
112. J. Fernández-Recio, A. Vásquez, C. Civera, P. Sevilla and J. Sancho, *J. Mol. Biol.*, 1997, **267**, 184.
113. E. Lacroix, A. R. Viguera and L. Serrano, *J. Mol. Biol.*, 1998, **284**, 173.
114. V. Muñoz and L. Serrano, *J. Mol. Biol.*, 1995, **245**, 275.
115. M. Petukhov, V. Muñoz, N. Yumoto, S. Yoshikawa and L. Serrano, *J. Mol. Biol.*, 1998, **278**, 279.
116. A. L. Lomize and H. I. Mosberg, *Biopolymers*, 1997, **42**, 239.
117. J. Wojcik, K. H. Altmann and H. A. Scheraga, *Biopolymers*, 1990, **30**, 121.
118. S. Padmanabhan, E. J. York, L. Gera, J. M. Stewart and R. L. Baldwin, *Biochemistry*, 1994, **33**, 8604.
119. C. N. Pace and J. M. Scholtz, *Biophys. J.*, 1998, **75**, 422.
120. C. A. Rohl, W. Fiori and R. L. Baldwin, *Proc. Natl. Acad. Sci. USA*, 1999, **96**, 3682.
121. R. J. Kennedy, K.-Y. Tsang and D. S. Kemp, *J. Am. Chem. Soc.*, 2002, **124**, 934.
122. P. C. Lyu, H. X. X. Zhou, N. Jelveh, D. E. Wemmer and N. R. Kallenbach, *J. Am. Chem. Soc.*, 1992, **114**, 6560.
123. M. Petukhov, K. Uegaki, N. Yumoto, S. Yoshikawa and L. Serrano, *Prot. Sci.*, 1999, **8**, 2144.
124. H. X. X. Zhou, P. C. C. Lyu, D. E. Wemmer and N. R. Kallenbach, *J. Am. Chem. Soc.*, 1994, **116**, 1139.
125. M. Petukhov, K. Uegaki, N. Yumoto and L. Serrano, *Prot. Sci.*, 2002, **11**, 766.
126. D. N. Ermolenko, J. M. Richardson and G. I. Makhatadze, *Prot. Sci.*, 2003, **12**, 1169.
127. L. Szalik, J. Moitra, D. Krylov and C. Vinson, *Nat. Struct. Biol.*, 1997, **4**, 112.
128. S. Liehr and H. K. Chenault, *Bioorg. Med. Chem. Lett.*, 1999, **9**, 2759.
129. C. D. Andrew, J. Warwicker, G. R. Jones and A. J. Doig, *Biochemistry*, 2002, **41**, 1897.
130. N. Errington and A. J. Doig, *Biochemistry*, 2005, **44**, 7553.
131. A. Horovitz, L. Serrano, B. Avron, M. Bycroft and A. R. Fersht, *J. Mol. Biol.*, 1990, **216**, 1031.
132. G. Merutka and E. Stellwagen, *Biochemistry*, 1991, **30**, 1591.
133. E. Stellwagen, S.-H. Park, W. Shalongo and A. Jain, *Biopolymers*, 1992, **32**, 1193.
134. B. M. Huyghues-Despointes, J. M. Scholtz and R. L. Baldwin, *Prot. Sci.*, 1993, **2**, 80.
135. B. M. Huyghues-Despointes and R. L. Baldwin, *Biochemistry*, 1997, **36**, 1965.

136. B. M. Huyghues-Despointes, T. M. Klingler and R. L. Baldwin, *Biochemistry*, 1995, **34**, 13267.
137. B. J. Stapley and A. J. Doig, *J. Mol. Biol.*, 1997, **272**, 465.
138. S. Padmanabhan and R. L. Baldwin, *Prot. Sci.*, 1994, **3**, 1992.
139. S. Padmanabhan and R. L. Baldwin, *J. Mol. Biol.*, 1994, **241**, 706.
140. A. R. Viguera and L. Serrano, *Biochemistry*, 1995, **34**, 8771.
141. L. K. Tsou, C. D. Tatko and M. L. Waters, *J. Am. Chem. Soc.*, 2002, **124**, 14917.
142. C. D. Andrew, S. Penel, G. R. Jones and A. J. Doig, *Proteins: Struct. Funct. Genet.*, 2001, **45**, 449.
143. L. Mayne, S. W. Englander, R. Qiu, J. X. Yang, Y. X. Gong, E. J. Spek and N. R. Kallenbach, *J. Am. Chem. Soc.*, 1998, **120**, 10643.
144. C. A. Olson, E. J. Spek, Z. S. Shi, A. Vologodskii and N. R. Kallenbach, *Proteins: Struc. Funct. Bioinf.*, 2001, **44**, 123.
145. T. I. Iqbalsyah and A. J. Doig, *J. Am. Chem. Soc.*, 2005, **127**, 5002.
146. Z. Shi, C. A. Olson and N. R. Kallenbach, *J. Am. Chem. Soc.*, 2002, **124**, 3284.
147. Z. Shi, C. A. Olson, A. J. Bell and N. R. Kallenbach, *Biophys. Chem.*, 2002, **101–102**, 267.
148. J. W. Taylor, *Biopolymers*, 2002, **66**, 49.
149. R. M. Campbell, J. Bongers and A. M. Felxi, *Biopolymers*, 1995, **37**, 67.
150. M. Chorev, E. Roubini, R. L. McKee, S. W. Gibbons, M. E. Goldman, M. P. Caulfield and M. Rosenblatt, *Biochemistry*, 1991, **30**, 5968.
151. G. Osapay and J. W. Taylor, *J. Am. Chem. Soc.*, 1990, **112**, 6046.
152. M. Bouvier and J. W. Taylor, *J. Med. Chem.*, 1992, **35**, 1145.
153. A. Kapurniotu and J. W. Taylor, *J. Med. Chem.*, 1995, **38**, 836.
154. C. Bracken, J. Gulyas, J. W. Taylor and J. Baum, *J. Am. Chem. Soc.*, 1994, **116**, 6431.
155. S. T. Chen, H. J. Chen, H. M. Yu and K. T. Wang, *J. Chem. Res. (S)*, 1993, **6**, 228.
156. W. T. Zhang and J. W. Taylor, *Tet. Lett.*, 1996, **37**, 2173.
157. P. Z. Luo, D. T. Braddock, R. M. Subramanian, S. C. Meredith and D. G. Lynn, *Biochemistry*, 1994, **33**, 12367.
158. D. Y. Jackson, D. S. King, J. Chmielewski, S. Singh and P. G. Schultz, *J. Am. Chem. Soc.*, 1991, **113**, 9391.
159. J. R. Kumita, O. S. Smart and G. A. Woolley, *Proc. Natl. Acad. Sci. USA*, 2000, **97**, 3803.
160. D. G. Flint, J. R. Kumita, O. S. Smart and G. A. Wolley, *Chem. Biol.*, 2002, **9**, 391.
161. J. R. Kumita, D. G. Flint, O. S. Smart and G. A. Woolley, *Prot. Eng.*, 2002, **15**, 561.
162. M. A. Jiménez, V. Muñoz, M. Rico and L. Serrano, *J. Mol. Biol.*, 1994, **242**, 487.
163. N. R. Kallenbach and Y. X. Gong, *Bioorg. Med. Chem.*, 1999, **7**, 143.
164. M. Petukhov, N. Yumoto, S. Murase, R. Onmura and S. Yoshikawa, *Biochemistry*, 1996, **35**, 387.

165. A. R. Viguera and L. Serrano, *J. Mol. Biol.*, 1995, **251**, 150.
166. K. M. Armstrong and R. L. Baldwin, *Proc. Natl. Acad. Sci. USA*, 1993, **90**, 11337.
167. J. S. Smith and J. M. Scholtz, *Biochemistry*, 1998, **37**, 33.
168. D. J. Lockhart and P. S. Kim, *Science*, 1993, **260**, 198.
169. J. M. Scholtz, S. Marqusee, R. L. Baldwin, E. J. York, J. M. Stewart, M. Santoro and D. W. Bolen, *Proc. Natl. Acad. Sci. USA*, 1991, **88**, 2854.
170. G. Yoder, P. Pancoska and T. A. Keiderling, *Biochemistry*, 1997, **36**, 5123.
171. C. Y. Huang, J. W. Klemke, Z. Getahun, W. F. DeGrado and F. Gai, *J. Am. Chem. Soc.*, 2001, **123**, 9235.
172. S. T. Thomas, V. V. Loladze and G. I. Makhatadze, *Proc. Natl. Acad. Sci. USA*, 2001, **98**, 10670.
173. R. Luo, L. David, H. Hung, J. Devaney and M. K. Gilson, *J. Phys. Chem. B*, 1999, **103**, 366.
174. S. Marqusee and R. T. Sauer, *Prot. Sci.*, 1994, **3**, 2217.
175. P. Luo and R. L. Baldwin, *Proc. Natl. Acad. Sci. USA*, 1999, **96**, 4930.
176. Z. Shi, C. A. Olson, A. J. Bell and N. R. Kallenbach, *Biopolymers*, 2001, **60**, 366.

CHAPTER 2
Kinetics and Mechanisms of α-Helix Formation

URMI DOSHI

Department of Chemistry and Biochemistry, and Center for Biomolecular Structure and Organization, University of Maryland, College Park, MD 20742, USA

2.1 Introduction

How an amino acid sequence guides the formation of a unique three-dimensional structure defines the most intriguing problem of protein folding. Concomitant development of secondary and tertiary structure makes it difficult to determine the mechanisms and the timescales of individual events. Hence, it is simpler from both conceptual and experimental points of view to investigate these events in isolation; that is outside the context of the protein. Understanding the factors governing the formation and stabilization of secondary structural elements, which is simpler, allows us to determine the dynamic processes underlying protein folding. Among secondary structures, helical motifs, in particular, are adopted not only by biological macromolecules such as nucleic acids and proteins but also by non-biological synthetic polymers. Due to this fundamental nature of helices, their formation has been a subject of inter-disciplinary interest with a long-standing history.

Since the early 1950s several statistical mechanical theories were developed to explain the helix–coil transition observed in very long homopolypeptides.[1] Although these theories differed in their approach to evaluating the partition function they had the same basic form, *i.e.* describing helix formation as a nucleation–elongation process. Nucleation involves the formation of the first turn of the helix in which the dihedral angles of four consecutive residues need to be fixed in helical states. The nucleation step is difficult as the loss in conformational entropy is more compared to the gain in enthalpy on formation of backbone interactions (the only backbone interactions originally considered were the hydrogen bonds between the *ith* carbonyl oxygen and the $i + 4th$ amide hydrogen in the case of α-helix). Fixing an additional residue results in the growth of the existing helix, which is much easier than helix initiation and can occur on either end. The equilibrium constants for nucleation and elongation are described by the parameters of the widely accepted Zimm–Bragg and Lifson–Roig theoretical treatments.[2,3] Using host–guest methods the nucleation and elongation parameters, which are the measures of the intrinsic preference for helical conformation, were determined experimentally for individual amino acids.[4] These extensive studies prompted further investigation of the various factors that govern α-helix stability and modification of the helix–coil theory to incorporate the effects applicable to heteropolymers.[5] Thorough thermodynamic characterization for the past fifty years has resulted in the accumulation of the various energetic contributions arising from the backbone as well as the side chains, *viz.* interactions of the charged groups with the helix macro-dipole, electrostatic and hydrophobic interactions between side chains and interactions responsible for stabilizing the N- and the C-caps of the helices.[6,7] Compilation of this knowledge into a database, when implemented in an algorithm (AGADIR) based on the nucleation–elongation model, has permitted successful prediction of the helical content of peptides from their amino acid sequences.[8–11] This notable ability to predict helical behavior of peptides at a given temperature, pH and solvent composition serves as a gauge of our understanding of the thermodynamics of α-helix formation. More details on helical structure, stability, and design are covered in Chapter 1 by Doig and in the references therein.

α-Helix formation represents a classic prototype of conformational transition in polymers. Usually such processes are fast, ranging from nanoseconds to milliseconds, thus making their measurements difficult. The common experimental approach to measure the dynamics of such fast events is to induce a shift in the equilibrium by a rapid perturbation and to monitor the ensuing relaxation to the new equilibrium. Analyses of the resulting kinetics are not straightforward due to the complexity involved in the measurements and in providing a theoretical description to explain them quantitatively. The $i, i + 4$ hydrogen bonding pattern in an α-helix implies that the conformation of each residue is affected by at most $i, i + 4$ (and/or $i, i - 4$) nearest neighbors. This renders quasi-one-dimensionality to the α-helix. Unlike proteins where different parts of the chain that are far apart in sequence are brought together, the α-helix is primarily stabilized by local interactions. Hence, one would expect formation of an α-helix to be faster than folding of a protein. But the one-dimensional networking in the α-helix may also be responsible for dampening its rate of formation when compared to proteins that are expected to have myriads of folding paths.

If the helix–coil transition were a first-order phase transition as implied by its name, a single relaxation between all-coil and all-helix states should be expected as observed in the two-state behavior of many small, single-domain proteins.

In fact, several microstates having more than one helix segment of varying lengths are also populated, indicating that α-helix formation is far from being an all-or-none transition. These microstates, in principle, may inter-convert with each other leading to a distribution of relaxation times. Nonetheless, these relaxation times may be clustered together into one or several separate timescales manifesting into single- or multi-phasic kinetics respectively. This continuum of relaxation times could not be easily determined theoretically or related to the global relaxation observed experimentally forty years back when the earliest studies on helix–coil kinetics were carried out. Due to the limited computational power of the time, finding the solution of several linear differential equations was considered very cumbersome. Hence, attempts were made to describe the problem in terms of more accessible quantities such as the mean relaxation times that could be obtained from experiments. This required treating the helix–coil transformation as a chemical relaxation process with a single relaxation time that could be obtained from the slope of the relaxation curve at time $t=0$.[12,13]

The seminal theoretical treatment by Schwarz[12] gave the relation between the fundamental parameters of the helix–coil theory, *i.e.* the Zimm–Bragg nucleation parameter σ and elongation parameter s, the rate constant for helix propagation k_f and the mean relaxation time τ^*. $\tau^* = 1/(4\sigma + (s-1)^2 k_f)$, where s, the elongation parameter, represents the degree of transition. At the transition mid-point $s \sim 1$ and τ^* is maximum and equal to $(4\sigma k_f)^{-1}$. From the earlier experimental findings[14,15] τ^* was reported to be ~ 1 microsecond and k_f was estimated to be on the order of $\sim 10^8 \, s^{-1}$. However, the analysis of these experiments was complicated, and the applicability of their results to protein folding was debatable due to the use of very long homopolypeptides that resembled protein helices neither in length nor in sequence. Hence, as compared to its thermodynamics, the kinetic aspects of the helix–coil transition remained mostly uncharacterized for decades.

It is only in the last decade that helix–coil kinetics has received renewed interest mainly due to developments in ultra-fast kinetic techniques and availability of short alanine-based peptides exhibiting considerable helical content in solution.[16,17] This new generation of kinetic studies suggested that folding of the α-helix is a fast event occurring on the sub-microsecond timescale.[17–20] Increased temporal resolution in relaxation experiments using more complex protein-like sequences revealed rich kinetic behavior.[21] The apparent relaxation times were found to depend on the magnitude of perturbation, *i.e.* size of the T-jump, and also on the specific regions of the peptides.[22] It was shown that nucleation–elongation theory, which is well established in describing the thermodynamic behavior of helix–coil transition (see Chapter 1 by Doig), is equally adequate in explaining the observed kinetics.[23] Besides, limitations in understanding the mechanistic details of α-helix formation by dynamic atomistic simulations were alleviated by the tremendous improvement in computational power, force field and sampling methods[24–31] (see Chapter 8 by Pande for more on computer simulations of protein folding). Consequently, direct comparison between experiment and theory became possible. More recently, in an effort to characterize helix–coil kinetics on a quantitative basis, several issues such as the dependence of relaxation times on chain length, sequence, and stability have been explored.[32–34]

This chapter builds on these recent experimental and theoretical developments to discuss what we have learned about helix–coil kinetics so far, and how this understanding is related and can be applied to the protein folding problem.

2.2 Experimental Techniques Employed to Study Helix–Coil Kinetics

Much of the recent progress in studying helix–coil kinetics can be attributed to the advances in modern laser techniques that led to improved T-jump instrumentation. In a general laser-induced T-jump setup, water (D_2O) containing the peptide is heated by a near-infrared nanosecond pulse in a frequency overlapping with the first vibrational overtone of water (D_2O). Water is then excited and relaxes back to the ground state in a non-radiative process that takes place in picoseconds, resulting in dead times limited by the duration of the pump laser pulse (most typically a few nanoseconds). The near infrared pulse is either generated by Raman-shifting the fundamental mode of a Nd:YAG laser or by coupling the Nd:YAG laser with a frequency difference mixing module or optical parametric oscillator. Faster T-jumps have also been produced by excitation of certain dyes with visible lasers, which then transfer the heat to the surrounding water molecules.[17,18,21] The near-infrared pulses are ideal because peptides and proteins do not absorb in the 1.5–2 µm range. Laser-induced T-jump techniques have also been instrumental for the development of fast-folding experiments (see Chapter 6 by Gruebele). The change in temperature is calibrated by monitoring the changes in the infrared absorption of water with temperature or the changes in fluorescence emission of a dye with known temperature-dependent quantum yield.

The subsequent relaxation of the peptide is followed by different time-resolved optical spectroscopic techniques such as:

1) *Fluorescence.* Helical content is indicated by the changes in or the quenching of fluorescence of a probe that interacts with the peptide backbone or side chain in an α-helical conformation.[18,19]
2) *Infrared (IR) spectroscopy.* The vibrational spectra of peptides are sensitive to their three-dimensional structures. The amide I band in the IR spectra arises predominantly from the stretching vibrations of the $C=O$ bond of the peptide backbone. From the modifications in the amide I IR absorbance and frequency shifts any changes in the secondary structure can be determined. Infrared spectroscopy when coupled with isotope editing (^{13}C) techniques provides an opportunity to obtain site-specific information. When backbone carbonyls are labeled with ^{13}C, the amide I band of ^{13}C residues is shifted by $\sim 40\,\text{cm}^{-1}$ from the amide I band of ^{12}C residues.[22,35,36]

3) *Ultraviolet resonance Raman spectroscopy (UVRS)*. When the sample is excited at a wavelength coinciding with a particular electronic absorption band (UV) and if the probing laser frequency resonates with the electronic excited state, certain amide vibrational modes are enhanced selectively in the Raman spectrum without any interference from water. The variation in the intensities at frequencies at which the enhanced modes are observed and the downshifts in frequencies provide a direct reporter of secondary structure. In addition, UVRS can also examine the exposure of aromatic residues to solvent.[20,37]

4) *Circular Dichroism (CD)*. CD measures the difference in the absorption of the right- and left-circularly polarized light. In the far-UV spectral region the peptide bond chromophore gives rise to a characteristic signal for each secondary structure. CD spectroscopy is widely used to determine the equilibrium α-helical content and can be coupled with rapid mixing methods such as stopped flow in which helix formation is initiated by dilution of peptides from a high denaturant concentration.[38] It is, however, not suitable for faster kinetic experiments (*i.e.* microsecond and sub-microsecond timescale).

5) *Vibrational Circular Dichroism (VCD)*. VCD in the amide I region probes for the changes in secondary structure by combining the features of both CD and IR. It results in variations in the band shapes as well as frequency shifts and potentially picosecond time-resolution. This provides an opportunity to monitor the relaxation occurring in the selective isotopically labeled regions of peptides.[39,40]

6) ^{15}N *NMR relaxation experiments* are also used to study conformational processes in peptides. Motions resulting from the global folding/unfolding contribute to the transverse relaxation rate that arises predominantly from the chemical exchange process.[41] Also NMR techniques are limited to processes slower than 100 microseconds.

The kinetic experiments in α-helical peptides are thus typical perturbation experiments (see Chapter 6 by Gruebele), in which the sample is usually subjected to a sudden increase in temperature such that the equilibrium shifts to the one at higher temperature. At the higher temperature the peptide relaxes to a new conformational ensemble corresponding to a decreased number and/or length of helical segments. The relaxation time to the new equilibrium is obtained from fits to the experimental time-dependent spectroscopic signal decay (Figure 2.1).

Helix–coil kinetics is expected to exhibit bi-exponential behavior due to the separation in timescales between nucleation and elongation (see below). In phenomenological fitting of a signal decay, such bi-exponential behavior can only be resolved when the fast and the slow phases are well separated in time. However, due to the heterogeneous nucleation and elongation rates based on the peptide sequence the two phases could overlap giving rise to an apparent stretched exponential relaxation. At some instances the fast phase, which could occur in a few nanoseconds, may not be resolved due to technical limitations resulting in an apparently mono-exponential behavior. Furthermore,

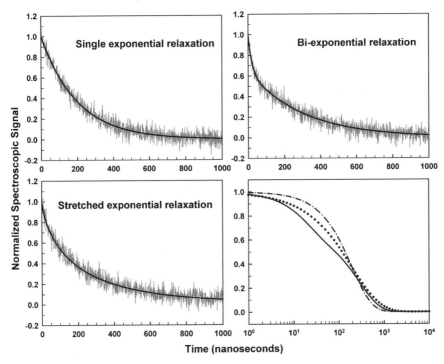

Figure 2.1 Simulated decays of spectroscopic signal showing different behaviors observed in kinetic experiments of α-helical peptides. Depending on the amino acid sequence of the peptide and the temporal resolution of the instrument, the spectroscopic signal decay following a T-jump can be fitted to a mono-exponential (top-left panel), bi-exponential (top-right panel), or sometimes even multi-exponential function (not shown). Recent kinetic experiments have reported relaxation fitted to a stretched exponential function (bottom left): $A.\exp(-t/\tau)^\beta$, where β is the measure of deviation from single exponential behavior, A is the amplitude and t = time. For clarity the fits to the mono-, bi-, and stretched exponential relaxation traces are shown in logarithmic scale (dashed dotted line: single exponential; solid line: double exponential; dotted line: stretched exponential) in the bottom-right panel.

sometimes in relaxation kinetics probed by IR spectroscopy or fluorescence the potential fast phase appears accompanied with an "instantaneous" component. This component is thought to result from the temperature-induced shift of the equilibrium IR spectra to higher frequency arising from the changes in solvation of the peptide at high temperatures (or the temperature-induced decrease in fluorescence quantum yield) without any contribution from the actual helix unfolding process.[17,22] Interpretation of kinetic data based on such phenomenological fits can lead to uncertainty in determining the mechanisms of α-helix formation. For example, single exponential relaxations are widely considered a signature for a two-state process (see Chapter 6 by Gruebele for more on kinetic decay curves), which leads to an unrealistic description of the process of α-helix formation. Hence, to better understand the origin of the observed relaxation, it is essential to model the kinetics using a theoretical backbone.

2.3 Theoretical Approaches to Explore Helix–Coil Kinetics

Simple analytical models developed essentially using the conceptual framework of nucleation–elongation theory have been used to explore the dynamics of α-helix formation. These models are helpful in examining the general physical features of helix–coil kinetics such as the dependence of relaxation rates on peptide length, temperature and stability. The equilibrium properties of the helix–coil transition are expressed only as a function of nucleation and elongation parameters, σ and s, which are assumed to be sequence independent (see Chapter 1 by Doig for a description of the thermodynamic features of helix/coil nucleation-elongation theory). Parameters for these models are either carefully guessed or obtained from free-energy surfaces generated from equilibrium simulations on short alanine peptides. To simplify the calculations, usually certain simplifying assumptions are made such as allowing helix elongation only from a single nucleus at a given time, and considering a single value of helix propensity for all amino acids. Kinetics is modeled by numerical integration of differential equations or, alternatively, as diffusive dynamics over multiple square potential energy barriers for formation of hydrogen bonds (*i.e.* for each helix propagation step).[42,43] Another approach utilizes a mean field approximation to evaluate mean first passage times.[44] This approach is more suitable for modeling longer peptides as it allows nucleation at multiple sites and bi-directional helix propagation.

Due to their fast formation rates (*i.e.* sub-microsecond) and short lengths, α-helices are suitable candidates for computer simulations using both minimalist and detailed atomistic models. While coarse-grained models are less intensive computationally, all-atom molecular dynamic simulations provide a detailed picture of solvent and side-chain dynamics. Improvement in computational power has permitted extended simulations in explicit solvent reaching up to microseconds and at temperatures comparable to experimentally relevant ones. Although use of replica exchange methods allows for exhaustive sampling of conformational space, it is now possible to obtain complete convergence to equilibrium between fully folded and coiled ensembles even in atomistic simulations with the help of distributed computing[27,30] (see Chapter 8 by Pande). These advances in simulation permit longer (and multiple) trajectories and thereby statistics for the direct comparison of helix-folding timescales with those obtained from experiment. However, one must be careful while employing force fields to simulate α-helix formation, as the overall mechanism may be skewed due to the different heliophilic character of commonly used force fields (*i.e.* force fields that selectively (de)stabilize helical conformations).[45]

Although simple analytical models capture the underlying physics of α-helix formation and all-atom molecular dynamic simulations provide detailed mechanistic information, there is a gap between these theoretical approaches and experimental observations. In order to directly analyze experimental data it is necessary to simulate the spectroscopic signals and take into account the complexity of the amino acid sequence of the peptide used in the study. In the absence of such treatment the free-energy surfaces resulting from analytical models or computer simulations have no straightforward connection with empirical observations.

2.4 General Observations in Helix–Coil Kinetics

Based on previous equilibrium experiments, most of the recent T-jump studies employ short alanine-based peptides ($\sim$15–25 residues) with only a few lysine or arginine substitutions at $i, i + 5$ positions (see Chapter 1 by Doig). For some cases the relaxation observed after temperature perturbation was described with a single exponential having a rate constant of $\sim$1/200 nanosecond^{-1},[19,20] whereas in others the relaxation followed a biphasic trace with the fast and the slow phases having respective lifetimes of $\sim$20 and $\sim$140–220 nanoseconds.[17] This timescale was six orders of magnitude faster than that reported from stopped flow experiments.[38] In order to explain this discrepancy it was argued that T-jump experiments probe only the local perturbations (*i.e.* local winding and unwinding of helices) that are much faster than the global folding/unfolding event that would then occur in $\sim$100 milliseconds detected by stopped flow experiments. However, this argument is disputable because in T-jump experiments the equilibrium amplitudes are reached within a few microseconds at most, indicating that there are no events occurring slower than microsecond timescale.

The relaxation for the Fs-MABA peptide (having the sequence (A)$_5$-(AAARA)$_3$A-NH$_2$ with a fluorophore MABA (4-methylaminobenzoic acid) attached at the N-terminus) was much faster than that obtained by Williams *et al.*[17] using IR spectroscopy for the same Fs peptide (having the same sequence Suc-(A)$_5$-(AAARA)$_3$A-NH$_2$ succinylated at the N-terminus) but without the N-terminal probe.[18] The changes in the fluorescent intensity of MABA measured the N-terminal helical content and corresponded to the relaxation occurring at the N-terminus of the peptide. Hence a much faster relaxation corresponding to the local helix unfolding was observed compared to the one in the IR study, which is expected to measure the average change in the helical content of the peptide. The relaxation time for the Fs-MABA peptide provided the first direct experimental estimate for helix propagation/depropagation rate ($\sim$1/20 nanosecond^{-1}).

One of the strengths of molecular simulations is that they can give an idea of the timescales for the elementary processes of helix initiation and propagation. Although earlier computer simulations were run for 200 picoseconds without reaching equilibrium, the elementary process of each residue transitioning from helical to coil states and *vice-versa* was assumed to be at equilibrium. From residence times in helical and coil states averaged over all the residues and the entire period of simulation it was concluded that propagation and initiation take place at the picosecond timescale.[25,26] Later on, Brooks and co-workers performed equilibrium simulations of short ends-protected polyalanine peptides. Using umbrella sampling free-energy surfaces were calculated as a function of dihedral angles and $i,i+4$ hydrogen bonds for a single propagation step (*i.e.* formation of a hydrogen bond) either at the N- or the C-terminus at 27 °C.[46] From the free-energy barriers of ~ 3 kcal mol^{-1} for each hydrogen-bond formation, helix propagation times were estimated to be ~ 100 picoseconds. The mean folding times were on the same order (~ 20–70 nanoseconds) as those found by T-jump experiments. Similarly simulations on Ala$_5$ and A$_2$-G-A$_2$ peptides blocked at both ends by Hummer *et al.* showed that formation of the first helical turn, *i.e.* helix nucleation, takes place within 0.1–1 nanoseconds.[29,47] Analysis of experimental relaxation kinetics of Ac-WAAAH$^+$-(AAAR$^+$A)$_3$A-NH$_2$ peptide (Ac: acetylated at the N-terminus) using a statistical mechanical model also suggested nucleation to occur in nanoseconds.[19]

Helix–coil theory predicts that for peptides with intermediate degrees of helicity this is concentrated in the central region with ends frayed.[1] In other words, the probability of forming helices in the middle of the peptide is greater than that at the termini (see Chapter 1 by Doig). In support of this prediction recent studies have shown that relaxation rates seem to vary for different regions of the peptide. Peptides having the same sequence (Ac-YGSPEAAA (KAAAA)(KAAAA)-D-Arg-NH$_2$) were ^{13}C labeled at the carbonyls of alanine residues placed at different locations in the molecule: N-terminal, middle, or C-terminus.[22] The ^{13}C-targeted relaxation of the C-terminus-labeled peptide was found to be faster than those of the N-terminus- or the middle-labeled peptides. Unexpectedly, the N-terminus-labeled peptide showed less extent of fraying with apparent relaxation times very close to the middle-labeled peptides. Moreover, the time courses for all three peptides were fitted to stretched exponential functions. Another intriguing and counter-intuitive observation was made for middle-labeled peptides in which the relaxation time decreased when the temperature before the T-jump to the same final temperature was lowered. A T-jump of ~ 14 K resulted in a relaxation ~ 1.5 times faster than a 4 K jump to the same final temperature. It was revealed for the first time that even simple short α-helical peptides could exhibit such complex behaviors.

Analytical models have predicted the dependence of folding times on the lengths and the stability of peptides. Using a sequential kinetic model Brooks found that for short peptides having lengths up to ~ 16-residues helix folding and unfolding times increased linearly with chain length. For lengths greater than 16, helix folding times did not change much whereas the unfolding times decreased dramatically.[25,26,42] Buchete and Straub found that mean first passage times estimated with an analytical model based on a mean field approximation showed an initial sharp increase followed by a much slower one with increasing chain length.[44]

The length dependence of relaxation times was then investigated empirically by Gai and co-workers using peptides with varying numbers of a repeating unit (Ac-YGSPEAAA(KAAAA)$_n$-D-Arg-NH$_2$, where $n = 2$ to 5 repeating units).[32] They showed that the relaxation times at low temperature decreased monotonically up to four repeating units. For greater numbers of repeating units a linear but rather flatter increase of relaxation times was observed. However, relaxations at higher temperatures exhibited a linear and monotonic dependence on chain length. Further, as the stability of the peptide was increased (*i.e.* using stronger N-caps (SPE instead of AKA) as those found in protein helices) the relaxation became faster.[33] On similar lines Gooding *et al.* explored the effects of a single amino acid substitution on the folding dynamics of the peptide Ac-(AAXAA)$_4$-GY-NH$_2$, where X represents R, K, E, or Q.[34] In contrast to the results of Gai and co-workers, the relaxation became slower for peptides with higher stability.

Summarizing, how fast is the helix/coil relaxation to the new equilibrium after a T-jump seems to depend on: a) peptide length, sequence, and stability; b) initial and final temperatures involved in the T-jump; c) solvent conditions such as pH, ionic strength, and viscosity; and d) the specific site probed within the molecule. These results exemplify how complex is the process of helix formation. Clearly, to fully understand the details of this essential process requires being able to account for all of these observations in a quantitative manner. In doing so one must be careful while comparing the results of different experiments that were carried out under various experimental conditions because the apparently discrepant observations may be misleading. Detailed kinetic analysis of these experiments emerges as an essential step before any conclusions regarding the characteristics and mechanisms of helix formation are drawn.

2.5 Kinetic Theory of the Helix–Coil Transition

It is well established that classical nucleation-elongation theory is valid in interpreting thermodynamic experiments on helix–coil transition (see Chapter 1 by Doig). The question is whether this theory is also suitable for explaining the observed experimental findings in the most recent kinetic studies that were summarized in Section 2.4. Indeed some of the results have been successfully interpreted with models that are based on a simple kinetic formulation of the nucleation-elongation theory of α-helix formation[23] (see Section 2.7).

Following a simple statistical mechanical treatment any basic unit, whether it be a residue or a peptide bond, can be assumed to be in one of two states: helical or coil. The state of each basic unit is defined according to the conformation of the pair of dihedral angles flanking the unit. When the dihedral angles of a single residue are fixed in α-helical conformation there is loss in conformational entropy. The loss in conformational entropy increases linearly upon incorporation of s_n successive residues into the helical conformation ($s_n \sim 4$–5 depicts the size of the nucleus). This unfavorable and rate-limiting process is the nucleation of a nascent helix, and gives rise to a free-energy barrier that is entropic in nature. As the helical segment extends from the nucleus, stabilizing interactions are formed that slightly over-compensate (in an α-helix-forming peptide) the entropy loss. The size of the nucleus is determined by the geometry of the helix: in an α-helix s_n pairs of dihedral angles must be fixed before hydrogen bonds between the carbonyl oxygen of *ith* residue and amide hydrogen of *i* + 4*th* residue are formed. However, each subsequent hydrogen bond is realized by fixing only one more pair of dihedral angles. This process leads to the elongation of the nucleated helix.

Apart from the $i,i + 4$ hydrogen bonds, the favorable interactions that are responsible for holding the helix together are van der Waals, dipole–dipole backbone interactions; $i,i + 3$ and $i,i + 4$ side-chain interactions; and the stabilizing effects arising from the N and C caps and the interactions of charge residues with the helix macrodipole (see Chapter 1 by Doig). With binary states allowed for each residue (or peptide bond) there can be 2^N possible combinations or species for a peptide of length N. These combinations can be drastically reduced ~ 100 times for a peptide with $N = 20$ by assuming that at most only two non-overlapping stretches of α-helix conformation can coexist at any given time (*i.e.* the double-sequence approximation). This approximation allows helix breaking in the middle of a larger helical segment as well as merging of two smaller helical segments. 2^N combinations can be reduced ~ 5000 times (for $N = 20$) if a single-sequence approximation is taken, which allows only a single stretch of helix at any given time. However, in contrast to the double-sequence, the single-sequence approximation has profound mechanistic implications because in this case helix breaking can take place only from the ends of an existing helical segment.

The kinetic connectivity is simple since species inter-convert by rotation of single peptide bonds. At a given time, for example, a species with four helical peptide bonds such as . . . . cccchhhhccc . . . . can be directly converted to species having five helical peptide bonds such as . . . . cccchhhhhcc . . . . or . . . . cchchhhhccc . . . . by a single flip but not to species . . . . cccchhhhhhc . . . . or . . . . chhchhhhccc . . . . that require double flips. The rate matrix can then be built from the on and off rates for each possible conversion. The idea behind the single- or double-sequence approximations is to reduce the complexity involved in solving the master equation and also to obtain the simplest possible model capable of explaining the experimental results. T-jumps can then be simulated by using the probability distribution of all species at the temperature before the jump as the initial condition. The kinetic decay is simulated either by solving the eigenvectors and eigenvalues of the rate matrix, or by numerical integration of the several thousand differential equations (*e.g.* for a peptide with $N = 20$ and with double-sequence approximation it involves solving as many as 52 596 differential equations).

In the kinetic zipper model[18] a residue was defined as helical only if it forms an $i,i + 4$ hydrogen bond, which meant that all conformations having helical segments of length less than four residues were ignored. This is equivalent to the Zimm–Bragg equilibrium treatment of the helix–coil transition (see Chapter 1 by Doig). Besides, the same free-energy cost was assumed in fixing the dihedral angles for each individual amino acid. Although a reasonable approximation for modeling homopolypeptides, it is not adequate for peptides with heterogeneous sequence. A statistical mechanical model used earlier to analyze the kinetics of Ac-WAAAH$^+$-(AAAR$^+$A)$_3$A-NH$_2$ peptide incorporated two different values for the entropy cost in fixing the dihedral angles, one for Ala and Arg and another for Trp and His.[19] In addition to the stabilizing effects of backbone hydrogen bonds, energetic contributions from favorable and unfavorable side-chain interactions were also included in the model.

However, the use of more complex sequences such as Ac-YGSPEAAA (KAAAA)$_2$-D-Arg-NH$_2$ in recent experiments demanded inclusion of additional parameters that describe more completely the thermodynamics of α-helix formation. This improvement was achievable by using the full set of sequence dependent parameters directly from AGADIR.[5,8,9] The AGADIR algorithm is parameterized with free-energy values for helical propensities of individual amino acids as well as for all possible backbone and side-chain interactions between any pair of amino acids known to date to affect the stabilization of α-helices. These values are obtained directly from empirical analyses and statistical distribution of the protein structure database[5,8–10] (see Chapter 1 by Doig). A detailed kinetic model that combines the empirical "force field" of AGADIR with the statistical mechanical description mentioned above has recently been developed.[23] Unlike earlier kinetic models, this model explicitly includes the species with short helical segments (with fewer than s_n helical residues), an approximation that allows breaking helices in the middle and a more realistic treatment of the molecular coil state (*i.e.* the ensemble of conformations with no significant α-helix content). For the conversion of each peptide bond from coil to helical state a transition state is assumed which is entropically destabilized without any specific enthalpic interactions. This was done for simplicity. The pre-exponential factor for the microscopic transition between helix and coil (see Chapter 6 by Gruebele) is an adjustable model parameter that defines the rate of the elementary event in helix formation, and perhaps even in protein folding, *i.e.* rotation of single peptide bonds. While there are no direct experimental estimates of this rate so far, the value of the pre-exponential factor obtained from modeling kinetic experiments gives us implicit information about the rates of elementary processes, and can be used to compare with those obtained from molecular dynamic simulations.

2.6 Free-Energy Landscape for α-Helix Formation

Using the nucleation-elongation model combined with the AGADIR "force field", single dimensional free-energy surfaces can be calculated as a function of order parameters. For α-helix formation the most natural order parameter is the number of helical units. Figure 2.2 shows the free-energy profiles at different temperatures calculated for a peptide of length $N=21$ using a double-sequence approximation and projected onto the number of helical peptide bonds (H). The free-energy profile produces a small barrier ($\sim 2RT$) separating two broad

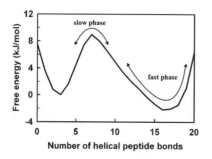 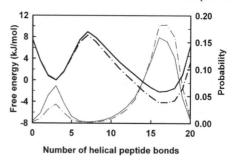

Figure 2.2 One-dimensional free-energy profiles. Free-energy surface calculated as a function of a single order parameter, *i.e.* number of helical peptide bonds, is shown in the left panel. The one-dimensional free-energy profile exhibits two minima corresponding to helical and coil ensembles that are separated by a small barrier. The slow phase corresponds to the equilibration between the helical and coil distribution whereas the fast phase arises from the local winding and unwinding processes in the helical ensemble. The right panel shows free-energy profiles at low (dotted-dashed line) and high (solid line) temperatures. The gray lines show the corresponding probability distribution. At high temperature not only does the increase in the coil distribution accompany the decrease in the helical one but the maximum of the helical peak also shifts towards the left indicating a redistribution of helical lengths.

basins corresponding to two ensembles – one with coil conformations and very short helices (with lengths $\sim < s_n$ residues) and another one with long helices (with lengths $\gg s_n$).

One of the predictions of helix–coil theory is that elongation-shortening of already existing helices is more favorable than helix initiation at new sites.[1] This means that the free-energy profiles at higher temperatures will show the helical ensemble shifted towards lower numbers of helical peptide bonds as a result of the changes in the distribution of helical lengths upon temperature-induced destabilization. From a kinetic standpoint the formation of shorter helices from longer ones is a much faster process than the nucleation-limited formation of new helices from coil conformations. Hence according to this nucleation-elongation mechanism one should expect to see two processes separated in time in a T-jump relaxation experiment – the barrier crossing event *i.e.* equilibration between coil and helical ensembles and re-equilibration between helices of varying lengths within the helical well. It is also important to notice that the difference in timescale of the two macroscopic phases is directly connected to the relative magnitude of the nucleation barrier, thus providing an experimental procedure to estimate it.

Recently, Sorin and Pande[45] carried out extensive all-atom MD simulations using world-distributed computing to predict a free-energy landscape projected onto two order parameters for a 21-residue helical peptide (see Chapter 8 for details on their simulation methodology). The free-energy landscape so obtained exhibited two broad and shallow basins corresponding to two macrostates – helical and coil state. A small free-energy barrier separates the helical and the coil basins, each containing a diverse population of microstates with different helical content and radii of gyration. The similarity in the basic features of the one-dimensional free-energy profile predicted from the nucleation-elongation model and the two-dimensional free-energy landscape produced from atomistic simulations supports the validity of the nucleation-elongation mechanism for α-helix formation.

2.7 Mechanisms of α-Helix Formation

Some kinetic helix–coil experiments on short peptides showing single exponential relaxations have been interpreted with a two-state model.[17] This assumption, however, ignores the complexity of helix formation and implies the existence of a much larger free-energy barrier ($\gg 2RT$) than the estimates discussed in the previous section. Molecular dynamic (MD) simulations of alanine-based peptides have suggested that helix formation corresponds to a diffusive search in the coil region of the phase space that leads to a barrierless transition into the helical state.[28,47] This conformation diffusion search model predicts non-exponential kinetics and dependence of the relaxation time on the T-jump width (*i.e.* ΔT). Since the relaxation kinetics observed by Huang *et al.*[22] for ^{13}C-labeled peptides were reported as non-exponential and showed T-jump width dependence it was concluded that helix formation kinetics is determined by downhill diffusion in the absence of a barrier. At first sight, this explanation could appear not to be in consensus with the nucleation–elongation model, which predicts a free-energy barrier. However, these views are not that different. Rather than diffusion occurring in the coil basin the fast phase in the nucleation–elongation model arises due to diffusion in a helical basin that needs to re-equilibrate to a new ensemble distribution after the perturbation. Since the free-energy barrier to helix formation is only $\sim 2RT$ the slow phase is not merely the crossing of the barrier but it also includes the diffusive motions in the helical basin. In support of this argument, Sorin and Pande pointed out that helix–coil kinetics resulting from atomistic simulations is better represented as a conformation diffusion process rather than a barrier-limited one.[45]

In contrast to nucleation-elongation or downhill diffusion mechanisms, Duan and co-workers reported three-phase kinetics for Ac-YG(AAKAA)$_2$AAKA-NH$_2$ peptide from all-atom MD simulations.[48] In the first phase helix initiation along with hydrophobic collapse takes place in less than 0.1 nanosecond. In a second nanosecond phase helix propagation occurs to develop a folding intermediate with very short helices comprising two independent turns, which in turn unfolds to form a transition state having a helix–turn motif. The breaking of the hydrophobic interactions in the helix–turn–helix motif is suggested to be the rate-limiting step rather than helix initiation. An MD study on the Fs peptide reported the folding process to occur in two phases with the helix–turn–helix motif as the dominant population at 300 K instead of the full helix state.[49] The common feature in these simulations was the implicit generalized Born (GB) continuum solvent model. It was demonstrated by Nymeyer and Garcia that the implicit GB solvent model predicts a helical bundle as the native state for the Fs peptide rather than the full helix state obtained from the explicit solvent treatment.[50] In a comprehensive all-atom MD study in explicit solvent using distributed computing the performance of several variants of the AMBER force field were compared in a quantitative manner.[45] The AMBER-94 variant, the force field used earlier by Hummer and co-workers[47] (kinetics modeled as barrierless one-dimensional diffusion), was assessed to overstabilize the helical conformations. Hence it is not surprising that simulation using this force field showed downhill folding towards a helix conformation. This indicates that results from atomistic simulations could heavily depend on the choice of solvent model as well as the nature of the force field.

Molecular interpretations proposed from computer simulations provide only a qualitative picture of helix formation. These models can be validated if they can reproduce the complex kinetic behavior seen in experiments (see Chapter 8 by Pande on how to quantitatively compare extensive simulations and experiment). To this end, the relaxation kinetic experiments on ^{13}C-labeled peptides probed by IR spectroscopy[22] were analyzed with a detailed kinetic model based on the nucleation-elongation description[23] (Section 2.5). This exercise should also provide a stringent test of nucleation-elongation mechanism in α-helix-formation.

In order to directly compare the results of the model with the equilibrium and kinetic experiments performed with any spectroscopic techniques, it is essential to model the signal decay as a function of time. In this case, the Fourier transformed IR signal was calculated theoretically from the weighted average of the amide I basis spectra of both labeled and unlabeled peptide bond chromophores and from the time-dependent probability distribution arising from the detailed kinetic model (Section 2.5). All the spectral features of the equilibrium Fourier transformed amide I spectra, *viz*. shifts in frequency of the ^{13}C-labeled peptide bonds and the decrease in the amide I band intensities with increase in temperature, were successfully reproduced. The model predicted a biphasic relaxation for each of the peptides labeled at the N-terminus, middle region or C-terminus. Furthermore, the fast phase of the C-terminus-labeled peptide had relatively larger amplitude and faster relaxation rate as compared to the N-terminus- or middle-labeled peptides, resulting in the same kind of kinetics observed experimentally. Inspection of the theoretical results indicated that it was the intricate balance between sequence effects (*i.e.* the different energetics arising from changes in amino-acid sequence) and the phenomenon of end fraying that gave rise to changes in the relative amplitudes of the fast and slow phases. These changes in relative amplitudes appeared as varying relaxation rates for different regions of the peptides when the decays were fitted to a stretched exponential.

In agreement with the experiments of Huang *et al.*[22] the model also reproduced the observation of faster apparent relaxation kinetics as the magnitude of perturbation becomes larger. At first this result appeared counter-intuitive, as one would expect longer relaxation times as the difference between the initial and final temperatures (ΔT) becomes greater. However, this dilemma was elegantly

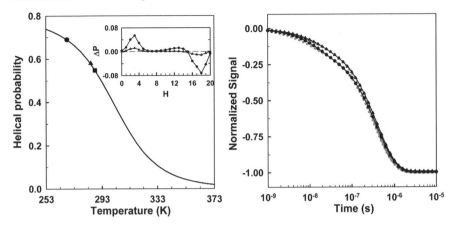

Figure 2.3 Simulation of a T-jump experiment. Left panel shows the calculated probability of forming helices as a function of temperature for a 21-residue peptide. This compares with the experimental temperature denaturation profile for α-helical peptides. The circle and triangle show the location of two different initial conditions before the T-jump and the square denotes the final temperature. The inset shows the difference in the distribution of probabilities between the initial and final temperatures as a function of the number of helical peptide bonds. For a shorter T-jump (line with triangle) the flux of molecules from the helical to the coil basin is reflected in the negative peak in the helical region and a positive peak of equal amplitude in the coil region. For a larger jump (line with circles), an increase in the positive coil peak intensity is not compensated for by an equal increase in the helical negative peak. Instead a positive shoulder appears in the helical region. The magnitude of the positive shoulder reflects the amount of redistribution in helical lengths that takes place after the T-jump and hence the amplitude of the fast phase. The right panel shows the predictions of the detailed model for the two T-jumps. A fast phase with relatively larger amplitude and shorter relaxation time is seen for a larger T-jump. The gray dotted lines are the relaxation traces predicted by the diffusive model.

resolved by the nucleation-elongation model (right panel of Figure 2.3). The mechanistic explanation is simple: as the magnitude of the T-jump increases, a larger amount of redistribution of helical lengths occurs with a fast diffusive rate that does not involve crossing a free-energy barrier.

The detailed nucleation-elongation kinetic model clearly demonstrated that all the complexities observed in α-helix formation are consequences of the inherent characteristics of the classical helix–coil transition on heterogeneous amino-acid sequences.

2.8 Reaction Coordinates for α-Helix Formation

The energy landscape theory suggests that folding kinetics can be determined from low-dimensional free-energy projections onto a few appropriate order parameters (the energy landscape, see Chapter 3 by Wolynes) and from the stochastic motions on such energy landscape.[51] These motions can be described to a first approximation as diffusion on a potential of mean force corresponding to an equilibrium population. Since protein folding is essentially a multi-dimensional problem involving innumerable degrees of freedom the question is whether it is possible to describe all the dynamic processes with a simple projection of the free energy. If yes, then how can we determine those few order parameters that are sufficient to characterize the folding kinetics? As seen from the above sections we now have a better understanding of the mechanistic details of α-helix formation. Hence it is possible to investigate whether diffusion on a low-dimension free-energy surface can capture all the complex kinetic behavior explored theoretically and experimentally for α-helix formation.

One-dimensional potentials of mean force for helix-formation can be straightforwardly calculated based on nucleation-elongation theory and using the number of helical peptide bonds H as order parameter (Section 2.6, Figure 2.2). The idea was then to simulate the kinetic decays for the ^{13}C-labeled peptides of Huang *et al.* (see Section 2.4) as diffusion on such a one-dimensional free energy profile and compare the results with those from the detailed kinetic treatment that reproduces the experimental observations. One-dimensional diffusion was computed using the analytic treatment of Szabo, Schulten, and Schulten.[52] This treatment had been used earlier in calculating the folding rates of proteins from statistical mechanical models as well as the rate of loop formation from the distribution of chain end-to-end distances.[53,54] The obvious advantage of the diffusive calculation is that it drastically simplifies the calculation (*e.g.* for a peptide with length $N=20$ the number of differential equations reduces from $\sim 50\,000$ to a mere $N \times N = 400$). The diffusion coefficient was assumed to be independent of temperature as well as the reaction coordinate.

Interestingly, the predictions of the diffusive calculation were in close agreement with the calculations of the detailed model (Figure 2.3, right panel). Relaxation kinetics was bi-exponential with the ratios of fast and slow amplitudes remarkably matching those calculated from the detailed model. All the complex kinetic behavior, *i.e.* dependence of relaxation times on the initial and final temperatures and different regions of the peptide, that was observed in the experiments and captured by the detailed model was also reproduced by the diffusive model.[55] This study showed that the number of helical peptide bonds was a valid reaction coordinate for α-helix formation, and constitutes one of the first direct empirical tests of the performance of simplified free-energy projections.

2.9 The Nature of the Diffusion Coefficient for α-Helix Formation

The rate of α-helix formation depends on the height of the nucleation barrier, the effective diffusion coefficient, which reflects the timescales of the dynamic motions, and the distance in reaction coordinate space between the coil and helical minima of the surface (*i.e.* the difference in number of helical peptide bonds). Hence, once a quantitative analysis of helix formation kinetics is available, together with a one-dimensional diffusion model that performs similarly to the detailed model, it is possible to examine the nature and timescale of the motions involved in determining the diffusion coefficient for helix formation. From the diffusive calculation of the ^{13}C-labeled peptides the diffusion coefficient was found to correspond to a timescale of ~ 2 nanoseconds.[55] This was in very good agreement with the timescale of ~ 4 nanoseconds for the elementary peptide bond rotation obtained from the detailed kinetic nucleation-elongation model.[23] For both the models the temperature dependence of the pre-exponential factor and the diffusion coefficient were derived from the temperature dependence of viscosity of water. Therefore, there is a very close connection between the peptide bond rotation rate and the effective diffusion coefficient for helix formation, supporting the idea that the number of helical peptide bonds is an appropriate reaction coordinate for helix formation.

Since the peptide bond rotation rate involves crossing an energetic microbarrier (which is estimated to be ~ 4–6 kcal mol^{-1} from the kinetic analysis of Ac-WAAAH$^+$-(AAAR$^+$A)$_3$A-NH$_2$ peptide using a statistical mechanical model[19]) arising from steric clashes as well as from solvent friction, the agreement between peptide bond rotation rate and one-dimensional diffusion coefficient emphasizes that diffusion coefficients in peptide conformational dynamics are likely to include activated terms (see Chapter 6 by Gruebele). Eaton and co-workers have suggested that these high energetic barriers to the elementary steps in α-helix formation result in motions against the solvent that are well separated from the barrier-crossing motions.[56,57] As a result, the overall barrier crossing during helix formation is less enslaved to the surrounding solvent, resulting in a fractional dependence of the kinetic relaxation rate on solvent viscosity, as has been observed experimentally.[56,57]

2.10 Implications for Protein Folding

The success of the diffusive calculations for α-helix formation, which is arguably the simplest protein-folding related process, gives support to the idea that protein folding kinetics can also be determined from simple free-energy projections. However, in contrast to α-helix formation free-energy projections for protein folding may require more than one order parameter (see Chapter 6 by Gruebele). Besides, in contrast to the results in α-helices, the effective diffusion coefficient for protein folding may also be dependent on the reaction coordinate. Protein folding accompanies a large increase in chain compactness and a significant degree of coupling between different parts of the chain as compared to α-helix formation. As protein folding progresses, formation of native interactions becomes more difficult because it may require breaking of existing interactions (both non-native and native). This may result in diffusion coefficients that decrease as the protein becomes more compact along the reaction coordinate.

As opposed to α-helix formation, we still lack a complete understanding of the various energetic factors needed to precisely model the free-energy surfaces for proteins and to obtain absolute barrier heights. Efforts have recently been made to estimate barrier heights from probability density extracted directly from the analysis of equilibrium data (differential scanning calorimetry thermogram)[58] (see Chapter 5 by Ibarra-Molero and Sanchez-Ruiz) as well as from using protein-length-scaling properties of thermodynamic parameters.[59] For many single-domain proteins the free-energy barriers are estimated to be large (~ 5–$10RT$) resulting from incomplete cancellation of interaction enthalpy and chain entropy (see Chapter 3 by Wolynes). In such cases the diffusion coefficient strictly corresponds to the dynamics on the top of the folding barrier.

However, when the free-energy barriers are low (close to RT as seen in the case of the α-helix) the effective diffusion coefficient reflects dynamic motions along the whole reaction coordinate. This also applies to fast-folding proteins that have marginal barriers and where the timescales of the barrier-crossing event approach those of diffusive motions in the unfolded and folded well. On the other hand, diffusion coefficients of proteins with negligible barriers provide direct estimates of the folding speed limit. Fast folding kinetics are then close to probing the intrinsic dynamics in protein folding.[60] A detailed discussion on fast folding can be found in Chapter 6 by Gruebele.

Studies on the formation of secondary structures in isolated peptides continue to provide important clues and to empirically validate simple approaches for analyzing protein-folding experiments. Recently a one-dimensional free-energy approach, similar to the one that has proved successful in simplifying the analysis of complex kinetic experiments on α-helical peptides (see Section 2.7), has been applied to proteins. Due to the lack of a precise force field and knowledge of entropic factors in protein folding, one-dimensional free-energy profiles for proteins were generated using a mean field approach and formulating simple mathematical functions that model the evolution of stabilization energy (*i.e.* enthalpy) and conformational entropy as folding progresses.[61] Such simple free-energy surfaces predict the overall thermodynamic properties (*i.e.* conformational ensembles and presence/absence of barriers), and diffusive kinetics on these surfaces allows calculation of folding rates. Moreover, this one-dimensional free-energy model has been successfully applied in the analysis of temperature dependence of relaxation rates of fast-folding proteins.[62] Ongoing efforts will allow this one-dimensional free-energy surface model to be used as a direct analytical tool for new equilibrium and kinetic experiments on protein folding as in the case of α-helix formation.

References

1. D. Poland and H. A Scheraga, *Molecular Biology: An International Series of Monographs and Textbooks. Theory of Helix–coil Transitions in Biopolymers Statistical Mechanical Theory of Order-Disorder Transitions in Biological Macromolecules*, Academic Press, New York, 1970.

2. B. H. Zimm and J. K. Bragg, *J. Chem. Phys.*, 1959, **31**(2), 526–535.
3. S. Lifson and A. Roig, *J. Chem. Phys.*, 1961, **34**(6), 1963–1974.
4. H. A. Scheraga, *Pure Appl. Chem.*, 1978, **50**(4), 315–324.
5. V. Muñoz and L. Serrano, *Nat. Struct. Biol.*, 1994, **1**(6), 399–409.
6. J. M. Scholtz and R. L. Baldwin, *Annu. Rev. Biophys. Biomol. Struct.*, 1992, **21**, 95–118.
7. A. Chakrabartty and R. L. Baldwin, *Adv. Protein Chem.*, 1995, **46**, 141–176.
8. V. Muñoz and L. Serrano, *J. Mol. Biol.*, 1995, **245**(3), 275–296.
9. V. Muñoz and L. Serrano, *J. Mol. Biol.*, 1995, **245**(3), 297–308.
10. V. Muñoz and L. Serrano, *Biopolymers*, 1997, **41**(5), 495–509.
11. E. Lacroix, A. R. Viguera and L. Serrano, *J. Mol. Biol.*, 1998, **284**(1), 173–191.
12. G. Schwarz, *J. Mol. Biol.*, 1965, **11**(1), 64–77.
13. D. Poland and H. A. Scheraga, *J. Chem. Phys.*, 1966, **45**(6), 2071–2090.
14. B. Gruenewald, C. U. Nicola and A. Lustig *et al.*, *Biophys. Chem.*, 1979, **9**(2), 137–147.
15. R. Zana, *Biopolymers*, 1975, **14**(11), 2425–2428.
16. S. Marqusee, V. H. Robbins and R. L. Baldwin, *Proc. Natl. Acad. Sci. USA*, 1989, **86**(14), 5286–5290.
17. S. Williams, T. P. Causgrove and R. Gilmanshin *et al.*, *Biochemistry*, 1996, **35**(3), 691–697.
18. P. A. Thompson, A. Eaton and J. Hofrichter, *Biophys. J.*, 1997, **72**(2), WP377–WP377.
19. P. A. Thompson, V. Muñoz and G. S. Jas *et al.*, *J. Phys. Chem. B*, 2000, **104**(2), 378–389.
20. I. K. Lednev, A. S. Karnoup and M. C. Sparrow *et al.*, *J. Am. Chem. Soc.*, 1999, **121**(35), 8074–8086.
21. C. Y. Huang, J. W. Klemke and Z. Getahun *et al.*, *J. Am. Chem. Soc.*, 2001, **123**(38), 9235–9238.
22. C. Y. Huang, Z. Getahun and Y. J. Zhu *et al.*, *Proc. Natl. Acad. Sci. USA*, 2002, **99**(5), 2788–2793.
23. U. R. Doshi and V. Muñoz, *J. Phys. Chem. B*, 2004, **108**(24), 8497–8506.
24. Y. Duan and P. A. Kollman, *Science*, 1998, **282**(5389), 740–744.
25. V. Daggett, P. A. Kollman and I. D. Kuntz, *Biopolymers*, 1991, **31**(9), 1115–1134.
26. V. Daggett and M. Levitt, *J. Mol. Biol.*, 1992, **223**(4), 1121–1138.
27. M. Shirts and V. S. Pande, *Science*, 2000, **290**(5498), 1903–1904.
28. P. Ferrara, J. Apostolakis and A. Caflisch, *J. Phys. Chem. B*, 2000, **104**(20), 5000–5010.
29. G. Hummer, A. E. Garcia and S. Garde, *Proteins: Struct. Funct. Genet.*, 2001, **42**(1), 77–84.
30. A. E. Garcia and K. Y. Sanbonmatsu, *Proc. Natl. Acad. Sci. USA*, 2002, **99**(5), 2782–2787.
31. M. Karplus and J. A. McCammon, *Nat. Struct. Biol.*, 2002, **9**(9), 646–652.
32. T. Wang, Y. J. Zhu and Z. Getahun *et al.*, *J. Phys. Chem. B*, 2004, **108**(39), 15301–15310.
33. T. Wang, D. G. Du and F. Gai, *Chem. Phys. Lett.*, 2003, **370**(5–6), 842–848.

34. E. A. Gooding, A. P. Ramajo and J. W. Wang et al., Chem. Commun., 2005, **48**, 5985–5987.
35. S. M. Decatur and J. Antonic, J. Am. Chem. Soc., 1999, **121**(50), 11914–11915.
36. C. Y. Huang, Z. Getahun and T. Wang et al., J. Am. Chem. Soc., 2001, **123**(48), 12111–12112.
37. I. K. Lednev, A. S. Karnoup and M. C. Sparrow et al., J. Am. Chem. Soc., 2001, **123**(10), 2388–2392.
38. D. T. Clarke, A. J. Doig and B. J. Stapley et al., Proc. Natl. Acad. Sci. USA, 1999, **96**(13), 7232–7237.
39. G. Yoder, P. Pancoska and T. A. Keiderling, Biochemistry, 1997, **36**(49), 15123–15133.
40. R. Silva, J. Kubelka and P. Bour et al., Proc. Natl. Acad. Sci. USA, 2000, **97**(15), 8318–8323.
41. I. Nesmelova, A. Krushelnitsky and D. Idiyatullin et al., Biochemistry, 2001, **40**(9), 2844–2853.
42. C. L. Brooks, J. Phys. Chem., 1996, **100**(7), 2546–2549.
43. B. Jun and D. L. Weaver, J. Chem. Phys., 2000, **112**(9), 4394–4401.
44. N. V. Buchete and J. E. Straub, J. Phys. Chem. B, 2001, **105**(28), 6684–6697.
45. E. J. Sorin and V. S. Pande, Biophys. J., 2005, **88**(4), 2472–2493.
46. W. S. Young and C. L. Brooks, J. Mol. Biol., 1996, **259**(3), 560–572.
47. G. Hummer, A. E. Garcia and S. Garde, Phys. Rev. Lett., 2000, **85**(12), 2637–2640.
48. S. Chowdhury, W. Zhang and C. Wu et al., Biopolymers, 2003, **68**(1), 63–75.
49. W. Zhang, H. X. Lei and S. Chowdhury et al., J. Phys. Chem. B, 2004, **108**(22), 7479–7489.
50. H. Nymeyer and A. E. Garcia, Proc. Natl. Acad. Sci. USA, 2003, **100**(24), 13934–13939.
51. P. G. Wolynes, Proc. Natl. Acad. Sci. USA, 1997, **94**(12), 6170–6175.
52. K. Schulten, Z. Schulten and A. Szabo, J. Chem. Phys., 1981, **74**(8), 4426–4432.
53. L. J. Lapidus, P. J. Steinbach and W. A. Eaton et al., J. Phys. Chem. B, 2002, **106**(44), 11628–11640.
54. V. Muñoz and W. A. Eaton, Proc. Natl. Acad. Sci. USA, 1999, **96**(20), 11311–11316.
55. U. Doshi and V. Muñoz, Chem. Phys., 2004, **307**(2–3), 129–136.
56. G. S. Jas, V. Muñoz and J. Hofrichter et al., Biophys. J., 1999, **76**(1), A175–A175.
57. G. S. Jas, W. A. Eaton and J. Hofrichter, J. Phys. Chem. B, 2001, **105**(1), 261–272.
58. V. Muñoz and J. M. Sanchez-Ruiz, Proc. Natl. Acad. Sci. USA, 2004, **101**(51), 17646–17651.
59. A. N. Naganathan and V. Muñoz, J. Am. Chem. Soc., 2005, **127**, 480–481.
60. A. N. Naganathan, U. Doshi and A. Fung et al., Biochemistry, 2006, **45**(28), 8466–8475.
61. U. Doshi and V. Muñoz, Unpublished data, 2007.
62. A. N. Naganathan, U. Doshi and V. Muñoz, J. Am. Chem. Soc., 2007, **129**(17), 5673–5682.

CHAPTER 3
The Protein Folding Energy Landscape: A Primer

PETER G. WOLYNES

Department of Chemistry and Biochemistry, University of California, San Diego, 9500 Gilman Drive, La Jolla, CA, 92093-0332, USA

3.1 Energy Landscape: Metaphor and Math

Protein folding unites the complexity of the huge phase space of even the smallest protein molecule with the combinatorial complexity of the evolution of that molecule through the ages. Surprisingly these difficulties largely cancel. The kinetics of natural protein folding is easier to predict in many ways than the kinetics of typical small molecule chemical reactions. These predictions can be made using the framework provided by the energy landscape theory of folding, an approach with its roots in the statistical mechanics of glasses and phase transitions.[1-4] This theory shows that while polymers made with randomly chosen sequences of amino acids should exhibit the very complex multi-exponential kinetics associated with glasses, natural proteins must have evolved to avoid this kinetic complexity. Instead folding proceeds fairly directly to the native structure being opposed only by chain entropy. While the landscapes for polymers with a randomly chosen order of amino acids are rugged, the energy landscapes of natural proteins have been smoothed to resemble a funnel.[5]

The mathematical basis for the ideas exposed in the opening paragraph, energy landscape theory, is a branch of statistical mechanics that focuses on the statistics of the energy landscapes for *finite size systems*. This article reviews the primary elements of this energy landscape theory, and how one can use the framework to understand folding kinetics in the laboratory.

3.2 Random Sequences – Prehistoric Proteins (Possibly), but Not Most Modern Proteins

We expect a random heteropolymer to associate with itself in a complex manner. A completely random heteropolymer would be either a random coil rarely making three-dimensional contacts or would collapse to form some adventitious contacts that are stabilizing in a free-energy sense. (We use the term "free energy" here to emphasize that the degrees of freedom of the solvent water are averaged over.) These adventitious contacts would largely be hydrophobic, but other "solvophilic" interventions are also possible. If sufficient stabilizing contacts can be made (perhaps by the molecule having a higher average hydrophobicity) the protein molecule will collapse but would remain fluid. While collapsed, the molecule would exist in many states: entropy still favors the protein making use of contacts in a variety of ways. Such a protein would resemble, at the nanoscale, something like what at a macroscopic scale we would call a gel. We know that gels generally have complex kinetics (cooking relies on manipulating these gel characteristics for the protein solutions of food). Individual gossamer biomolecules resembling nanoscopic gels may well exist. In fact, it is possible that some sequences which are thought to be disordered on the basis of sequence analysis are these kinds of gossamer molecules, possessing transient and fragile structures. Such transient organization is not, however, characteristic of how the workhorse globular proteins, enzymes, receptors, *etc.*, with which we are familiar, act. Most working proteins are more compact and are fairly rigid. If still more stabilizing contacts were to be made in a gel-like self-interacting biopolymer, without careful placement, the timescales of rearrangement would continue to lengthen. The resulting more strongly cross-linked molecule would exhibit the characteristics of artificial rubbers or polymer glasses. For a nanoscale *glass*, a small number of possible conformations can take advantage of the numerous adventitious contacts to form low energy structures. The resulting candidate low-energy configurations will be structurally quite distinct. By converting between these species a compact collapsed *random* heteropolymer will therefore exhibit the complex kinetics displayed by glasses. The kinetics will have highly multi-exponential decays that change rapidly with temperature.

3.3 The Statistical Energy Landscape

Purely random heteropolymers have been hard to study in the laboratory. This difficulty arises because strongly interacting random heteropolymers will not be soluble and will "crash" out of solution. On the other hand, the sketch of the behavior of random heteropolymers presented above is well established by simulation. Simulations of simple lattice models of protein folding have confirmed the basic ideas of energy landscape theory. While not doing justice to the stereochemistry intrinsic to polypeptides, in these lattice or "minimalist" models, a heteropolymer is captured in its essence, as merely a necklace of beads.[6-10] Nevertheless intellectual analysis of lattice polymers is much simpler than for real polymers because each bead is positioned on a specific location of a crystal lattice. Much as chess is simpler to analyse than real warfare, lattice proteins are easier to study than more realistic models. The discreteness of possibilities allows specific counting of states when necessary. The types of beads have differing interactions and thus can model the heterogeneity of real protein chains. Generally a random lattice heteropolymer will be "frustrated." This term was introduced in the field of spin glasses, a kind of alloy in which magnetic spins interact in conflicting ways, so the system does not know how to order.[11] For most randomly chosen sequences the conflict necessarily arises because it is impossible to simultaneously satisfy the desire of each residue to be surrounded by its most stabilizing partners. The covalent connections of the chain cannot be broken. These unbreakable connections generally prevent locally optimal arrangements of the three-dimensional partners. Finding the lowest-energy state of a random heteropolymer resembles the situation encountered by many married couples in American culture. The partners in each couple can get along quite well with each other. But family gatherings involving many couples can be rather painful because of the incompatibility of the spouses of various siblings. The spouses were generally not chosen for their compatibility with the rest of the family.

In thinking about the physical consequences of heteropolymeric frustration using a variety of models and approximations, many years ago Joe Bryngelson and I came to the conclusion that the energy landscape of a random heteropolymer would resemble the extreme case of a rugged energy landscape, the random energy model introduced by Derrida to model spin glasses.[12] Owing to the many conflicting intersections, each protein conformation would have a seemingly unpredictably varying random energy. In fact, the actual non-pairwise additive nature of the solvent-averaged forces exaggerates this trend towards uncorrelated randomness. While an extreme caricature of the nature of a random heteropolymeric energy landscape, the random energy model nevertheless does capture the universal aspects of the low-energy state of the heteropolymer's landscape. The lack of correlation in the model makes the analysis of the model rather straightforward. The random energy model is characterized by the total number of configurations, W, which is related to the configurational entropy through the equation

$$\Omega = e^{S_o/k_B} \tag{3.1}$$

The only other parameter characterizing this most rugged energy landscape is the mean square fluctuation in energy states (ΔE^2). This quantity scales with the protein's length if the protein is compact.

The probability distribution of the energy of any individual state should be Gaussian because it is the sum of many potentially conflicting terms,

$$P(E) = \frac{1}{\sqrt{2\pi\Delta E^2}} e^{-E^2/2\Delta E^2} \tag{3.2}$$

Since each state's energy is statistically independent in this model, the distribution of configurations at equilibrium is given by a Boltzmann factor further weighting this Gaussian

$$P = \frac{1}{Z_e} - E/k_B T_e^{-E^2/2\Delta E^2} = \frac{1}{\sqrt{2\pi\Delta E^2}} e^{-(E-\bar{E})^2/2\Delta E^2} \tag{3.3}$$

The resulting Boltzmann-weighted population is itself a shifted Gaussian about the averaged thermal energy

$$\bar{E} = \frac{-\Delta E^2}{2k_B T} \tag{3.4}$$

According to this result as the system cools, ever deeper states in the landscape, farther out in the Gaussian distribution, should be sampled. But, eventually, a problem must arise. The thermally sampled states rapidly drop in number. The entropy drops with temperature as does the energy. The entropy corresponding to the most probable thermal energy can be evaluated by taking the logarithm of the number of configurations with the thermal averaged energy.

$$S = k_B \log \Omega P(E) = S_o - \frac{\Delta E^2}{2k_B T^2} \tag{3.5}$$

This result quantitatively captures the fall of configurational entropy upon cooling. How far can this drop in entropy go? We see that, finally, a problem will occur at the temperature, T_0, given by

$$T_o = \sqrt{\frac{\Delta E^2}{2S_o}} \tag{3.6}$$

At this temperature the entropy would vanish on such a random landscape. The formula would appear to give negative entropy below T_0! This possibility contradicts the Third Law of Thermodynamics. This catastrophe will actually be avoided by the system undergoing a thermodynamic transition at T_0, becoming trapped in at most a few states. The vanishing of the entropy shows that the real histogram of states begins to show its "graininess" at this temperature. Only a few states (a number polynomial in the length of the chain) are to be found in the tail of the distribution. The actual energy distribution can resemble a Gaussian until we sample its very-low-energy states. For a significant fraction of the sequences, in fact, only thermally occupied states might be found below T_0. Since a single state would be thermodynamically occupied, it seems the molecule would have self-organized, merely by obeying statistical mechanics. One might therefore think that the random energy model phase transition is perhaps itself a model of the folding cooperative transition. For a time some authors held this view.[13,14] The situation is, however, not this simple, when we examine laboratory kinetics or, more importantly, when we think about molecular evolution! The problem with the random energy model for the kinetics becomes clear when we recognize that the thermodynamic transition of the random energy model describes the phenomenology of the liquid glass transition quite well.[15] The apparent vanishing of configurational entropy is a generic phenomenon in supercooled liquids, known as the Kauzmann paradox.[16] Laboratory experiments show this impending entropy crisis is correlated with the ever-slowing dynamics of a viscous liquid when strongly cooled.

We can also see the kinetic consequences of the entropy crisis. We can theoretically impose a set of rules that describe the connections between the configurational states of the heteropolymers allowed by elementary chain motions.[17-19] The slow dynamics of the random energy landscape comes about because it is globally unpredictable which specific configurations will be stable enough to act as traps. As the temperature is lowered, the system will only explore lower energy states, but these configurations have very few structural elements in common. The different low-energy structures are compromises – they satisfy the frustrating conflicts as best as can be done but in very different ways. Thus if at one moment the protein is found in a satisfactorily low-energy state, there is no way that the molecule can tell that there isn't a slightly deeper still more stable state which it would rather be in, but which is rather far off in the landscape. Through undirected Brownian motion the molecule will keep trying to find lower energy states, largely unsuccessfully, but eventually the still lower state may be found. Search on the rugged energy landscape resembles a blindfolded golfer on a golf course: diffusion on a rugged landscape is just like diffusion on an absolutely flat but high-dimensional-energy landscape with a single deep minimum.

The kinetics of a globally random energy landscape is complex in detail. But the rates can be described in a statistically simple way when the density of states is high enough. Models using several connection rules for describing which states can go to which others have been studied.[18,19] When the polymer is in a particularly low energy state, to escape from it the molecule must jump to other states which are more typical in their energy. Consequently, the activation barrier for trap escape increases just as the average energy $\bar{E}$ goes down with temperature. This increasing barrier upon cooling leads to a typically super-Arrhenius temperature dependence for the rate of diffusing in configuration space. The average rate is given by the equation

$$R = R_o e^{-\Delta E^2/2(k_B T)^2} \qquad (3.7)$$

The super-Arrhenius decrease of the rate will plateau once the bottom of the landscape is reached but, at the transition temperature, the rate reflects the time to search through all the states

$$R = R_e \, e^{-S_o/k_B} \tag{3.8}$$

In globally connected models where even distant configurations are connected (by unfolding!) when only a few states are thermally accessible, the time it takes to escape from whatever minimum the molecule is in currently becomes typically of the same order as the time needed to explore all of the configuration space if the landscape were flat. (This is really a worst-case analysis. For a *correlated* landscape the rate depends on the number of basins, not the number of configurations, but it is still exponentially small.[20,21]) While a typical random heteropolymer may be able to thermodynamically occupy a small part of its configuration space at a low enough temperature, it will be very difficult to reach this part of phase space kinetically at those temperatures by means of Brownian motion.

As we have just said, the random energy model is a caricature. Clearly most energy landscapes are correlated. Polymer configurations which look similar have fairly similar values of the energy. This degree of correlation is actually not obviously relevant for polymers where even changing a single dihedral angle can bring together very distant and large parts of the sequence that were not in contact before. This topological constraint can break structural correlations, making the problem worse. Simply correlated energy landscapes in which pairs of energy levels are correlated lead to a search that is somewhat easier because the basins of attraction are larger.[20,21] The search problem still remains difficult at the transition temperature having a rate scaling exponentially with N. The locality of interaction that correlates the landscape[22,23] can lower the search time to scale exponentially with a fractional power of N instead of having a rate scaling exponentially with N.

A second problem emerges with assuming that a random energy landscape applies to real protein when we consider the evolution of random heteropolymers. Suppose a protein had, despite the typical kinetic difficulties, efficiently evolved to fold and function but still possessed a random, rugged landscape. The statistics of related, local landscapes shows this to be a possibility, but also that such a route to foldability is not probable.[24] Such landscapes can be called "buffed" since their specific irregularities have been removed (by evolution!). On a buffed landscape there are traps but they can easily be escaped from or bypassed. However, a mutation in such a protein would give a protein with new ground state structure. The new structure would be unlikely to function as the old one did. By calculating their probability of occurrence, Plotkin and Wolynes have suggested that such "buffed" energy landscapes while possible will be eliminated by evolution.

3.4 The Energy Landscape of Long Evolved Proteins

Rather than using the buffing strategy to evolve sophisticated folding pathways that avoid the typical traps for frustrated systems, nature has taken a different route. There is much evidence that natural proteins simply are not as highly frustrated as typical random heteropolymers would be. Proteins do not fold by gradual loss of entropy until you run out of states, as they would in the random energy model. The crystal structures of folded proteins that show native structures do not exhibit obvious energetic conflicts in the way pairwise interactions are satisfied. Major compromises of the rules of structural chemistry are generally absent. Evolved protein structures are not so highly frustrated as the ground states of random sequences structures should be. Instead, examining structures reveals many local themes of consistency and symmetry between a given sequence and the structure it adopts. One specific way of achieving consistency was highlighted early in the work of Go.[25] Go postulated there was a self-consistency principle that ensured compatibility between those local secondary structure forces local in sequences giving rise to helices and sheets and the higher order forces acting between distant residues that gave rise to the particular packings.[26]

Bryngelson and Wolynes generalized this consistency idea to allow more general sorts of self-consistency (not just between secondary and tertiary structure[1]). They christened the more general idea the "minimal frustration principle" and gave the principle a quantitative formula. They argued the mechanism of consistency need not involve secondary structure specifically. Instead different tertiary interactions by themselves should also be consistent with each other if they are sufficiently numerous. Structural biologists can tell you there are many consistent aspects of the distance in sequence contacts found in natural proteins. The core of a protein is largely hydrophobic. Furthermore, if hydrophilic residues are found in the core, they are usually compensated in some kind of salt bridge. Assuming tertiary structure consistency alone is, as we shall see, usually sufficient to predict the folding route although secondary structure biases must also play a role. Saven and Wolynes suggest local signals give about one third of the bias in any event.[27] We need a mathematical definition of consistency in order to understand the kinetics and evolution of proteins quantitatively. To quantify the notion of consistency, one must introduce the idea of a "stratified" statistical energy landscape.[2]

In contrast to a glass or a random heteropolymer the energy landscape of a minimally frustrated system must be *stratified*. That is, the landscape can be naturally divided into layers with common energetic properties. Within each layer the states have a specified degree of similarity to the ground state. There are many choices for this similarity measure. The similarity measure might be chosen to quantify how many dihedral angles are in the correct configuration. Alternatively, one can stratify the landscape *vs.* quantifying how many pair contacts are correctly made, that is, how many are the same as in the native structure. The stratification of the landscape of a minimally frustrated system implies that the energy decreases the deeper the layer is. The energy decreases faster as the global minimum is approached than would be expected for a random heteropolymer. Quantitatively, the average slope of the landscape towards the native structure is larger than the energetic slope found approaching any typical local minimum on a random landscape.

Knowing the depth of the layer where a configuration is located in the landscape does not perfectly predict the energy of a configuration. Non-native contacts will make a random contribution. The latter random energy will be distributed in a way that can be predicted from the random energy model. Even with non-native (partially) frustrating interactions the extra slope of the landscape towards the native structure leads to an energy landscape resembling a still rough, multi-dimensional funnel. For a real protein, this folding funnel will be pockmarked with many overlapping mini-funnels reflecting small traps that can be rapidly escaped from. It is not necessary for evolution to have completely eliminated landscape ruggedness for the protein to fold. Clearly when mutations disrupt the folding of the naturally evolved protein some features of the random landscape may still be present but in addition there will be enough of a stabilizing slope to guide the protein to a properly folded native structure. The entropy and the ruggedness energy parameter characterize the landscape of a random heteropolymer. In addition, a landscape for a *foldable* protein must be characterized by a third parameter, the size of the stability or, better, the "specificity" gap measuring the difference in energy of the global minimum from the typical random states. Owing to this extra stability, the similarity between the ground state and any other state can be used as a collective coordinate to describe the protein folding reaction. This is shown in Figure 3.1. Even when some energetic ruggedness remains, knowing the similarity to the native structure of a given candidate configuration gives us an approximation to its specific energy.

Bryngelson and Wolynes showed that kinetics on a stratified energy landscape can be treated in a simple way. First one groups together all those states in a single layer. They will have a common value of the collective folding coordinate. The average flow of probability between different strata in the folding funnel is then determined not just by the energy but by the gradient of a total free energy as a function of the collective coordinate measuring the depth of the layer. The total free energy includes now an average over all the states of a structure. Each stratum can be approximately described by a separate random energy model appropriate to the non-native interactions that can be formed at this level. The free energy profile as a function of the collective coordinates can be estimated by using the random energy model results to average over the states in a layer. The resulting free energy profile is

$$F(n) = \overline{E}(n) - \frac{\Delta E^2(n)}{2k_B T} - TS_0(n) \tag{3.9}$$

where n measures the depth of the layer, *i.e.* similarity to the native structure. The configurational entropy at any level of the funnel depends on the temperature much as the entropy for the completely random heteropolymer. The

The Protein Folding Energy Landscape: A Primer

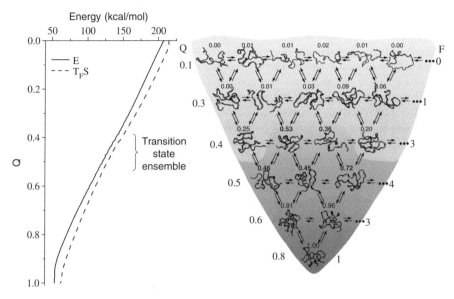

Figure 3.1 The stratified energy landscape of a foldable protein is shown in two ways. On the left are plotted the average energy of a structure at a given stratum (value of Q, the fraction of native contacts) and the entropy of that stratum. On the right is a corresponding "funnel" diagram. Depth in the funnel corresponds to energy (and for this ideal model is identified with Q, after scaling). The width of the funnel diagram represents again the configurational entropy of that stratum. The numbers labeling the different sampled structures represent their probability to complete folding before unfolding, as computed in a native-contacts-only model. Clearly intermediate depths correspond to intermediate values of probability of fold completion. Entropy favors the set of states near the funnel's top, the stabilization free energy (F) of individual structures favors the bottom.

configurational entropy decrease is offset by the average energy gradient of the funnel. By trading off entropy for energy, the Brownian motion of the system will cause it to descend to the global minimum if the temperature is low enough. If there were no energy gradient or if the energy gradient remained near zero until a final stabilizing crash nearly at the folded structure (as some have recently argued![28]), this free energy function would have a high entropic barrier giving again very slow search rates. Instead the free energy gradient provided by the funnel offsets much of this huge entropy barrier. This leads to the possibility of very rapid folding. Indeed, the overall free energy gradient may become nearly downhill below the folding temperature.

Only a small residual activation barrier typically remains due to the incomplete cancellation of entropy and energy through the folding process. In fact such a barrier may be absent in some natural cases[29] or may be removable by protein engineering.[30] While the mean flow of probability may be downward towards the global minimum below T_F, the average flow of an ensemble of proteins may again be rather slow because of trapping in the mini-funnels whose local coordinates are orthogonal to the folding coordinate. These trapping events act as a source of friction on the folding motions. These trapping events involve disentangling motions like those expected in the gel-like random heteropolymer. The Bryngelson–Wolynes (BW) theory shows that the escape from these transient traps determines the effective mobility or diffusion constant of the collective reaction coordinate for folding if it takes place through collapsed structures. Accounting for this diffusive process, energy landscape theory yields a folding time which reflects the mobility effects which get worse at low temperature and the thermodynamic effects which generally favor organization at low temperatures.

$$\tau_F = D(T)^{-1} e^{\Delta F^{\neq}/K_B T} \tag{3.10}$$

The diffusion constant here has a temperature dependence like Equation (3.7). If there were no friction the thermodynamic effects alone would give rise to a folding time *versus* temperature resembling a rectifier's response to a current[31] (see Figure 3.2). At high temperature there will be a big entropy barrier slowing folding but folding becomes nearly downhill in a thermodynamic sense and very fast at low temperature. On the other hand, the frictional slowing due to trapping prevents the rate from actually becoming very fast at the low

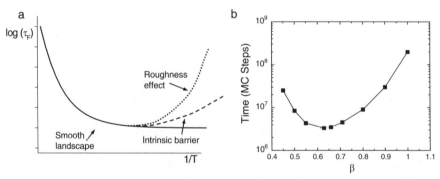

Figure 3.2 On the left is plotted the logarithm of the folding time *versus* $1/T$. At high T, T_F becomes essentially the time to search all the states. This entropy barrier is more effectively canceled by the funnel's energy bias at lower temperature, speeding folding. There may be some residual energy barriers from water expulsion, *etc.*, at low T for real proteins. The behavior in this case is indicated by the dotted curve assuming no glassy trapping. Off-lattice models with only attraction become downhill folders at low T. There will be a bigger barrier at low T if the landscape is rugged, *i.e.*, if non-native contacts lead to trapping. Off-lattice models based on most currently used empirical, sequence-dependent energy functions show this behavior. The low temperature slowing shows these potentials are to some extent frustrated with natural sequences. This increase of the folding time reflects glassy dynamics among the compact state. Panel b shows the same plot for a lattice 27-mer with a modest level of frustration meant to give a T_F/T_G ratio of 1.6 as calculated by Socci, Onuchic, and Wolynes.

The Protein Folding Energy Landscape: A Primer

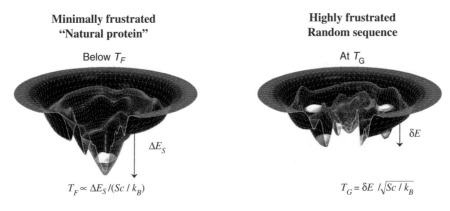

For fast reliable folding must have $T_F > T_G$

Figure 3.3 The left panel shows the landscape of a minimally frustrated protein (entropy is the radial coordinate, energy the depth). At T_F only states with a great deal of structural similarity to the lowest energy are occupied (shown as a single cloud). The right panel shows the landscape of a random heteropolymer. At T_G a small number of discrete traps (shown as multiple distinct clouds of states) are occupied. Minimal frustration is characterized by T_F being bigger than the T_G of the compact but non-native ensemble of structures. The folding transition temperature T_F depends on the specificity gap ΔE, and the configurational entropy. The glass transition temperature T_G depends on the root mean square fluctuation of non-native collapsed structural energies δE and (with a different power!) the configurational entropy. T_G represents the temperature at which a few misfolded and structurally distinct states would dominate the Boltzmann weighted population. Escape from these traps (sometimes called "topomers") would limit the rate.

temperatures even when it is thermodynamically favorable. Folding slows again because of trapping. This characteristic non-monotonic behavior, obtained analytically by BW, has also been seen in many simulations. Socci, Onuchic, and Wolynes showed the validity of the Bryngelson–Wolynes picture for a protein folding funnel reaction coordinate for one of the simplified lattice models.[32] The parabolic temperature dependence they found was characteristic of a partially rough folding funnel. In order to fold much faster than the glassy limit, the temperature at which the energy gradient can overcome the entropy gradient must exceed the glass transition temperature. That is, the folding temperature must be greater than the glass transition temperature of a random heteropolymer of the same composition. The landscapes of frustrated and funneled polymers are sketched in Figure 3.3. Because of the statistical mechanical analysis of the energy landscape, we do not need to argue about what is meant by "consistency," "harmony," *etc*. The quantitative form of the minimal frustration principle can simply be stated "T_F exceeds T_G." By evaluating these temperatures for realistic energy functions the hypothesis of minimal frustration can be directly verified or refuted. One can also try to infer these temperatures from measurements of residual structure and dynamics of collapsed, but denatured, protein, which was done more than ten years ago.[33] The estimated ratio of temperatures was $T_F/T_G = 1.6$. Chan has recently argued the ratio is still larger.[34]

This mathematical form of the "principle of minimal frustration" can be used to find energy functions suitable for protein structure prediction[35–37] and to design proteins that fold in the laboratory.[38] There is now widespread agreement that minimizing frustration by considering sequences and energy functions which yield T_F over T_G ratios bigger than 1 defines what is special about proteins as opposed to random heteropolymers.[39–41]

Because of the minimum frustration principle, the main degrees of freedom that are needed to characterize partially folded ensembles of natural proteins should measure locally whether the native structure is formed or not. Most partially folded molecules that have some fraction of their native interactions formed will have only a few non-native interactions formed, only enough to induce a weakly cross-linked gel – not a quenched glass. This weak cross linking slows folding, but only by a small amount. If the denatured state is strongly collapsed the friction effect increases. This is seen in the so-called "salt-induced detour."[42] Even when non-native trapping effects are highlighted, the native contacts still must fight entropy to complete the organization of the molecule. Nevertheless, because of minimal frustration this fight can go on both locally and, more importantly, quasi-independently throughout the molecule. On a minimally frustrated funnel-like landscape there are numerous possible folding routes. These routes are not equally likely; those routes that gain free energy of stabilization quickly while at the same time paying a low entropy cost will dominate. Many of the dominant folding patterns will pass through a common region of configurational space. This small region of phase space represents a bottleneck for the folding process. This region of configuration space is called the Transition State Ensemble (TSE). The free energy of the TSE relative to the denatured ensemble determines the rate of folding *via* these routes.[43]

Sometimes multiple bottlenecks will occur in sequence, with low free energy "intermediate" ensembles in between. These ensembles can be kinetically blocked from completing their folding or unfolding because the entropy/stabilization trade-off is not uniform. Such ensembles may sometimes be detected as kinetic intermediates.

The minimal frustration principle and the funnel concept, therefore, suggest that the main features of folding kinetics can be predicted by knowing the stabilization energies of elements of the native structure and the entropic costs of bringing together parts of the scaffold. If these two scales suffice to determine the dominant folding route the problem simplifies because these ingredients are functions of the contact map of the native protein – alone. The minimal frustration principle thus has a consequence that the "topology" of a protein structure largely determines the mechanism of folding to that structure. Different natural sequences with the same endpoint structure will have similar mechanisms or a small set of mechanisms under different conditions. This seems to be true experimentally.[44] This pattern is borne out, as we shall see, when it comes to predicting the structural details of folding mechanisms.

3.5 Minimal Frustration, Capillarity, and Protein Topology

When we analyse folding mechanisms we must recognize that the energy landscape analysis so far tells us *globally* what makes a protein a protein and not just a random heteropolymer – the relative lack of energetic frustration (precisely $T_F > T_G$). To describe the mechanisms of protein folding we must appreciate that these globally consistent, only minimally frustrated interactions still must act in real three-dimensional space: they must act locally. Because of the local character of the interactions, usually, only some key parts of the protein need to be assembled before the rest will inevitably fall into place. This is the essence of the capillarity picture of protein – a local version of the energy landscape theory which in its original form concentrated on global collective coordinates.

Even with a local picture of the interactions much of the landscape paradigm can be followed. One can still write the free energy in terms of a collective coordinate for the protein as it folds, the number of residues folded, N_f.

The expression for the profile contains a linear "bulk" term and an interfacial term scaling like $N^{2/3}$.

$$F(N_f) = (f_f - f_u)N_f + \gamma N \quad (3.11)$$

The bulk term that scales with N reflects the free-energy difference per particle, Δf, between folded (f_f) and unfolded (f_u) protein; the small value of $\Delta f = f_f - f_u$ under folding conditions reflects the near cancellation of the entropy of unfolded state and the stabilizing energy of the native structure that is so familiar. In an early paper the "interface" term γ was taken by Bryngelson and Wolynes to be largely energetic and independent of the stability.[45] Putting in the typical stabilization of proteins under physiological conditions ($10k_BT$), BW incorrectly concluded that the order of ~ 100 residues would need to be ordered at the folding transition state, a number comparable to a protein domain size. The error of their analysis was rather basic and apparent to those skilled in thermodynamics of small systems and can be traced back to Kelvin's work on the evaporation of small drops. The bulk transition temperature at which Δf vanishes does not coincide with the transition temperature of the cluster T_F unlike what Bryngelson and Wolynes thought. The freezing temperature is depressed by the surface contribution, as was known to Lord Kelvin. We know that the initial and final full free energies must balance in folding – not their hypothetical bulk values of the free energies per particle! This was pointed out in the folding context by Finkelstein and Badretdinov.[46,47] Taking account of this fact to rewrite Equation (3.1) one can show explicitly the free energy profile referenced near the folding temperature

$$F_{id}(\tilde{N}_f) = (-\tilde{\gamma} + \Delta H(T - T_F)/T_F)\tilde{N}_f + \tilde{\gamma}\tilde{N}_f^{2/3} \quad (3.12)$$

where we have written $\gamma = \gamma N^{2/3}$ and scaled N_f by the chain length N. At T_F, this is a crudely universal form for the "ideal" free-energy profile since there is only one parameter present. (This is a special case of the general rule that at T_F only one basic energy scale should enter a completely unfrustrated model!) The temperature dependence of the stability depends on the enthalpy of unfolding ΔH. The normalized folded fraction $\tilde{N}_f = N_f/N$ now can serve as a collective coordinate for the folding process as in other theories of the funnel based on the fraction of native contacts. The specific numeric coefficients used assume that the protein is nearly spherical, so the curvature of the front is equally limited by all dimensions of the protein (see Figure 3.1). The coefficients would be quite different for helical bundles or other nearly one-dimensional structures such as modular repeat proteins.

As emphasized before,[48] at T_F the free-energy profile obtained by the "spherical cow model" above is quite universal. Also it indicates that there is a rather broad barrier for folding (Figure 3.4) where we superimpose the

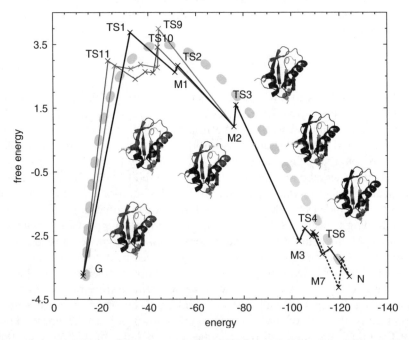

Figure 3.4 The folding free energy profile of UIA calculated *via* a specific topology-based variational method and the ideal "spherical cow" capillarity model. The overall shape but not the bumps and wiggles ("fine structure") are quite well described by capillarity.

capillarity prediction on a detailed calculation for a specific topology. The maximum occurs at $N_f = 8/27\ N$. The barrier can be ascribed to the interface term and is given by $\Delta F = 4/27\ \gamma N^{2/3}$. See Figure 3.4. This same barrier scaling was obtained by Finkelstein and Badretdinov,[47] who used a more elaborate treatment of the interface contributions and a more careful treatment of the protein shape.

How broad is the region that corresponds with the transition state ensemble? Structures within $k_B T$ of the barrier top should be included. Denoting the breadth of the free-energy profile as δN_{TST}, defined as the range over which $F(N)$ changes by $k_B T$, we find $\delta \tilde{N}_{TST} = \sqrt{k_B T/(\delta^2 F/\delta^2 \tilde{N}^2)}$. At T_F, the number of residues displaced in moving over the transition region is thus approximately $\delta N_{TST} \cong 2.\sqrt{2}/3 N^{2/3} \sqrt{k_B T/\gamma}$.

The result indicates the barrier becomes broader with increasing chain length, thus justifying a collective diffusive treatment of the chain motions for larger proteins. An elementary move of the chain in folding must typically displace a loop whose length will scale like $N^{1/3}$. Because of the different N scaling crossing the broad variable will take many elementary moves and will be expected to at least be at the border of diffusive behavior for proteins of the size of 100 resident domains.

Real proteins may well have an intricate interface between the "folded" and "unfolded" regions. Thermal fluctuations, variations in protein connectedness and topology, and heterogeneity of native contacts roughen the interface between the parts of the protein considered folded or not. A metastable compact liquid or even liquid crystalline state[49] of proteins is possible. Conceivably a topologically correct structure lacking side-chain ordering may also exist as has been theoretically suggested. Such intermediately ordered phases will partially wet the interface between completely folded and unfolded states. Wetting reduces the interface energy γ. The resulting thermodynamic activation barrier may thus be smaller than anticipated. The absolute magnitude of the barrier for folding will be hard to predict.

Despite the difficulty of predicting the absolute magnitude of the folding terrain, the location of the interface should follow with only a small error from the locality of the interactions. The structure of the transition state ensemble depends on how strong local contacts can be made given the topology of the final native structure. Theory thus can answer: Where exactly will the nucleation front reach the critical size to allow folding to continue downhill? Experimentally, this can also be answered by protein engineering and kinetic φ value analysis.[50,51]

Unlike the pathway paradigm, energy landscape theory leads to a quasi-thermodynamic theory for folding based on a few reaction coordinates. This picture works because many, many elementary barrier-crossing steps describing the individual dihedral angle isomerization of the chain must come to a steady state before the bottleneck region for folding is reached. Only a globally determined bottleneck ensemble remains to limit the speed of assembly. Due to this quasi-equilibrium, on an overall funnel-like surface, changes in protein stability are directly reflected in the observed kinetics.[49–52] In small-molecule chemistry, such extra-thermodynamic relations only hold for small perturbations. For folding, the collective nature of the reaction coordinate smoothes out the relationship between kinetics and thermodynamics and thus allows one to apply to much bigger changes in stability than in $k_e T$.

If the landscape of evolved proteins were very rugged, intermediates would possess many specific non-native interactions. These intermediates would be found in the tail of the Gaussian distribution expected for a glass. Changing any of the contacts in these unpredictably structured intermediates would stabilize or destabilize minima at random and qualitatively change the folding mechanism, making kinetic predictions impossible.

It is only because of the funnel-like nature of the folding landscape that it is possible to develop a structural interpretation of the φ values. Such an interpretation was already clear in the pioneering experiments of Fersht.[53] Simulations show that the structural interpretation of φ values does indeed break down for frustrated rough landscapes.[54] Simulations show that on highly frustrated surfaces, mutations do indeed stabilize idiosyncratically specific traps on a very rugged landscape.[55,56] Empirically, it seems, perfect funnel models based on the native structure of the protein do a quite adequate job at predicting the location of the key residues in the transition state for folding and the location of the capillary interface between folded and unfolded regions. Thus, one must conclude that ruggedness is a small effect and that proteins do obey the principle of minimal frustration.

3.6 Delightful Prediction of Many of the Devilish Details of Folding

The energy landscape theory provides an organized strategy for picking apart folding kinetics. Uncovering the details of folding requires today a strong collaboration between theory and experiment.[57] The first step of the strategy, according to the minimal frustration principle, is to analyse perfect funnel models. The dynamics of unfrustrated models depends on topology alone, which encodes the balance of changes in entropy and stabilization energy reflecting the average value of native interactions. Since many native contacts must form, the heterogeneity of the energies of such contacts is, to a crude approximation, averaged over. If an evolved protein is minimally frustrated the non-native interactions will certainly be present but these non-native interactions will act non-specifically – they may act to encourage collapse, thereby lowering the entropy barrier.[58] But there is a limit to their help: increasing the strength of non-native contacts leads to trapping, *i.e.* it increases the friction on the motion of funnel reaction coordinate.

Such non-specific effects of landscape ruggedness can be changed by adding denaturants, which again act nonspecifically. Even with non-native interactions present, so long as they are weak most structural features of transition state and intermediate ensembles can be predicted from the native structure alone. This strategy works quite well, but apparently is not perfect.

A wide range of studies has established that the dominant structures of transition state ensembles and of the partially folded ensembles corresponding to intermediates are in fact determined by perfect funnel landscape models much like the spherical cow capillarity model. A perfect funnel landscape depends on the detailed protein topology since the overall entropy and stabilization energies balance so closely.[59-64] Folding mechanism is not nearly as robust to mutation as the final structure is to sequence changes but the entropy/energy imbalances intrinsic to any given protein topology often can provide all the information needed to determine the number of intermediates and which regions of the protein transition state ensembles are natively ordered. This realization has powered an explosion of studies using energy functions based on perfect funnels. These models do a good job of predicting the presence and structures of partially folded intermediates when they are observed.[65] The predictions of φ values from the simple off-lattice models generally agree well in gestalt with experiment.[66,67] At the same time achieving precision predictions of φ values and of absolute barrier heights requires the model to contain non-additive forces that arise from solvation and side-chain placement.[68] Even with non-additivity the landscape is a funnel, although a somewhat narrower one. Variational techniques of polymer statistical mechanics that directly balance chain-entropy and contact stabilization are a useful complement to simulations. The predicted folding intermediates and transition states for U1A[67-70] are shown in Figure 3.4. The overall profile corresponds well with the capillarity form ("spherical cow model") but the fine structure wiggles on the profile reflect details of the protein contact map.

Recently several seeming exceptions to the simple funnel picture have actually served to confirm our confidence in the energy landscape paradigm. An often cited possible exception to the landscape picture has been the folding of cytochrome c.[71] This protein exhibits discernible fine structure in unfolding, as has been explored by Englander's group using H/D exchange.[72] The nature of the folding subunits is currently predicted, however, in detail by a perfect funnel model in simulations.[73] The quasi-independence of the different "foldons" arises because the heme separates many parts of the chain. The mutual contacts between foldons and the heme provide the first key elements of the folding nucleation process. The cooperation between different parts of the polypeptide chain usually needed in proteins without co-factors is less important than these first interactions with the heme for cytochrome.

The mysterious kinetics of the ROP dimer has also challenged the native funnel paradigm.[74,75] ROP dimer folding exhibits a very strong disconnect between stability and folding speed, which do not track each other. Engineered mutations in this dimer speed up both wings of the kinetic chevron and unfolding. How can this be reconciled with a quasi-equilibrium theory dominated by native-like interactions? In fact non-native interactions appear necessary. Simulations suggest this violation of the expected funnel behavior comes about because the symmetry of the dimer allows two near-degenerate topologically distinct structures to compete in some mutants.[76]

Interestingly, although all the engineered mutants bound their target RNA *in vitro* they do not all function *in vivo*. The higher concentrations of RNA in the test tube allow binding to pull the equilibrium over to the competent binding structure. The *in vivo* studies suggest such frustrated mutants, if they had not come about naturally, would likely lose out in natural selection. This explanation of the strange case of ROP time limits is not iron-clad. Other explanations are possible. Nevertheless it seems likely the ROP system may well be the felicitous "exception that proves the rule" – proteins need not have simple funnel landscapes – if they actually do not function.

Many biological functions probably invite frustration: multiple states of a protein are needed for switches (allostery) or to accommodate multiple substrate conformations along the mechanistic pathway of an enzyme transformation. Violations of the minimal frustration principle will often signal such functional constraints. Folding of proteins with complex functions may thus bring new challenges to energy landscape theory. Perhaps it is good to end by emphasizing that it is through challenging paradigms that scientific progress is made. Energy landscape theories of folding can flexibly accommodate many phenomenological observations but deviations may tell us much more about how the molecules of life evolved than has been anticipated.

References

1. J. D. Bryngelson and P. G. Wolynes, *Proc. Natl. Acad. Sci. USA*, 1987, **84**, 7524–7528.
2. J. D. Bryngelson and P. G. Wolynes, *J. Phys. Chem.*, 1989, **93**, 6902–6915.
3. J. Bryngelson, J. Onuchic, N. Socci and P. G. Wolynes, *Proteins: Struct. Funct. Genet.*, 1995, **21**, 167–195.
4. P. G. Wolynes, *Philos. Trans. R. Soc. London, Ser. A*, 2005, **363**, 453–464.
5. P. E. Leopold, M. Montal and J. N. Onuchic, *Proc. Natl. Acad. Sci. USA*, 1992, **89**, 8721–8725.
6. P. G. Wolynes, J. N. Onuchic and D. Thirumalai, *Science*, 1995, **267**, 1619–1620.
7. M. Levitt and A. Warshel, *Nature*, 1975, **253**, 694–698.
8. Y. Ueda, H. Taketomi and N. Go, *Biopolymers*, 1978, **17**, 1531–1548.
9. E. Shakhnovich and A. M. Gutin, *Proc. Natl. Acad. Sci. USA*, 1993, **90**, 7195–7199.
10. K. A. Dill, S. Bromberg, K. Yue, K. M. Fiebing, D. P. Yee, P. D. Thomas and H. S. Chan, *Protein Sci.*, 1995, **4**, 561–602.
11. G. Toulouse, *Helv. Phys. Acta*, 1984, **57**, 459–469.
12. B. Derrida, *Phys. Rev. Lett.*, 1980, **45**, 79–82.
13. E. Shakhnovich, G. Farztdinov, A. M. Gutin and M. Karplus, *Phys. Rev. Lett.*, 1991, **67**, 1665–1668.
14. A. Gutin, A. Sali, V. Abkevich, M. Karplus and E. I. Shakhnovich, *J. Chem. Phys.*, 1998, **108**, 6466–6483.

15. P.G. Wolynes, in *Symposium Proceedings on "40 Years of Entropy and the Glass Transition", J. Res. Natl. Inst. Stand. Technol.*, 1997, **1022**, 187–194.
16. W. Kauzmann, *Chem. Rev.*, 1948, **43**, 219–256.
17. J. Wang, J. G. Saven and P. G. Wolynes, *J. Chem. Phys.*, 1996, **10524**, 11276–11284.
18. J. Wang, S. Plotkin and P. G. Wolynes, *J. Phys. I Fr.*, 1997, **73**, 395–421.
19. J. G. Saven, J. Wang and P. G. Wolynes, *J. Chem. Phys.*, 1994, **10112**, 11037–11043.
20. S. S. Plotkin, J. Wang and P. G. Wolynes, *Phys. Rev. E*, 1996, **536**, 6271–6296.
21. S. S. Plotkin, J. Wang and P. G. Wolynes, *J. Chem. Phys.*, 1997, **1067**, 2932–2948.
22. D. Thirumalai, *J. d. Phy.*, 1995, **5**, 1457–1467.
23. P. G. Wolynes, *Proc. Natl. Acad. Sci. USA*, 1997, **94**, 6170–6174.
24. S. S. Plotkin and P. G. Wolynes, *Proc. Natl. Acad. Sci. USA*, 2003, **100**, 4417–4422.
25. N. Go, *Annu. Rev. Biohys. Bioeng.*, 1983, **12**, 183–210.
26. H. Taketomi, Y. Ueda and N. Go, *Int. J. Petp. Protein Res.*, 1975, **7**, 445–459.
27. J. Saven and P. G. Wolynes, *J. Mol. Biol.*, 1996, **257**, 199–216.
28. O. Schueler-Furman, C. Wang, P. Bradley, K. Misura and D. Baker, *Science*, 2005, **310**, 638–642.
29. M. M. Garcia-Mira, M. Sadqi, N. Fischer, J. M. Sanchez-Ruiz and V. Muñoz, *Science*, 2002, **298**, 2191–2195.
30. H. R. Ma and M. Gruebele, *Proc. Natl. Acad. Sci. USA*, 2005, **102**, 2283–2287.
31. R. P. Feynman, R. B. Leighton and M. Sands, *The Feynman Lectures in Physics, II*, Addison-Wesley, Reading, 1964.
32. N. D. Socci, J. N. Onuchic and P. G. Wolynes, *J. Chem. Phys.*, 1996, **10415**, 5860–5868.
33. J. N. Onuchic, P. G. Wolynes, Z. Luthey-Schulten and N. D. Socci, *Proc. Natl. Acad. Sci. USA*, 1995, **92**, 3626–3630.
34. H. S. Chan, S. Shimizu and H. Kaya, *Methods Enzymol.*, 2004, **380**, 350–379.
35. R. Goldstein, Z. Luthey-Schulten and P. G. Wolynes, *Proc. Natl. Acad. Sci. USA*, 1992, **89**, 4918–4922.
36. Y. Fujitsuka, S. Takada, Z. A. Luthey-Schulten and P. G. Wolynes, *Proteins: Struct. Funct. Genet.*, 2004, **54**, 88–103.
37. G. A. Papoian, J. Ulander, M. P. Eastwood, Z. Luthey-Schulten and P. G. Wolynes, *Proc. Natl. Acad. Sci. USA*, 2004, **101**, 3352–3357.
38. W. Z. Jin, O. Kambara, H. Sasakawa, A. Tamura and S. Takada, *Structure*, 2003, **11**, 581–590.
39. I. A. Hubner, E. J. Deeds and E. I. Shakhnovich, *Proc. Natl. Acad. Sci. USA*, 2005, **102**, 18914–18919.
40. J. G. Saven, *Chem. Rev.*, 2001, **101**, 3113–3130.

41. C. M. Dobson, A. Sali and M. Karplus, *Angew. Chem. Int. Ed.*, 1998, **37**, 868–893.
42. D. E. Otzen and M. Oliveberg, *Proc. Natl. Acad. Sci. USA*, 1999, **96**, 11746–11751.
43. J. N. Onuchic, N. D. Socci, Z. Luthey-Schulten and P. G. Wolynes, *Fold. Des.*, 1996, **16**, 441–450.
44. V. Grantcharova, E. J. Alm, D. Baker and A. L. Horwich, *Curr. Opin. Struct. Biol.*, 2001, **11**, 70–82.
45. J. D. Bryngelson and P. G. Wolynes, *Biopolymers*, 1990, **30**, 177–188.
46. A. V. Finkelstein and A. Badretdinov, *Fold. Des.*, 1997, **2**, 115–121.
47. A. V. Finkelstein and A. Y. Badretdinov, *Mol. Biol.*, 1997, **31**, 391–398.
48. M. Sasai and P. G. Wolynes, *Phys. Rev. A*, 1992, **4612**, 7979–7997.
49. Z. Luthey-Schulten, B. E. Ramirez and P. G. Wolynes, *J. Phys. Chem.*, 1995, **997**, 2177–2185.
50. A. R. Fersht, *Curr. Opin. Struct. Biol.*, 1997, **7**, 3–9.
51. V. Daggett and A. Fersht, *Nat. Rev. Mol. Cell Biol.*, 2003, **4**, 497–502.
52. M. Oliveberg and P.G. Wolynes, *Q. Rev. Bio. Phys.*, submitted.
53. L. S. Itzhaki, D. E. Otzen and A. R. Fersht, *J. Mol. Biol.*, 1995, **254**, 260–288.
54. C. Clementi, H. Nymeyer and J. N. Onuchic, *J. Mol. Biol.*, 2000, **298**, 937–953.
55. J. Chahine, H. Nymeyer, V. B. Leite, N. D. Socci and J. N. Onuchic, *Phys. Rev. Lett.*, 2002, **88**, 168101.
56. H. Nymeyer, N. D. Socci and J. Onuchic, *Proc. Natl. Acad. Sci. USA*, 2000, **97**, 634–639.
57. C. L. Brooks, M. Gruebele, J. N. Onuchic and P. G. Wolynes, *Proc. Natl. Acad. Sci. USA*, 1998, **95**, 11037–11038.
58. C. Clementi and S. S. Plotkin, *Protein Sci.*, 2004, **13**, 1750–1766.
59. B. A. Shoemaker, J. Wang and P. G. Wolynes, *J. Mol. Biol.*, 1999, **287**, 675–694.
60. Y. Levy, J. N. Onuchic and P. G. Wolynes, *Proc. Natl. Acad. Sci. USA*, 2003, **101**, 13786–13791.
61. V. Muñoz and W. A. Eaton, *Proc. Natl. Acad. Sci. USA*, 1999, **96**, 11311–11316.
62. E. Alm, A. V. Morozov, T. Kortemme and D. Baker, *J. Mol. Biol.*, 2002, **322**, 463–476.
63. N. Koga and S. Takada, *J. Mol. Biol.*, 2001, **313**, 171–180.
64. C. Clementi, A. E. Garcia and J. N. Onuchic, *J. Mol. Biol.*, 2003, **326**, 933–954.
65. C. Clementi, P. A. Jennings and J. N. Onuchic, *Proc. Natl. Acad. Sci. USA*, 2000, **97**, 5871–5876.
66. M. R. Ejtehadi, S. P. Avall and S. S. Plotkin, *Proc. Natl. Acad. Sci. USA*, 2004, **101**, 15088–15093.
67. B. A. Shoemaker, J. Wang and P. G. Wolynes, *Proc. Natl. Acad. Sci. USA*, 1997, **94**, 777–782.

68. J. J. Portman, S. Takada and P. G. Wolynes, *Phys. Rev. Lett.*, 1998, **81**, 5237–5240.
69. J. J. Portman, S. Takada and P. G. Wolynes, *J. Chem. Phys.*, 2001, **114**, 5082–5096.
70. T. Shen, C. P. Hoffman, M. Oliveberg and P. G. Wolynes, *Biochemistry*, 2005, **44**, 6433–6439.
71. Y. Bai, T. R. Sosnick, L. Mayne and S. W. Englander, *Science*, 1995, **269**, 192–197.
72. H. Maity, M. Maity and S. W. Englander, *J. Mol. Biol.*, 2004, **343**, 223–233.
73. P. Weinkam, C. Zong and P. G. Wolynes, *Proc. Natl. Acad. Sci. USA*, 2005, **102**, 12401–12406.
74. M. A. Willis, B. Bishop, L. Regan and A. T. Brunger, *Structure*, 2002, **8**, 1319–1328.
75. M. Munson, K. S. Anderson and L. Regan, *Fold. Des.*, 1997, **2**, 77–87.
76. Y. Levy, S. S. Cho, T. Shen, J. N. Onuchic and P. G. Wolynes, *Proc. Natl. Acad. Sci. USA*, 2005, **102**, 2373–2378.

CHAPTER 4

Hydrogen Exchange Experiments: Detection and Characterization of Protein Folding Intermediates

YAWEN BAI

Laboratory of Biochemistry, National Cancer Institute, NIH, Building 37, Room 6114E, Bethesda, MD 20892, USA

4.1 Introduction

Protein folding is the last step in the central dogma of molecular biology, which involves the transfer of genetic information from protein sequences to protein structures. The accepted rule for protein folding has been the thermodynamic principle established in the 1960s by Anfinsen and coworkers.[1] It states that the native structure of a protein is at the most stable thermodynamic state and is determined by the amino acid sequence of the polypeptide chain in the given physiological environment. Since then, protein folding studies have focused on the understanding of the process of folding, including the characterization of intermediate steps and seeking the physical rules that control them. The finding that partially unfolded intermediates could be the precursors for the formation of amyloid fibers[2] also makes protein folding studies important to the understanding of the causes of various amyloid diseases (see Chapter 10 by Serrano for more details in protein aggregation). Since the thermodynamic principle of protein folding suggests that one should, in principle, be able to predict protein structures by minimizing the energy of the conformation of a polypeptide chain, prediction of protein structures has been another important issue in the field of protein folding and has been spurred by the increased power of computers and the Genome Project. The Genome project has accumulated a huge number of DNA and protein sequences. However, in order to understand the functional information encoded in these sequences, one needs to know the structures of these proteins. Thus, it is highly desirable to understand the relationship between protein sequences and protein structures so that one can predict protein structures by computational methods based on protein sequences alone.

This chapter describes the experimental methods for studying folding intermediates by taking advantage of the hydrogen exchange reaction between protein amide groups and water molecules of the surrounding solvent. To structurally characterize the process of protein folding at the residue level, two major experimental methods have been developed: amide hydrogen exchange (HX) and protein engineering.[3,4] Between the two methods, amide hydrogen exchange plays a dominant role in detecting and characterizing partially unfolded intermediates, whereas protein engineering is targeted at obtaining structural information of the transition state. Direct hydrogen exchange measurements have revealed the structures and stability of partially unfolded states under acidic conditions.[5,6] The hydrogen exchange pulse-labeling method has been used to determine the structures and stability of folding intermediates that populate transiently during folding.[7,8] The native-state hydrogen exchange method is capable of detecting folding intermediates that exist as infinitesimally populated high-energy states under native conditions,[9] leading to the determination of their high-resolution structures.[10] A number of excellent reviews on various aspects of these studies have been published.[11–18] This chapter illustrates the basic principle underlying HX methods for detecting and characterizing the structures of partially unfolded intermediates. Typical examples are provided for each method.

4.2 Intrinsic Exchange Rates for Unfolded Polypeptides

Amide protons in polypeptides are chemically labile and can exchange with hydrogen isotopes in solvent water such as:

$$>N-H + D_2O \rightarrow >N-H + DOH \tag{4.1}$$

Because of the extreme pK_a values of main chain amides, the exchange of their hydrogen atoms with solvent is relatively slow and catalyzed only by the strongest of aqueous acids and bases (hydronium and hydroxide ion). Thus, the exchange rate is pH-dependent. Figure 4.1 illustrates the exchange rate constants as a function of pD ($pD = pH_{read} + 0.4$)[19] for amide protons in unstructured poly D/L alanine. Here pH_{read} is the reading value from the pH

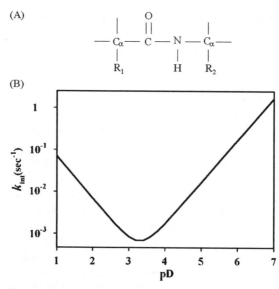

Figure 4.1 Amide hydrogen exchange in the unfolded state. (A) Chemical structure of a peptide. The exchange rate of an amide hydrogen is dominantly affected by the two nearest neighbor side chains. (B) Intrinsic exchange rate constant as a function of pD* (pD* = pH$_{read}$ + 0.4) at 20 °C for amide protons in PDLA.

meter. At pD 7.0 and 20 °C, the exchange rate constant of an amide proton in an unfolded peptide, k_{int}, is affected mainly by the side chains of its two nearest amino acid residue neighbors. Both inductive[20] and steric blocking effects[21] are apparent. These effects have been characterized for all 20 amino acids using dipeptides as models.[22] Accordingly, k_{int} can now be predicted among a broad range of pHs and temperatures.[22,23] An online program for calculating k_{int} is available (http://www.fccc.edu/research/labs/roder/sphere/). For unfolded polypeptides, the predicted k_{int} is normally within a factor of 2 of the measured k_{ex}.[22,24]

4.3 Linderstrøm–Lang Model for Amide Hydrogen Exchange in Folded Proteins

In folded proteins, many amide protons are protected from exchange with water molecules due to hydrogen bonding and burial within the native structure. Therefore, the experimentally measured exchange rate constant of a given protein amide proton (k_{ex}) relative to that of a reference molecule, like an unstructured peptide (*i.e.* the protection factor: PF = k_{int}/k_{ex}) provides information on the native structure and stability. Linderstrøm-Lang and his co-workers assumed a 2-state situation and that amide hydrogens can exchange with solvent hydrogens only when they are transiently exposed to solvent in some kind of closed-to-open reaction, as indicated in Equation (4.2).[25]

$$\text{NH(closed)} \underset{k_{cl}}{\overset{k_{op}}{\Leftrightarrow}} \text{NH(open)} \overset{k_{int}}{\Rightarrow} \text{exchanged}; \quad K_{op} = k_{op}/k_{cl} \quad (4.2)$$

Here, k_{op} is the kinetic opening-rate constant; k_{cl} is the kinetic closing-rate constant. Under steady-state conditions, the exchange rate, k_{ex}, of the above scheme is given by Equation (4.3).

$$k_{ex} = k_{op} k_{int}/(k_{op} + k_{cl} + k_{int}) \quad (4.3)$$

For conditions in which the native structure is stable ($k_{op} \ll k_{cl}$), this equation can be simplified under two extreme cases. (i) The closing reaction is much faster than the intrinsic exchange rate constants ($k_{cl} \gg k_{int}$). In this case, termed EX2 regime, the exchange rate of any hydrogen (k_{ex}) is determined by its chemical exchange rate in the open form multiplied by the equilibrium opening constant, K_{op}.

$$k_{ex} = K_{op} \times k_{int} \quad (4.4)$$

This leads to an empirical expression for the free energy change in the dominant opening reaction, as represented by the equation:

$$\Delta G_{HX} = -RT \ln K_{op} = -RT \ln(k_{ex}/k_{int}) = -RT \ln(1/\text{PF}) \quad (4.5)$$

In this equation, R is the gas constant and T is the temperature. As seen above, k_{int}/k_{ex} is defined as the protection factor (PF). The free energy defined in Equation (4.5) represents a combination of opening transitions from both structural unfolding and local fluctuations. (ii) The closing reaction is much slower than the intrinsic exchange rate constant ($k_{cl} \ll k_{int}$). In this case, termed EX1 regime, the exchange rate is approximately equal to the opening-rate constant (k_{op}). For amide protons that can only exchange through global unfolding, the k_{op} will correspond to the global unfolding-rate constant of the protein. A more general pre-steady-state solution for reaction scheme (2) without any assumptions about the relative magnitudes of k_{op}, k_{cl}, and k_{int} was also solved.[26,17]

4.4 Characterization of Acid Denatured States by Hydrogen Exchange

Some proteins become partially unfolded under mild denaturing conditions such as low pH and high concentration of salt. In some cases, it has been found that such intermediates are similar to those identified in kinetic folding experiments by the pulse-labeled H/D exchange method (see below). The structural features of the partially unfolded intermediates under acidic conditions can be characterized directly by measuring the hydrogen exchange rates of amide protons and comparing them with those of an idealized unfolded state, predicted based on short unstructured peptide models. The obtained protection factors can be used to determine which region of the protein has a folded-like environment in the partially unfolded intermediates. Amide protons with PF >1 are in folded-like environments; whereas amide protons with PF ~ 1 are unfolded.

In these hydrogen exchange experiments, the exchange process is allowed to proceed at low pH, and is quenched subsequently at different time points by changing the pH to a value at which proteins fold to the stable native state quickly. The extent of exchange for each amide proton as a function of exchange time is determined by measuring the peak intensities of the amide protons using two-dimensional NMR methods. A quench procedure is used because partially unfolded intermediates usually do not have well-dispersed chemical shifts in the two-dimensional NMR spectra and are less stable, leading to fast exchange of amide protons. A direct measurement of the exchange rates from the intermediates using NMR is usually not possible. More recently, it has been shown that DMSO quenches the hydrogen exchange process even in unfolded polypeptide chains. Furthermore, the problem of poor chemical shift dispersion for unstructured proteins in proton NMR can be greatly alleviated using two-dimensional amide ^{1}H and ^{15}N correlation spectroscopy. Typical examples for the application of this method are discussed below.

4.4.1 Apomyoglobin (AMb)

AMb is an α-helical protein with 158 amino acids. It has eight α-helices named from A to H. At near neutral pH, all helices are folded except helix F. An equilibrium partially unfolded intermediate is found at pH 4.2.[5] Exchange rates of amide protons in this AMb intermediate have been determined at pD 4.2.[5] The exchange was allowed to proceed in D$_2$O and was quenched by the addition of heme while adjusting to higher pH (~ 8), leading to a reconstitution of stable holo-myoglobin. A high pH solution was used for the reconstitution of the holo-protein since heme tends to aggregate at low pH. AMb is not very stable even under native conditions and tends to aggregate at the high protein concentrations that are needed for NMR measurement. The measured protection factors indicate that helices A, G, and H are folded in this intermediate whereas the remaining helices are unfolded. The structural features of this intermediate are very similar to those observed in the kinetic pulsed-H/D labeling experiment (see below). Recently, this experiment has been reexamined using DMSO to quench the exchange reaction at low pH using ^{15}N-labeled proteins. The results are similar to the earlier experiments but include additional amide proton exchange rates.

4.4.2 Cytochrome c (cyt c)

Cyt c is a protein with a 104 amino acid sequence and a covalently linked heme. It is primarily α-helical: N-terminal helix, C-terminal helix, and the helix including residues from 60 to 70. Oxidized cyt c (Fe^{3+}) at low pH in the presence of high concentration of salt forms a partially unfolded structure, with a molten-globule-like property.[6] To determine the structure of the intermediate, hydrogen exchange was performed at pD 2 and 1.5 M NaCl at 4 °C.[6] The samples are quenched at pD 5.5 at different time points of exchange with the addition of reducing agent (50 mM ascorbate). Under these conditions, cyt c refolds completely within seconds. Ascorbate was included in the quench buffer because cyt c is substantially more stable in the reduced form (Fe^{2+}) and thus the exchange for most amide protons is slower. Proton NMR spectra were recorded at 30 °C by taking the ^{1}H COSY spectra in the magnitude mode. The most slowly exchanging amide protons in this intermediate state were found in the three major helical segments of cyt c.

4.4.3 Ribonuclease H (RNase H)

RNase H has 155 amino acids. The structure of RNase H consists of several α-helices (A to E) and four β-strands (1 to 4). The hydrogen exchange experiment was performed at pD 1.26 with 50 mM KCl and 4 °C.[27] The exchange was quenched at different exchange times by diluting the sample to native conditions (pD 5.5), under which the protein can fold rapidly to the native state. Amide protons in helices A, D, E and strand 2 show varying degrees of protection in the acid state, while amide protons in other regions of the structure show an almost complete lack of protection (PF ~ 1). This experiment was performed using ^{15}N-labeled protein samples with the collection of a series of ^{1}H–^{15}N HSQC spectra.

4.5 Pulsed-Amide H/D Exchange Method

Amide hydrogen exchange can also be used to characterize the structure and measure the stability of folding intermediates in a pulsed-H/D exchange experiment as illustrated in Figure 4.2.[8,7] In a typical experiment, the protein is initially unfolded in D_2O in concentrated chemical denaturant or at low pH. Amide NH groups then exchange to ND. Refolding from the fully deuterated form is initiated by millisecond dilution into a folding buffer in H_2O under the same conditions where folding experiments are normally performed, for example, at pH 5.0–7.0 and 5–20 °C (see Chapter 6 by Gruebele for sub-millisecond folding methods). Formation of early-folding intermediates under these conditions is commonly faster than the exchange rate constants. After some folding time t_f, a brief H-labeling pulse (t_p) is applied by mixing with a high pH buffer (*i.e.* fast exchange conditions), for example at pH 10 and 20 °C. The exchange time will be ~0.3 ms under such conditions. Amide deuterons

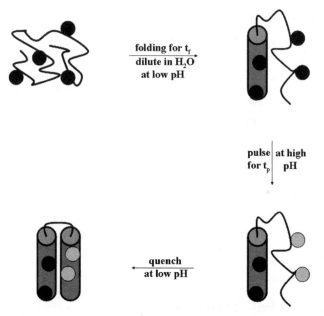

Figure 4.2 Illustration of the pulse-labeling procedure for detecting early-folding intermediates. Grey and black balls represent D and H respectively. t_f is the time for folding. t_p is the time for the high pH pulse.

that are not protected in the populating intermediates will exchange to NH but those in already-formed structures are protected remaining as ND. In practice, multiple time points are taken for either t_f or t_p. t_f may be in the range from milliseconds to several seconds. t_p may be in the range from 5 to 100 ms. A third step involves mixing to low pH to terminate the labeling. Within seconds the protein folds into its native state, which practically freezes the existing H–D labeling profile. In this experiment protein samples are highly concentrated. The exchange can be monitored by either 2D NMR or mass spectrometry.[7,8,28] The NMR experiments provide site-specific information. The quenched sample can also be analysed with the use of mass spectrometry. Although it has low resolution, mass spectrometry has the advantage of rendering the statistical distribution of exchanged protons, which is useful to identify folding processes involving a few parallel folding pathways.[28] A resolution at the level of peptide fragments can also be obtained by mass spectrometry coupled with controlled proteolysis.[29] Another advantage of using mass spectrometry is that large proteins (>20 kD) can be studied.[30]

For a simple case in which the folding intermediate forms in submillisecond timescales and folds to the native state in the timescale of seconds, the fraction of the proton labeled in a pulse-labeling experiment is described by Equation (4.6):

$$H_{label} = [1 - \exp(-PF \times k_{int} \times t_p)] \times \exp(k_f \times t_f) \quad (4.6)$$

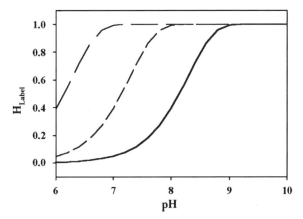

Figure 4.3 Illustration of the results for an early-folding intermediate populated in the submillisecond timescale from a pulse-labeling experiment. The parameters used are: $t_f = 10$ ms, $t_p = 50$ ms, $k_{int} = 100$ ms at pH 6.0. It also assumes that the conversion from the intermediate to the native state is much longer than t_p at the pH of the pulse. X-axis is the pH value of the pulse. The solid line represents an amide proton that is fully unfolded. The short and long dashed lines represent the amide protons that have protection factors of 10 and 100 respectively.

H_{label} can be measured at a series of different folding times from the start of folding (ms) to its completion (s). The folding kinetics of the protein therefore can be monitored at multiple structural sites. Since k_{int} is different for different amide protons at a given pH, H_{label} is also different for different amide protons for a given pulse length t_p. As a consequence, the H_{label} alone is not always sufficient to determine which amide proton is protected in the intermediate. This problem is normally solved by changing the pH of the pulse while fixing t_f and t_p to measure the protection factor (PF) for each amide proton.[31,32] Figure 4.3 illustrates the results from such an experiment.

A simplified version of the pulse-labeling method is the hydrogen exchange-competition method,[33] in which the high pH pulse is omitted. Folding and exchange occur concurrently. Although it is limited to the cases where the folding rate constant, k_f, is close to the intrinsic exchange rate constant (k_{int}), it is very useful to test whether a stable submillisecond intermediate exists owing to its simplicity. The proton occupancy (H_{occ}) as a function of k_{int}, k_f, and t_f is described by Equation (4.7).

$$H_{occ} = [PF \times k_{int}/(PF \times k_{int} + k_f)] \times [1 - \exp(-k_f \times t_f)] \quad (4.7)$$

Typical examples that have been characterized by the pulse-labeling method include cyt c, AMb, lysozyme, and ribonuclease H.

4.5.1 Cytochrome c

Cyt c is the first protein that was studied by the pulse-labeling experiment.[8] In this experiment, the protein is unfolded in a D_2O-denaturant solution. All exchangeable NH sites become deuterated. Refolding, initiated by rapid dilution of the denaturant, is allowed to proceed for variable time periods before the partially refolded protein is exposed to a 50-ms H_2O labeling pulse. Under conditions chosen for the pulse (pH 9.3, 10 °C, 40–60 ms), the free-peptide H-exchange time constant is about 1 ms, so that amide sites in still unstructured parts of the protein become fully protonated. On the other hand, the proton label is excluded from sites where exchange is retarded more than 50-fold by prior formation of (H-bonded) structure. The labeling pulse is terminated by a rapid change to slow-exchange conditions and refolding is allowed to proceed to completion. To quantify the individual proton occupancies, the intensities of well-resolved NH–H_α cross peaks were measured using 2D J-correlated (COSY) spectra.

The degree of protection acquired in the early phase is greatest for amide protons of the N-terminal and C-terminal helices; their amide proton occupancy drops to about 40% in 30 ms. At the same time, other protons located throughout the intervening polypeptide segment remain almost completely accessible for H-labeling; these include sites in two helical segments other than the N- and C-terminal helices, as well as some protons involved in tertiary hydrogen bonds. These results indicate that an intermediate involving the formation of the N- and C-terminal helices is formed in the early stage of the folding process of this protein. The amide proton protection pattern is more complicated at later times, largely due to the intermolecular interactions.

4.5.2 Apomyoglobin

Pulse-labeling experiments were performed at 5 °C in a rapid-mixing device.[34] Fully protonated AMb was unfolded in 6 M urea and 10 mM acetate, pH 6.1, at 5 °C. Refolding was initiated by rapid dilution (1:7.5) into acetate buffer (10 mM, pH* 6.1, in D_2O) for variable time periods before being pulse-labeled (pD 10.2 in D_2O); the final pH was 10.2. Labeling was quenched after 20 ms by dilution into buffer pH 1.9 in D_2O to pH* 5.6. This solution was injected into a reservoir that contained a 1:1 molar excess of bovine hemin to apoprotein. Reconstituted protein was concentrated into a small volume, equilibrated with CO and reduced with sodium dithionite. Double quantum and NOESY spectra were collected for each sample on a 600-AMX Bruker spectrometer.

The amide protons of residues from A, G, and H helices are fully protected within 6.1 ms after exposure to refolding conditions. Some of the amide protons in the B-helix also exhibit complete protection within 6.1 ms while others become fully protected within 1 second after initiation of refolding. Protons in other regions of the protein are protected later.

4.5.3 RNase H

^{15}N-labeled RNase H was dissolved at 2 mg ml^{-1} in deuterated buffer containing 20 mM potassium acetate, pD 5.5, 50 mM KCl, and 7 M deuterated urea, and allowed to unfold and exchange for at least 1 h.[35] Pulse-labeling hydrogen exchange was carried out in a Biologic SFM4/Q quench flow instrument. Refolding was initiated by 1:10 dilution into protonated refolding buffer (20 mM potassium acetate, pH 5.5, 50 mM potassium chloride). A 43.5 ms labeling pulse was applied by a two-fold dilution into pulse buffer (200 mM trisHCl for final pH 9.0 and 9.5, 200 mM glycine for pH 10 and 10.5). A second two-fold of dilution into quench buffer (300 mM potassium acetate) was used to adjust the pH to 5.5. The sample was then concentrated and the buffer was exchanged to 3 mM d$_3$-sodium acetate in H$_2$O, pH 5.5. The final sample was lyophilized and stored at −70 °C until the day of the NMR experiment. The lyophilized samples were resuspended at 10–20 mg ml^{-1} in D$_2$O, followed by the addition of d$_3$-sodium acetate, pD 5.5. ^{1}H–^{15}N HSQC spectra were recorded at 25 °C on a Bruker DMX 600 MHz spectrometer.

The pulse-labeling hydrogen exchange experiment shows that amide proton probes in helices A and D and in strand 4 are well protected (low proton occupancy), and hence well structured within the 14 ms of refolding. Helix E also shows slight protection in this time period. Probes located in the remaining beta-strands are unprotected at the earliest time point, and become protected with an average rate constant of 1 s^{-1}. Thus this kinetic intermediate resembles the acid state.

4.5.4 Hen Egg White Lysozyme (HEWL)

HEWL has 129 amino acids and two sub-domains. One sub-domain contains α-helical structures (α-domain) and the other one has β-structure (β-domain). The pulse-labeling experiment was done at 20 °C using a biologic QFM5 rapid mixing quench flow apparatus.[36] Lysozyme (20 mg ml^{-1}) was initially dissolved in 6 M GdmCl deuterochloride in D$_2$O and pH 6.0, leading to complete denaturation and substitution of all exchangeable hydrogen atoms by deuterium. Refolding was initiated by 10-fold dilution of this solution into 20 mM sodium acetate pH 5.5 in H$_2$O. At the resulting pH of 5.2 the half-life for amide exchange is about 1.6 s so that negligible labeling occurred during this phase. After variable refolding times (3.5 ms to 2000 ms) the solution was diluted again with a volume of 5 times that of the initial protein solution of 0.2 M sodium borate, pH 10.0. This step initiated labeling at a pH of 9.5. After 8.4 ms the labeling was terminated by further dilution, with a volume again 5 times that of the initial volume of protein solution, of 0.5 M acetic acid in H$_2$O. The final pH was about 4.0, at which exchange of the 49 amides studied from the native structure is very slow. Protein samples were concentrated and the buffer exchanged for 40 mM deuterated sodium acetate, pH 3.8 in D$_2$O by ultrafiltration at 4 °C. A phase-sensitive COSY spectrum of each sample was recorded on a 500 MHz GE/Nicolet spectrometer at 35 °C. The intensities of the C$_\alpha$H–NH cross-peaks were measured.

The labeling curves are different for different protons. In addition, for a given amide proton, the exchange curve is not monophasic. Each curve was modeled well by a sum of two exponentials. The rates of the fast phase show no clear pattern but those of the slower phase fall quantitatively into two groups differing in their average time constant by a factor of 4. The more rapidly protected group comprises amide protons in the four α-helical segments, the 3^{10} helix close to the C-terminus of the protein, and three amides, Try 63, Cys 64, and Ile 78, that lie in the loop region in the native enzyme. With the exception of the loop region, these structural elements all occur in one of the two lobes of the native conformation (*i.e.* α-domain). In contrast, amides that become protected more slowly are located, with the single exception Asn 27, in the β-domain, which comprises a short-stranded and a longer triple-stranded beta-sheet, a 3^{10} helix, and a long loop. These results could be interpreted as arising from a non-sequential assembly process that involves two parallel alternative pathways. Thus, most of the molecules would fold through the fast track involving the earlier formation of the α-helical domain. A smaller fraction of molecules are side-tracked onto the slower pathway involving the earlier formation of the β-domain. This parallel folding behavior was subsequently confirmed by a pulse-labeling experiment monitored by mass spectrometry.

4.6 Native-State Hydrogen Exchange Method

As seen above, hydrogen exchange pulse labeling is a kinetic method for characterizing folding intermediates. A native-state hydrogen exchange method was also developed to detect the equilibrium intermediates based on the exchange behavior of amide protons at low concentrations of chemical denaturant. Depending on the free-energy landscape of folding, amide protons can exchange through different processes. Figure 4.4 illustrates the three processes of exchange that occur for a putative three-state system. Amide protons that are not strongly protected or deeply buried in the native protein can exchange through local structural fluctuations that result in breaking one or two hydrogen bonds without significantly exposing solvent-accessible surface area.[37–39] If the protein has a partially unfolded state that is more stable in native conditions than the unfolded state, then the amide protons in the unfolded region of the intermediate can exchange from such partially unfolded states. Of course, all amide protons can also exchange from the fully unfolded state. The measured exchange rate constant is the sum of the exchange rate constants of the three processes, weighted by the unfolding equilibrium constants of the intermediate and the unfolded state under EX2 conditions:

$$k_{ex} = k_{loc} + K_{NI} \times k_{int} + K_{NU} \times k_{int} \qquad (4.8)$$

i.e. the exchange reactions from the intermediate and the unfolded state are modulated by the (partial) unfolding equilibrium constants. Here, k_{loc}

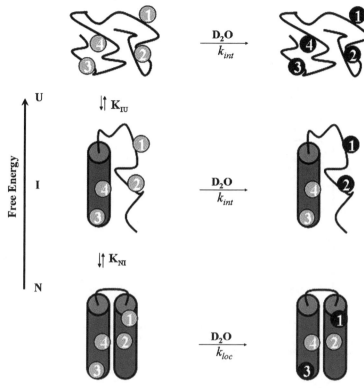

Figure 4.4 Three basic processes for hydrogen exchange.

represents the exchange process from the native structure. K_{NI} and K_{NU} are the equilibrium unfolding constants. In a native-state hydrogen exchange experiment, the hydrogen exchange rates are measured at different concentrations of denaturant. The denaturant is used to perturb the equilibrium constants and help to reveal the different exchange behavior for amide protons in different regions of the protein. When the values of ΔG_{HX} for different amide protons are plotted against the concentration of chemical denaturant, different patterns can be observed. The idea is that exchange arising from partial unfolding (e.g. folding intermediates) will result in ΔG_{HX} values smaller than the global ΔG_{NU}, and also in weaker chemical denaturant dependence. This is so because partial unfolding involves smaller changes in accessible surface. Accordingly, different structural segments that unfold cooperatively as structural units will converge to particular ΔG_{HX} values, which will then join the global ΔG_{NU} process at the concentration of chemical denaturant at which their sensitivity to denaturant crosses the global one. Figure 4.5 illustrates the hydrogen exchange pattern for a protein with a folding intermediate. Based on the exchange pattern, one can deduce the structure, stability, and relative exposed surface area of the intermediates from a native-state hydrogen exchange experiment.[9]

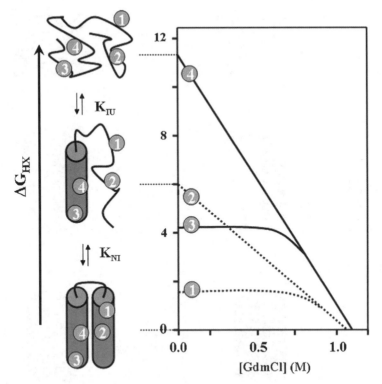

Figure 4.5 Illustration of native-state hydrogen exchange results for a protein with a partially unfolded intermediate (GdmCl = guanidinium chloride).

4.6.1 Cytochrome c

The native-state hydrogen exchange on cyt c was performed at pD 7 and 30 °C.[9] The exchange was rapidly initiated by changing the protein sample in H$_2$O to D$_2$O using a spin column. The exchange process was allowed for a given period of time and then quenched for NMR measurements. The quench was done by lowering the pH and adding ascorbate to reduce the sample, in similar fashion to the quenching experiment for characterizing the acid state. Four groups of amide protons were identified. The first group involves the amide protons in the N- and C-terminal helices. The second group involves the amide protons in the 60s helix and the 30s loop. The third group involves the amide protons in the loop from residues 36 to 60. The last group involves the amide proton from 70 to 85. This leads to the three partially unfolded intermediates. The first one has the N- and C-terminal helices folded. The second intermediate has the N- and C-terminal helices, the 60s helix, and the loop involving residues from 30s to 40s folded. The third intermediate involves the folding of all parts of the structure except the Omega loop from residue 71 to 85.

4.6.2 RNase H

Hydrogen exchange rates were measured for 53 amide protons dispersed throughout RNase H in D_2O at 12 different guanidinium chloride concentrations in the range 0–1.3 M.[40] Exchange of NH to ND was measured by recording two-dimensional ^1H–^{15}N HSQC spectra as a function of time, ranging from hours to months. The hydrogen exchange data suggested that RNAse H unfolds in three distinct regions defining two partially unfolded forms. The most stable region of the protein encompasses helices A and D. Amide protons in these helices exchanged with an average unfolding free energy of 10.0 kcal mol^{-1}. The second region encompasses helix B and strand 4 of the beta-sheet. The third region includes helices C and E, and strands 1, 2, 3, and 5.

4.6.3 Rd-apocytochrome b_{562}

Rd-apocyt b_{562} is a redesigned four-helix bundle protein based on apocytochrome c.[41] The native-state hydrogen exchange experiment was performed at pH 5.0 and 25 °C in GdmCl solution starting with deuterated proteins. Two partially unfolded forms have been found. One partially unfolded intermediate has the N-terminal helix unfolded. The other intermediate has the N-terminal helix and the C-terminal part of the C-terminal helix unfolded.

References

1. C. B. Anfinsen, *Science*, 1973, **181**, 223.
2. D. Canet, A. M. Last, D. B. Archer, C. Reddield, C. V. Robinson and C. M. Dobson, *Nat. Struct. Biol.*, 2002, **9**, 308.
3. C. R. Matthews and M. R. Hurle, *Bioessays*, 1987, **6**, 254.
4. A. Matouschek, J. T. Kellis Jr, L. Serrano and A. R. Fersht, *Nature*, 1989, **340**, 122.
5. F. M. Hughson, P. E. Wright and R. L. Baldwin, *Science*, 1990, **249**, 1544.
6. M. F. Jeng and S. W. Englander, *J. Mol. Biol.*, 1991, **221**, 1045.
7. J. B. Udgaonkar and R. L. Baldwin, *Nature*, 1988, **335**, 694.
8. H. Roder, G. A. Elove and S. W. Englander, *Nature*, 1988, **335**, 700.
9. Y. Bai, T. R. Sosnick, L. Mayne and S. W. Englander, *Science*, 1995, **269**, 192.
10. H. Q. Feng, Z. Zhou and Y. Bai, *Proc. Natl. Acad. Sci. USA*, 2005, **102**, 5026.
11. R. L. Baldwin, *Curr. Opin. Struct. Biol.*, 1993, **3**, 84.
12. S. W. Englander and L. Mayne, *Annu. Rev. Biophys. Biomol. Struct.*, 1992, **21**, 243.
13. R. Li and C. Woodward, *Protein Sci.*, 1999, **8**, 1571.
14. C. E. Dempsey, *Prog. Nucl. Mag. Reson. Spectrosc.*, 2001, **39**, 135.
15. J. Rumbley, L. Hoang, L. Mayne and S. W. Englander, *Proc. Natl. Acad. Sci. USA*, 2001, **98**, 105.

16. S. W. Englander, *Annu. Rev. Biophys. Biomol. Struct.*, 2000, **29**, 213.
17. M. M. G. Krishna, L. Hoang, Y. Lin and S. W. Englander, *Methods*, 2004, **34**, 51.
18. D. Wildes and S. Marqusee, *Methods Enzymol.*, 2004, **380**, 328.
19. P. F. Glasoe and F. A. Long, *J. Phys. Chem.*, 1960, **64**, 188.
20. R. S. Molday, S. W. Englander and R. G. Kallen, *Biochemistry*, 1972, **11**, 150.
21. Y. Bai and S. W. Englander, *Proteins*, 1994, **18**, 262.
22. Y. Bai, J. S. Milne, L. Mayne and S. W. Englander, *Proteins*, 1993, **17**, 75.
23. G. P. Connelly, Y. Bai, M. F. Jeng and S. W. Englander, *Proteins*, 1993, **17**, 87.
24. B. M. Huyghues-Despointes, J. M. Scholtz and C. N. Pace, *Nat. Struct. Biol.*, 1999, **6**, 910.
25. A. Hvidt and S. O. Nielsen, *Adv. Protein Chem.*, 1966, **21**, 287.
26. H. Qian and S. I. Chan, *J. Mol. Biol.*, 1999, **286**, 607.
27. J. Debora and S. Marqusee, *Biochemistry*, 1996, **35**, 11951.
28. A. Miranker, C. V. Robinson, S. E. Radford, R. T. Aplin and C. M. Dobson, *Science*, 1993, **262**, 896.
29. Z. Zhang and D. L. Smith, *Protein Sci.*, 1993, **2**, 522.
30. H. Pan, A. S. Raza and D. L. Smith, *J. Mol. Biol.*, 2004, **336**, 1251.
31. J. B. Udgaonkar and R. L. Baldwin, *Proc. Natl. Acad. Sci. USA*, 1990, **87**, 8197.
32. G. A. Elove and H. Roder, *ACS Symp. Ser.*, 1991, **470**, 50.
33. F. X. Schmid and R. L. Baldwin, *J. Mol. Biol.*, 1979, **135**, 199.
34. P. A. Jennings and P. E. Wright, *Science*, 1993, **262**, 892.
35. T. M. Raschke and S. Marqusee, *Nat. Struct. Biol.*, 1997, **4**, 298.
36. S. E. Radford, C. M. Dobson and P. A. Evans, *Nature*, 1992, **358**, 302.
37. S. L. Mayo and R. L. Baldwin, *Science*, 1993, **262**, 873.
38. H. Qian, S. L. Mayo and A. Morton, *Biochemistry*, 1994, **33**, 8167.
39. Y. Bai, J. S. Milne, L. Mayne and S. W. Englander, *Proteins*, 1994, **20**, 4.
40. A. K. Chamberlain, T. M. Handel and S. Marqusee, *Nat. Struct. Biol.*, 1996, **3**, 782.
41. R. A. Chu, W. H. Pei, T. Takei and Y. Bai, *Biochemistry*, 2002, **41**, 7998.

CHAPTER 5
Statistical Differential Scanning Calorimetry: Probing Protein Folding–Unfolding Ensembles

BEATRIZ IBARRA-MOLERO AND JOSE MANUEL
SANCHEZ-RUIZ

Facultad de Ciencias, Departamento de Quimica Fisica. Universidad de Granada, 18071-Granada, Spain

5.1 Differential Scanning Calorimetry (DSC) as a Tool for the Complete Energetic Description of Protein Folding/Unfolding Thermal Equilibria

Differential scanning calorimetry experiments lead to the determination of the heat capacity of a solution (a protein solution, in the case of interest here) as a function of temperature. The thermodynamic quantity known as "heat capacity" provides a measure of the system's capacity to store energy. That is, if the heat capacity value is large, the system will be able to store a significant amount of energy upon temperature increase. This means, in molecular terms, that some efficient energy-storage mechanism must be operative in the system. For instance, liquid water has a high heat capacity due to an efficient energy-storage mechanism that, in very simplistic terms, could be described as "breaking of hydrogen bonds." In a protein solution, the unfolding process provides an energy-storage mechanism, since the energy of the unfolded protein is (under most circumstances) higher than that for the native protein. Thus, in the temperature range in which protein unfolding occurs, the heat capacity value is high and unfolding processes are revealed by positive "peaks" in DSC

RSC Biomolecular Sciences
Protein Folding, Misfolding and Aggregation: Classical Themes and Novel Approaches
Edited by Victor Muñoz
© Royal Society of Chemistry 2008

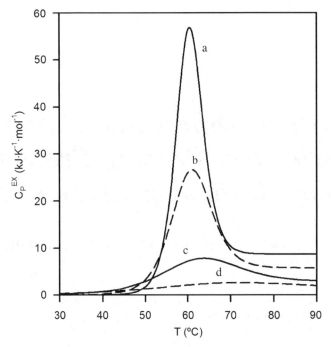

Figure 5.1 Theoretical excess heat capacity *versus* temperature profiles computed on the basis of the two-state model (Equations (5.5)–(5.11)). In all cases, the unfolding equilibrium constant is unity at 60 °C. The unfolding enthalpy values (kJ mol^{-1}) at that temperature are: a) 438, b) 292, c) 146, and d) 73. For the sake of illustration and simplicity, we assume constant (temperature-independent) unfolding heat capacity in the calculation. The values used (kJ K^{-1} mol^{-1}) are: a) 8.7, b) 5.8, c) 2.9, and d) 1.45. The unfolding enthalpy and heat capacity values are of the order expected for proteins of 150 (a), 100 (b), 50 (c), and 25 (d) residues, according to the correlations reported by Robertson and Murphy.[15]

thermograms (plots of heat capacity *versus* temperature; see Figure 5.1 for an illustrative example).

Scanning calorimeters used for diluted protein solutions (*i.e.* in the submillimolar range) are differential instruments: what they actually measure is the heat capacity difference between the protein solution and the pure solvent (buffer). Therefore, some mathematical manipulations are required to get the so-called absolute heat capacity of the protein. This and other relevant issues of the standard DSC data acquisition and analysis of protein unfolding reactions (baseline corrections, reversibility-related issues, model fitting, *etc.*) have been discussed in detail elsewhere.[1–13] Here, we focus on the main distinctive feature of the DSC approach to the study of protein folding/unfolding, that is, the fact that, unlike other techniques, differential scanning calorimetry has the potential to provide a complete energetic description of protein folding/unfolding equilibria. This important result was demonstrated by Ernesto Freire and Rodney Biltonen many years ago[14] by showing that *undistorted* DSC thermograms are essentially equivalent to the relevant protein partition function. This connection between DSC data and partition function is the basis of many of the most informative procedures of DSC data analysis. These include not only the fitting procedures used to ascertain the number and energetic features of the protein macrostates significantly populated during the thermally induced unfolding processes, but also the recently developed approaches to the determination of thermodynamic barrier heights in protein folding.

In this chapter, we will try to introduce the reader to the concept and uses of partition functions in the simplest and most intuitive manner. Subsequently, we will explain some of the partition-function-based approaches to DSC-data analysis and interpretation. In doing so, however, we will find ourselves discussing central issues, such as the size of barriers in protein folding, the existence of downhill folding, and the adequacy of the two-state model for small fast-folding proteins, and, finally, the implications of the essentially kinetic character of protein stability in biological conditions.

5.2 Partition Functions of Folding/Unfolding Processes

In the simplest possible way, a partition function can be regarded as a sum of statistical weights (more "sophisticated" interpretations can be found in textbooks on statistical thermodynamics, but these will not be required here). Consider, for instance, that a certain number of different "situations" ("states", "number of bound ligands", *etc.*) are possible for a given protein. We shall label these different situations with numbers: 0, 1, 2... For each situation, a statistical weight is defined in such a way that it is *proportional* (not necessarily equal) to the probability of finding the protein in that situation at equilibrium (defined as the situation in which probabilities do not change with time). The probability for a given situation (P_i) is actually calculated as the corresponding statistical weight (w_i) divided by the sum of the statistical weights for all situations ($w_0 + w_1 + w_2 + \cdots$):

$$P_i = \frac{w_i}{w_0 + w_1 + w_2 + \cdots} = \frac{w_i}{\sum_i w_i} \tag{5.1}$$

and the sum of the statistical weights is known as the partition function (Q):

$$Q = w_0 + w_1 + w_2 + \cdots = \sum_i w_i \tag{5.2}$$

It is customary to assign a given situation the status of reference, thus resulting in a statistical weight of unity (*e.g.* $w_0 = 1$). Then, we have $P_0 = 1/Q$, $P_1 = w_1/Q$, $P_2 = w_2/Q \ldots$, or, in general,

$$P_i = \frac{w_i}{Q} \tag{5.3}$$

It is now straightforward to write expressions for any quantity that can be expressed as an average over all the "situations" available to the protein. Thus, if the value of a given property (X) is X_0 for situation "0", X_1 for situation "1", and so forth, its average value is:

$$\langle X \rangle = X_0 P_0 + X_1 P_1 + X_2 P_2 + \cdots = \sum_i X_i P_i = \frac{1}{Q} \sum_i X_i w_i \qquad (5.4)$$

It must be noted that not all experimental quantities can be expressed as an average over the different possible situations. Although the enthalpy ($\cong$ energy, for the cases of interest here) can be thus expressed, quantities such as entropy and heat capacity cannot. In particular, expressions for the heat capacity are obtained as temperature derivatives of the average enthalpy.

The above formalism can be illustrated with the well-known two-state model for protein thermal unfolding. The model assumes that the protein can exist in two different macrostates, native and unfolded (the meaning of these macrostates and the apparent and "hidden" implications of the model will be discussed below). Here "situation 0" is the native state and "situation 1" is the unfolded state. We take the native state as reference and assign to it a statistical weight of unity ($w_0 = 1$). The statistical weight of the unfolded state must then be the unfolded to native concentration ratio, which equals the unfolding equilibrium constant: $w_1 = [U]/[N] = K$. Therefore, the partition function for this case is,

$$Q = w_0 + w_1 = 1 + K = 1 + \exp\left(-\Delta G/RT\right) \qquad (5.5)$$

where we have used the well-known thermodynamic relation between equilibrium constants and free energy changes ($\Delta G = -RT \ln K$) and ΔG is the free energy of the unfolded state with respect to the native state (the reference situation). This ΔG obviously has both enthalpic and entropic components:

$$\Delta G = \Delta H - T\Delta S \qquad (5.6)$$

In the cases of interest here, enthalpy can be interpreted essentially as energy, and entropy is related to the number of different ways ("number of microstates") compatible with a given situation ("macrostate"). Several reviews and scientific publications on the structural interpretations of these energetic parameters are available,[8,15-18] and the reader is referred to them for details.

The probabilities that a protein molecule is found in the native and unfolded states are now given by Equation (5.3),

$$P_0 = \frac{1}{Q} = \frac{1}{1+K} = \frac{1}{1 + \exp\left(-\Delta G/RT\right)} \qquad (5.7)$$

$$P_1 = \frac{K}{Q} = \frac{K}{1+K} = \frac{\exp\left(-\Delta G/RT\right)}{1+\exp\left(-\Delta G/RT\right)} \tag{5.8}$$

These probabilities are temperature dependent, since the unfolding equilibrium constant changes with temperature as given by the van 't Hoff equation: $\partial \ln K/\partial T = \Delta H/RT^2$ (see any elementary textbook on chemical thermodynamics).

The expression for the average enthalpy can be easily written from Equation (5.4),

$$\langle H \rangle = H_0 P_0 + H_1 P_1 = \Delta H \cdot P_1 = \Delta H \frac{K}{1+K} \tag{5.9}$$

where we have used the native state as reference and, therefore $H_0 = 0$ and $H_1 = \Delta H$ (the unfolding enthalpy change). Straightforward derivation of Equation (5.9) and use of the van't Hoff equation leads to an expression for the protein heat capacity,

$$C_P^{EX} = \frac{\partial \langle H \rangle}{\partial T} = \Delta C_P \frac{K}{1+K} + \frac{\Delta H^2}{RT^2} \frac{K}{(1+K)^2} \tag{5.10}$$

where ΔC_P is the unfolding heat capacity change (the heat capacity of the unfolded state with respect to that of the native state) and the protein heat capacity is labeled with a superscript "EX" (for excess), meaning again that it is measured with respect to the reference state (i.e. $C_P^{EX} = C_P - C_P(\text{native})$). Using Equations (5.7) and (5.8), Equation (5.10) can be written as,

$$C_P^{EX} = \Delta C_P \cdot P_1 + \frac{\Delta H^2}{RT^2} P_0 \cdot P_1 \tag{5.11}$$

The first term in the right-hand side of Equation (5.11) is simply the average heat capacity of the native and unfolded states and, in the parlance of the field, it is usually known as the "chemical baseline." The chemical baseline reflects the change in heat capacity due to the increase in the population of the unfolded state. The second term in the right-hand side of Equation (5.11) is significant when *both* the native and unfolded sates are significantly populated (the term contains the product $P_0 \cdot P_1 = P_0 \cdot (1-P_0)$ whose maximum value is reached when $P_0 = P_1 = 0.5$). This second term reflects an additional energy-storage capability in the system that is associated to the temperature-induced shift in the folding/unfolding equilibrium (i.e. in the temperature range of the unfolding transition, the system – protein solution – may store energy in increasing amounts of unfolded protein). Clearly, the second term corresponds to a "peak" in the DSC thermogram (see Figure 5.1 for illustrative examples).

The above analysis summarizes the treatment of the two-state model in the partition-function based formalism. The reader is referred to published work[7,19–21] for further discussion on other issues related with the two-state model, such as the calculation and interpretation of the protein stability curve (plot of unfolding free energy *versus* temperature) and the existence of cold denaturation. We proceed now to apply the formalism to a more general multi-state denaturation process. Assuming that the protein can exist in n number of states (macrostates) at equilibrium:

$$I_0 \leftrightarrow I_1 \leftrightarrow I_2 \leftrightarrow \cdots \leftrightarrow I_{n-1} \leftrightarrow I_n \qquad (5.12)$$

where I_0 and I_n are the native and "most-unfolded" states and I_1 to I_{n-1} could be viewed as states of intermediate degree of unfolding ("intermediate equilibrium states").

We take the native state as reference (statistical weight unity: $w_0 = 1$) and consequently the statistical weights (w_i) for all other states are given by the value of the corresponding $I_0 \leftrightarrow I_i$ equilibrium constant:

$$w_i = \frac{[I_i]}{[I_0]} = K_i = \exp\left(-\Delta G_i/RT\right) \qquad (5.13)$$

where ΔG_i is the free energy of the state I_i with respect to that of the native state (related to the corresponding changes in enthalpy and entropy by $\Delta G_i = \Delta H_i - T\Delta S_i$). From Equations (5.2), (5.3), (5.4), and (5.13), the partition function, the probability of state occupation, and the average enthalpy are:

$$\begin{aligned} Q &= 1 + \sum_{i=1}^{n} K_i = 1 + \sum_{i=1}^{n} \exp\left(-\Delta G_i/RT\right) \\ &= 1 + \sum_{i=1}^{n} \exp(\Delta S_i/R) \cdot \exp(-\Delta H_i/RT) \end{aligned} \qquad (5.14)$$

$$\begin{aligned} P_i &= \frac{K_i}{Q} = \frac{\exp\left(-\Delta G_i/RT\right)}{Q} \\ &= \frac{1}{Q} \exp(\Delta S_i/R) \exp(-\Delta H_i/RT) \end{aligned} \qquad (5.15)$$

$$\langle H \rangle = \sum_{i=1}^{n} \Delta H_i \cdot P_i = \frac{1}{Q} \sum_{i=1}^{n} \Delta H_i \cdot \exp\left(-\Delta G_i/RT\right) \qquad (5.16)$$

Straightforward differentiation of Equation (5.14) (and use of the van 't Hoff equation $\partial \ln K_i/\partial T = \Delta H_i/RT^2$) leads to the following relation between the average enthalpy and temperature derivative of the partition function,

$$\frac{\partial \ln Q}{\partial T} = \frac{\langle H \rangle}{RT^2} \qquad (5.17)$$

and differentiation in Equation (5.16), together with Equation (5.17), produces the following expression for the excess heat capacity,

$$C_P^{EX} = \frac{\partial \langle H \rangle}{\partial T} = \sum_{i=1}^{n} \Delta C_{P,i} \cdot P_i + \frac{\langle H^2 \rangle - \langle H \rangle^2}{RT^2} \quad (5.18)$$

where $\Delta C_{P,i}$ is the heat capacity of state I_i with respect to the reference state (the native state: I_0), and $\langle H^2 \rangle$ is the average value of the squared enthalpy (given by Equation (5.4) with $X_i = (\Delta H_i)^2$). The first term in the right-hand side of Equation (5.18) is the average heat capacity of all protein states (the "chemical baseline") and the second term represents the contribution from the temperature-induced shift in states populations (compare Equations (5.18) and (5.11)). However, the most interesting result from the analysis of the multi-state equilibrium is related to the fact that the average enthalpy is proportional to a temperature derivative of the partition function (Equation (5.17)) whereas the excess heat capacity equals a temperature derivative of the average enthalpy (first equality in Equations (5.10) and (5.18)).

From these results, it follows that the partition function is given by a double integral of the heat capacity:

$$\ln Q = \int_\Theta \left[\frac{1}{RT^2} \int_\Theta C_P^{EX} \cdot dT \right] \cdot dT \quad (5.19)$$

where integration starts at a temperature, Θ, low enough to ensure that the probability of occupation of the native state is essentially unity ($P_0 = 1$). Equation (5.19) shows that the experimental DSC thermograms are equivalent to the partition function and, therefore, provides a complete description of the process. This point was first noted, about 30 years ago, by Ernesto Freire and Rodney Biltonen,[14] who showed how the number of significantly populated states in the multi-state equilibrium (Equation (5.12)) and their energetic parameters can be obtained from the partition function using a simple "peeling-off" procedure. Nowadays, with the popularization of computers, a non-linear least-squares fit of Equation (5.18) to the experimental heat capacity data is the preferred approach.

The multi-state equilibrium discussed above (Equation (5.12)) assumes that the protein can exist in a given number of distinct and well-defined states (i.e., protein *macro*states). However, protein folding/unfolding is not a chemical reaction in the sense that it does not involve breaking and forming of one or a few strong covalent bonds, but the reorganization of myriads of weak non-covalent interactions. For certain kinds of analyses, therefore, it is preferable to view protein folding/unfolding as a continuous process that involves an ensemble of protein microstates. This is a more general approach in which the existence of well-defined macrostates (such as "native" and "unfolded") is not imposed *a priori* (as in the chemical model described by Equation (5.12)).

For this purpose, we consider enthalpy as a continuous variable and thus the partition function is written as an integral over enthalpy (instead of as a sum over discrete states):[22]

$$Q = \int \rho(H) \cdot \exp(-H/RT) \cdot dH \qquad (5.20)$$

where we have written H (instead of ΔH), since in Equation (5.20) we have not specified the reference state. $\rho(H)$ in Equation (5.20) plays a role analogous to $\exp(\Delta S_i/R)$ in Equation (5.14). Indeed, $\rho(H)dH$ gives the number of protein microstates in the infinitesimal enthalpy interval $\{H, H + dH\}$. In the statistical thermodynamics field, $\rho(H)$ is termed the "density of states" (see Goodstein[23] for several illustrative applications). Note that, in this formulation of the partition function, the enthalpy scale and the density of states are taken to be independent of temperature. That is, microstates are assigned constant energy (enthalpy) value.

Other properties such as entropy and heat capacity arise from the characteristic probability distribution of the ensemble of microstates. In particular, the heat capacity defines the temperature dependence of the average enthalpy. The probability of finding the protein in a microstate of enthalpy H at a given temperature is obtained with an expression analogous to Equation (5.15):

$$P(H) = \frac{1}{Q}\rho(H)\exp(-H/RT) \qquad (5.21)$$

Note that, since H is now a continuous variable, $P(H)$ is actually a probability density, so that $P(H) \cdot dH$ gives the probability of finding enthalpy values within the infinitesimal range $\{H, H + dH\}$. The average value of a quantity X (provided that it can be expressed as an average over protein microstates) is now,

$$\langle X \rangle = \int XP(H) \cdot dH \qquad (5.22)$$

and the excess heat capacity,

$$C_P^{EX} = \frac{\langle H^2 \rangle - \langle H \rangle^2}{RT^2} \qquad (5.23)$$

which is identical to Equation (5.18), except for the first term on the right-hand side of Equation (5.18) that is absent here. The reason obviously is that in an ensemble-based description heat capacities are properties of the complete ensemble and are not individually assigned to microstates. Equation (5.17), which connects the partition function with the calorimetric data, is still valid in the continuous case.

The formalism embodied in Equations (5.20)–(5.23) is simply the continuous analogue of the Freire–Biltonen approach (Equations (5.14)–(5.16) and (5.18)) and provides the basis of recent developments addressed at determining folding/unfolding barriers from calorimetric data (see further below).

5.3 The Two-state Equilibrium Model: A Historical Perspective

The two-state model is the most popular mechanism currently used to describe protein denaturation processes. It is also the simplest one as only native and unfolded (reversibly denatured) states are assumed to be significantly populated. According to the two-state model, denaturation is an all-or-none reaction in which the protein is in either the fully native (N or I_0 in the terminology of the preceding section) or the fully unfolded state (U or I_1), strictly corresponding to an infinite degree of cooperativity between protein interactions:

$$N \Leftrightarrow U \qquad (5.24)$$

The interpretation of protein conformational transitions *via* a two-state picture is deeply rooted in the current protein-folding literature. One of the reasons is that its simplicity allows for a complete thermodynamic characterization of the process in a very *easy* and *convenient* manner (provided, of course, that the model holds). However, the extensive use of this approximation in protein folding and stability studies is such that, quite often, it is taken for granted without strong experimental evidence.

The validity of the two-state scenario to describe reversible protein denaturation processes was hotly debated among early investigators in the protein-folding field. At the time, the amount of experimental data and its quality was obviously limited, giving rise to different views of the folding process. The two-state approximation for protein folding/unfolding was already used in the 1930s by Anson and Mirsky,[24] but different theoretical and experimental results published in the 1960s appeared to support a less-cooperative, multi-state, or gradual mechanism.[25–28] For example, by comparing experimental and theoretical values for the steepness of urea-induced unfolding curves for a number of small globular proteins Tanford concluded in 1964 that the existence of stable intermediate forms between native and unfolded states must be a general phenomenon.[28] A similar argument was proposed later on by Poland and Scheraga.[25] They used different approaches (theoretical arguments, model calculations and experimental data for several proteins) to propose a mechanism of gradual unfolding, rather that an all-or-none, two-state type of denaturation. Furthermore, based on theoretical simulations, Poland and Scheraga pointed out that a small finite degree of cooperativity between interactions can result in rather sharp transitions, as sharp as those that were being observed experimentally for many proteins.

In the present authors' view, a critical explanation of the meaning and implications of the two-state model, which is still useful today, can be found in a review paper published by Lumry, Biltonen, and Brandts in 1966.[29] In this work the native and unfolded states of the two-state model (and its extension to the multi-state equilibria as in Equation (5.12)) are considered macrostates corresponding to ensembles of protein microscopic states. These microstates could be formally defined in terms of given sets of protein/solvent conformations). It is also clear from the Lumry–Biltonen–Brandts analysis that the existence of well-defined macrostates implies a very low population for microstates of intermediate degree of unfolding. They illustrated the idea by using energy ($\cong$ enthalpy) as a measure of the degree of unfolding and plotting probability of microstate occupation versus energy (see Figure 5.2).

It is very important to note now that probability is related to free energy through a Boltzmann exponential [probability $\propto \exp(-G/RT)$]. Therefore, a lower probability for microstates of intermediate degree of unfolding means a *higher* free energy. In other words, the existence of well-defined macrostates implies significant free-energy barriers (*i.e.*, significantly higher than the thermal energy RT), as is shown in the illustrative examples of Figure 5.2. Moreover, it is important to keep in mind that protein folding is not a chemical reaction. While in an organic chemistry reaction typically one strong covalent bond is broken and formed, protein folding involves reorganizations of thousands of weak, non-covalent interactions (hydrogen bonding, van der Waals interactions, hydrophobic effect, *etc.*). We may expect an organic chemistry reaction to occur through very-well-defined intermediate states separated by high-energy barriers (since breaking a covalent chemical bond may require energies on the order of hundreds of kJ mol^{-1}). For protein folding, there is no physical reason for the barriers to be high (since covalent bonds are not formed/broken). Indeed, kinetic studies and theoretical analyses suggest that folding barriers for many natural proteins are small[30-32] (see Chapter 3 by Wolynes and Chapter 6 by Gruebele for more details on the magnitude of free-energy barriers to protein folding). Furthermore, computer-designed proteins have been found to fold faster than their natural counterparts, although no selection for folding efficiency was included in the design[33] and theoretical analyses of polymer models[34] indicate that it may be harder for proteins to achieve cooperativity (a large free-energy barrier and clear two-state behavior) than a stable folded structure.

All of the above suggests that the observation of significant folding barriers for natural proteins is not intrinsic to the folding process itself, but rather the result of natural selection (see Chapter 3 by Wolynes). Thus, as we discuss below, a significant barrier may be advantageous in order to guarantee kinetic stability of the native state when confronted with the destabilizing effect of irreversible alterations (such as those involved in misfolding diseases).[35] In any case, energy barriers must be necessarily rather small for single domain proteins that fold fast (otherwise it is not clear how they can fold so fast). Interestingly, if the thermodynamic barrier is of the order of the thermal energy (a few kJ mol^{-1}) microstates of all degrees of unfolding may become populated during

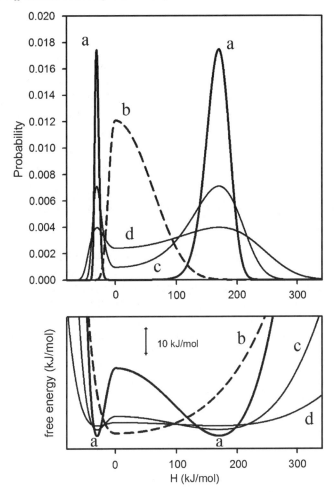

Figure 5.2 Interpretation of protein macrostates as ensembles of microstates. An enthalpy ($\cong$ energy) scale is used as a measure of the degree of unfolding and plots of probability and free-energy *versus* enthalpy are given in the upper and lower panels, respectively (calculated from Landau free-energy functionals, as explained in the text). The line labeled a in the upper panel corresponds to a two-macrostate situation, since microstates of intermediate degree of unfolding are not significantly populated; this implies a higher free energy for those microstates and, therefore, a thermodynamic barrier between the two macrostates, as is shown by line labeled a in the lower panel. A barrierless free-energy profile (line labeled b in the lower panel) produces a single macrostate (line labeled b in the upper panel). The thin lines labeled c and d correspond to barrier heights on the order of the thermal energy (marginal barriers): twice the thermal energy (c) and half the thermal energy (d). These profiles show clear deviations from the two-macrostates scenario and, in fact, the results for a barrier smaller than the thermal energy are almost identical to the strict barrierless case.

denaturation at equilibrium. For a sufficiently small barrier, the free energy *versus* degree of unfolding profile would essentially show a single minimum that would shift upon changing temperature and the folding–unfolding process would be continuous and involve a single macrostate (see Figure 5.2). Note also that, in a free-energy surface with only one minimum, the folding–unfolding relaxation is always downhill (since no barrier needs to be overcome in reaching the surface minimum). Experimental evidence for the existence of this global downhill regime in some cases has been recently reported[22,36–39] (see Chapter 6 by Gruebele).

5.4 Folding Free-energy Barriers from Equilibrium DSC Experiments

It is clear from the discussion in the preceding section that significant deviations from two-state behavior may perhaps occur often with small proteins, not due to the existence of additional significantly populated macrostates ("intermediate" states), but to a low thermodynamic folding/unfolding barrier. This situation may have gone unnoticed in many cases, since methods of analysis based on "chemical-like models" (Equations (5.12) and (5.24)) do not take such possibility into account. In addition, small proteins usually give rise to very broad unfolding transitions, with poorly defined low- and high-temperature heat capacity levels. Under these circumstances, good fits of the two-state model can always be achieved, provided that some "flexibility" in native and unfolded baseline tracing is allowed (see below for further discussion on this issue).

However, it has been recently shown[22] that the continuous formulation of the partition function formalism (Equations (5.20)–(5.23)) allows the analysis of DSC data to be posed in terms of a one-dimensional folding/unfolding free-energy surface, in the same spirit of the energy landscape approach to protein folding (see Chapter 3 by Wolynes). In this case, the presence of a large barrier separating the folded and unfolded minima on the surface is not implicitly assumed (as in the "traditional" two-state model), but rather estimated from the shape and broadness of the DSC thermogram.

Essentially, what is needed for this kind of analysis is a procedure to describe processes that, depending on conditions, behave as first-order (two clearly defined macrostates "separated" by a high barrier) or continuous (low or non-existing barrier and, therefore, a single macrostate). Fortunately, this problem also arises in a well-known branch of thermodynamics: the theory of critical transitions. For instance, the gas and liquid phases of a given substance can coexist at equilibrium (as two distinct phases) for temperatures and pressures in the liquid–vapor equilibrium line. As temperature and pressure are increased along that line, liquid and gas become similar and eventually merge into a single phase at the critical point (for a pictorial illustration, see Sengers and Sengers[40]). In the classical Landau theory of critical transitions (see Chapter 10 in Callen[41]) this is phenomenologically described with a free-energy functional expressed as a series expansion in powers of an "order parameter" (the thermodynamic quantity that exhibits large fluctuations near critical conditions) and truncating the expansion at the quartic level. The truncated expansion produces a free energy functional with one or two free-energy minima depending on the sign of the coefficient of the quadratic term.

A Landau free energy functional can be implemented in DSC data analysis by writing the probability density for enthalpy microstate occupation (Equation (5.21)) at a given characteristic temperature (T_0) in terms of the enthalpy dependence of the free energy,

$$P(H) = C \cdot \exp(-G_0(H)/RT_0) \qquad (5.25)$$

and expressing the free energy as,

$$G_0(H) = -2\beta \left(\frac{H}{\alpha}\right)^2 + |\beta| \left(\frac{H}{\alpha}\right)^4 \qquad (5.26)$$

which is actually the Landau functional with the coefficients of the H^2 and H^4 expressed in terms of two parameters, α and β. These parameters have a clear and intuitive meaning (see Muñoz and Sanchez-Ruiz[22] for details).

For $\beta > 0$, $G_0(H)$ has a maximum at $H=0$ and two minima at $H=\pm\alpha$. Therefore, for $\beta > 0$, there are two macrostates with minima separated by an enthalpy of 2α. In this case, β corresponds to the height of the free-energy barrier separating the two minima at the characteristic temperature. For $\beta \leq 0$, the free-energy profile shows only a minimum and there is only one macrostate. In this case, α and β are just convenient parameters that describe the shape of the free-energy functional. Therefore, it is just the *sign* of the parameter β that determines the observation of either two macrostates or a single macrostate at the characteristic temperature T_0. Of course, positive but very small values of β (of the same scale of the thermal energy, RT) are essentially equivalent to the single-macrostate, barrierless case. A final modification is introduced in the free-energy functional to take into account the asymmetry expected in a folding–unfolding process. Such modification is explained in some detail in Muñoz and Sanchez-Ruiz[22] and illustrated by the free energy and probability density profiles displayed in Figure 5.2.

The important point, however, is that the above approach permits the calculation of heat capacity *versus* temperature profiles for different values of the thermodynamic folding/unfolding barrier (β). Thus, the probability density at any temperature can be easily calculated from the one at the characteristic temperature (Equations (5.25) and (5.26)).The averages $<H^2>$ and $<H>$ are then obtained from Equation (5.22) and the heat capacity value from Equation (5.23) (see ref. 22 for further details). A practical implication of this exercise is that it becomes possible to set a non-linear least-squares procedure to fit experimental DSC data in which the height of the folding barrier is an experimentally obtainable parameter. Such procedure does not impose the existence of two well-defined macrostates but, instead, the two-macrostate or single-macrostate character of the thermal unfolding process is the outcome of the analysis. In fact, in the first application of this *variable-barrier* phenomenological model,[22] a significant barrier was determined for the two-state protein *E. coli* thioredoxin. On the other hand, essentially no barrier was found for the small, fast-folding protein BBL, which had been previously characterized as a downhill folder[36] (see Chapter 6 by Gruebele for more on downhill folding).

An important point to note is that, *a priori*, the barrier heights determined from the variable-barrier model do not necessarily have a kinetic meaning. That is, within the context of the model, the role of the barrier is to reduce the population of intermediate microstates. A kinetic meaning for this "thermodynamic" barrier would imply that the folding/unfolding process can be described to some acceptable extent by a single reaction coordinate and the enthalpy scale used in the variable-barrier analysis is an acceptable approximation to that reaction coordinate. However, a very recent analysis[39] including 15 proteins demonstrates an excellent correlation between the barrier heights derived from the variable-barrier analysis of the DSC transitions and the experimental folding rates. This result suggests the surprising and unexpected possibility of estimating folding *kinetic* barriers from equilibrium DSC data.

5.5 The van 't Hoff to Calorimetric Enthalpy Ratio Revisited

One of the main motivations for the original development of differential scanning calorimetry as a tool to study protein solutions was the need for reliable tests of the validity of the two-state model.[42] The reason is that finding "hard" evidence for two-state behavior is not as straightforward as it might seem at first. For a two-state denaturation, agreement between the experimental transition profiles obtained using different physical probes (fluorescence, far-UV CD, near UV-CD, *etc.*) is expected. However, while a clear disagreement between such profiles effectively rules out two-state behavior, the agreement does not constitute definitive evidence in its support, since there is always the possibility that an additional transition profile (based on a yet untested physical probe) could differ.

On the other hand, DSC leads to a straightforward two-state test based upon the fact that the unfolding enthalpy is determined in two different ways from the DSC experiment. The unfolding enthalpy can be directly calculated from the area under the heat capacity peak, resulting in the so-called calorimetric enthalpy. In addition, it can be obtained from the shape (width) of the transition, since the unfolding enthalpy determines the temperature dependence of the unfolding equilibrium constant. A high value for the unfolding enthalpy implies that the unfolding equilibrium constant shows a high-temperature-dependence. In other words, the transition from the native state ($\sim 100\%$ native) to the unfolded state ($\sim 100\%$ unfolded) occurs in a narrow temperature range. Conversely, when the unfolding enthalpy value is low, the unfolding equilibrium constant changes less abruptly with temperature and the DSC transition is wide (see Figure 5.1). The unfolding enthalpy value calculated from the shape of the DSC transition is known as the van 't Hoff enthalpy (after the van 't Hoff equation that gives the temperature dependence of any equilibrium constant). The calculation of the calorimetric enthalpy is essentially model-independent, while the calculation of the van 't Hoff enthalpy relies on the two-state model. Therefore, the agreement between the two values constitutes evidence for two-state behavior. Nowadays, this test is usually carried out on the basis of the non-linear, least-squares fitting of a pseudo-two-state equation, analogous to Equation (5.10) also including a native heat capacity baseline.

The calorimetric criterion based in the comparison between the calorimetric and van 't Hoff enthalpies has been successfully used in many studies to ascertain the validity of the two-state model. However, an important point of its application is often overlooked. Small proteins have necessarily low values of the denaturation enthalpy and, therefore, give rise to very broad DSC transitions in which the native and unfolded baselines (the approximately linear pre- and post-transition heat capacity levels) are not clearly apparent. In these cases, good fits to the experimental data with calorimetric enthalpy equal or close to the van 't Hoff enthalpy can almost certainly be obtained, provided that the native and unfolded baselines are adequately "chosen" by the person who analyses the data or by the fitting program.[43,44] Obviously, these fits are not proof of two-state behavior without evaluation of the baselines involved in the fitting. In particular, crossing of the native and unfolded baselines at a temperature in the middle of the transition is clearly unphysical (it would imply that the unfolding heat capacity change becomes negative at that temperature). Recently, such crossings have been detected in several experimental cases and interpreted indeed as evidence of deviation from two-state behavior.[36,38,45,46]

The above consideration is particularly relevant in light of the recent developments indicating that deviations from two-state behavior may often occur for small proteins due to marginal (or even non-existent) folding/unfolding thermodynamic barriers. In fact, it is not clear that the van 't Hoff to calorimetric enthalpy ratio can detect such situations, at least without evaluation of the fitted baselines. The point is illustrated in Figure 5.3, which shows several DSC profiles calculated on the basis of the continuous formulation of the Freire–Biltonen partition function formalism (Equations (5.20)–(5.23)) and the Landau free-energy functional (Equation (5.26)). Marginal values (smaller than the thermal energy) have been used for the barrier height (β), which produce a clear deviation from two-state behavior due to significant population of microstates of intermediate degree of unfolding. We also show the best fits of a pseudo-two-state equilibrium that includes van 't Hoff and calorimetric enthalpy values as independent fitting parameters. The result is that the van 't Hoff to calorimetric enthalpy ratios derived from the fittings are close to unity. However, the crossing of the fitted baselines in these cases (see Figure 5.3) provides a clear clue of the non-physical character of such two-state fits.

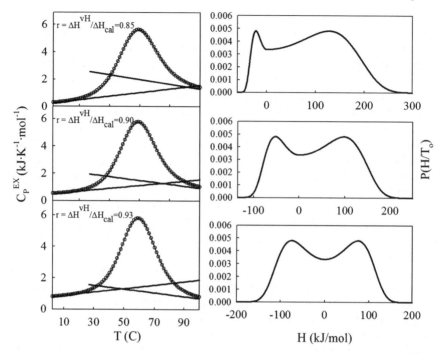

Figure 5.3 Profiles of excess heat capacity *versus* temperature (left) calculated from Landau free energy functionals, as explained in the text. The corresponding profiles of probability of microstate occupation versus enthalpy at the characteristic temperature are also shown (left). In all cases, a barrier significantly smaller than the thermal energy has been used in the calculation so that two distinct macrostates are *not* observed. However, it is possible to achieve good fits of a pseudo-two-state model (continuous lines in the heat capacity plots) with van 't Hoff to calorimetric enthalpy ratios reasonably close to unity. Such fits, however, involve fitting native and unfolded baselines (straight lines in the heat capacity plots) that cross at a temperature in the middle of the transition range. This would imply that the unfolding heat capacity depends strongly on temperature and changes sign in the transition temperature range, a physically unrealistic result.

Overlooking the fitted baselines would mislead us into assigning two-state character to processes which are closer to global downhill folding.[47]

5.6 Protein Kinetic Stability: Free-energy Barriers for Irreversible Denaturation from Scan-rate Dependent DSC

The possibility of determining folding kinetic barriers from equilibrium DSC experiments, as described in Section 5.4, may certainly come as a surprise to the biochemist. Nevertheless, kinetic barriers for irreversible protein denaturation have been studied for many years (see Plaza del Pino et al.[35] and references quoted therein) on the basis of the scan-rate effect of DSC transitions, an approach that reveals the possible kinetic-control character of the denaturation process. As in the preceding sections, some of the general features of this approach can be explained with a partition-function-based analysis.

Assuming a multi-state equilibrium, such as that depicted in Equation (5.12), and that the most unfolded state can undergo an irreversible alteration (aggregation, chemical modification of amino acids, etc.[35,48]) that leads to a "final" state that is unable to fold back the native state,

$$I_n \rightarrow F \tag{5.27}$$

and that the conversion is determined by a simple first-order rate constant,

$$\frac{d[F]}{dt} = k[I_n] \tag{5.28}$$

which, for the sake of simplicity and illustration, we take to be temperature-independent. It is straightforward to write the above rate equation as,

$$\frac{d[F]}{dT} = \frac{k}{\alpha} \cdot (C_t - [F]) \cdot P_n \tag{5.29}$$

where C_t is the total protein concentration (including the I_i states in equilibrium (Equation (5.12)) and the final, irreversibly denatured state), P_n is the probability of occupation of state I_n (as given by $(1/Q) \cdot \exp(-\Delta G_n/RT)$; see Equation (5.15)), and we have used the fact that in a DSC experiment temperature increases with time according to a constant scan rate ($\alpha = dT/dt$).

Separation of variables in Equation (5.29) followed by integration from a low temperature (T_0) at which essentially all the protein is in the native state ($P_0 \cong 1$ and $[F] \cong 0$) leads to,

$$\ln \frac{C_t - [F]}{C_t} = -\frac{k}{\alpha} \int_{T_0}^{T} P_n dT \tag{5.30}$$

with [F] equal to the concentration of final state at the temperature used as the upper integration limit. We now use as that limit the temperature ($T_{1/2}$) at which, for a given scan rate, half of the protein has denatured irreversibly (that is, for $T = T_{1/2}$, $[F] = C_t/2$) and solve for the integral in Equation (5.30) to obtain,

$$\int_{T_0}^{T_{1/2}} P_n dT = \alpha \cdot \tau \tag{5.31}$$

where $\tau [=(\ln 2)/k]$ is the half-life time for the irreversible denaturation step (Equation (5.27)). Equation (5.31) implies that for a sufficiently fast irreversible alteration step (very short half-life time) the integral on the right-hand side must necessarily be a very small number, which means a very low population of state I_n throughout the temperature range in which irreversible denaturation takes place. In other words, if the $I_n \to F$ step is sufficiently fast, irreversible denaturation occurs through a very small amount of state I_n, which in turn never gets significantly populated. It can be easily shown along the same lines that, if the irreversible alteration step is fast enough, irreversible denaturation will take place in a temperature range in which all I_i states (except the native state, I_0) are not significantly populated. The result is a denaturation process that can be phenomenologically described as the kinetic conversion from the native protein to the final, irreversible denatured state:

$$I_0 \to F \qquad (5.32)$$

This is a situation usually referred to as the *two-state irreversible model*,[7,49] since only states I_0 and F are significantly populated.

The two-state irreversible model is found as a limiting case in different theoretical analyses of protein denaturation that include irreversibility in a realistic manner.[6,7,35,49] More importantly, the *in vitro* irreversible thermal denaturation of many proteins (in particular, large complex protein systems) has been found to conform to this very simple model (see Plaza del Pino *et al.*[35] and references quoted therein). It is clear that, in these cases, the partition function cannot be obtained from the DSC data, since the states I_1 though I_n of the multi-state equilibrium (Equation (5.12)) never get significantly populated. For two-state irreversible denaturation (Equation (5.32)) the kinetic distortion introduced by the irreversible step eliminates all thermodynamic information about the equilibrium thermal unfolding mechanism.

From a general viewpoint, however, the compliance of many protein thermal denaturation processes with a two-state irreversible model indicates that protein stability (*in vitro*, as well as *in vivo*) is often of kinetic origin and determined by a free-energy barrier that can be characterized by the scan-rate dependence in DSC experiments. In other words, thermodynamic stability (a positive value for the unfolding free energy at physiological temperature) does not guarantee that the protein will remain in the native (biologically functional) state during the biologically relevant time period, since irreversible protein alterations (even if they occur from lowly populated unfolded or partially unfolded states) may deplete the native state in a time-dependent manner.[35] It appears likely then that many proteins, particularly complex protein systems, must have been naturally selected to have significant kinetic stability.[35,50]

The interest of understanding protein kinetic stability is emphasized by the fact that some emerging molecular approaches to the inhibition of amyloidogenesis focus on the increased kinetic stability of the protein native state,[51–53] as is in fact suggested by the simple Lumry–Eyring models of irreversible denaturation.[35] Furthermore, kinetic stability may be of considerable biotechnological importance. Researchers interested in fundamental aspects of protein folding may generally choose "model" proteins and solvent conditions in such a way that equilibrium folding–unfolding is observed. However, the proteins and/or solvent conditions employed in technological applications often involve irreversible denaturation and kinetic control of the stability. In fact, enhancing protein stability for biotechnological applications may in many cases mean enhancing protein kinetic stability. Scan-rate dependent DSC studies and the subsequent data analyses based on suitable kinetic models[6,7,35,49] certainly provide a convenient approach to the characterization of protein kinetic stability and the associated free-energy barrier.

Finally, it is worth noting that we have discussed here two different applications of the statistical analysis of differential scanning calorimetry that are complementary to one another. The variable-barrier analysis of equilibrium DSC data discussed in Section 5.4 is targeted to the determination of marginal folding/unfolding barriers in small, fast-folding proteins. On the other hand, scan-rate dependent DSC studies of irreversible protein denaturation processes provide information about the much larger barriers that determine kinetic stability in more complex protein systems.

References

1. P. L. Privalov, *Adv. Protein Chem.*, 1979, **33**, 167–241.
2. P. L. Privalov, *Adv. Protein Chem.*, 1982, **35**, 1–104.
3. P. L. Privalov, *Annu. Rev. Biophys. Biophys. Chem.*, 1989, **18**, 47–69.
4. E. Freire, *Methods Mol. Biol.*, 1995, **40**, 191–218.
5. E. Freire, *Methods Enzymol.*, 1994, **240**, 502–530.
6. E. Freire, W. W. van Osdol, O. L. Mayorga and J. M. Sanchez-Ruiz, *Annu. Rev. Biophys. Biophys. Chem.*, 1990, **19**, 159–188.
7. J. M. Sanchez-Ruiz, *Subcell. Biochem.*, 1995, **24**, 133–176.
8. G. I. Makhatadze and P. L. Privalov, *Adv. Protein Chem.*, 1995, **47**, 307–425.
9. V. V. Plotnikov, J. M. Brandts, L. N. Lin and J. F. Brandts, *Anal. Biochem.*, 1997, **250**, 237–244.
10. M. M. Lopez and G. I. Makhatadze, *Methods Mol. Biol.*, 2002, **173**, 121–126.
11. P. L. Privalov, *Pure Appl. Chem.*, 1980, **52**, 479–497.
12. G. I. Makhatadze, *Current Protocols in Protein Science*, Wiley, New York, 1998, 7.9.1–7.9.14.
13. B. Ibarra-Molero and J. M. Sanchez-Ruiz, *Emerging Techniques in Biophysics*, ed. J. L. Arrondo and A. Alonso, Elsevier, Amstadam, 2006, 27–48.
14. E. Freire and R. Biltonen, *Biopolymers*, 1978, **17**, 463–479.
15. A. D. Robertson and K. P. Murphy, *Chem. Rev.*, 1997, **97**, 1251–1268.
16. I. Luque and E. Freire, *Methods Enzymol.*, 1998, **295**, 100–127.

17. A. Cooper, *Biophys. Chem.*, 2005, **115**, 89–97.
18. M. S. Moghaddam, S. Shimizu and H. S. Chan, *J. Am. Chem. Soc.*, 2005, **127**, 303–316.
19. J. A. Schellman, *Annu. Rev. Biophys. Biophys. Chem.*, 1987, **16**, 115–137.
20. P. L. Privalov, *Crit. Rev. Biochem. Mol. Biol.*, 1990, **25**, 281–305.
21. C. R. Babu, V. J. Hilser and A. J. Wand, *Nat. Struct. Mol. Biol.*, 2004, **11**, 352–357.
22. V. Muñoz and J. M. Sanchez-Ruiz, *Proc. Natl. Acad. Sci. USA*, 2004, **101**, 17646–17651.
23. D. L. Goodstein, *States of Matter*, Prentice Hall, Englewood Cliffs, New Jersey, 1975.
24. M. L. Anson and A. E. Mirsky, *J. Gen. Physiol.*, 1934, **17**, 393–398.
25. D. C. Poland and H. A. Scheraga, *Biopolymers*, 1965, **3**, 401.
26. S. J. Gill and R. L. Glogovsky, *J. Phys. Chem.*, 1965, **469**, 1515–1519.
27. W. A. Klee, *Biochemistry*, 1967, **6**, 3736–3742.
28. C. Tanford, *J. Am. Chem. Soc.*, 1964, **86**, 2050–2059.
29. R. Lumry, R. Biltonen and J. F. Brandts, *Biopolymers*, 1966, **4**, 917–944.
30. J. Kubelka, J. Hofrichter and W. A. Eaton, *Curr. Opin. Struct. Biol.*, 2004, **14**, 76–88.
31. A. Akmal and V. Muñoz, *Proteins*, 2004, **57**, 142–152.
32. A. N. Naganathan and V. Muñoz, *J. Am. Chem. Soc.*, 2005, **127**, 480–481.
33. M. Scalley-Kim and D. Baker, *J. Mol. Biol.*, 2004, **338**, 573–583.
34. H. S. Chan, S. Shimizu and H. Kaya, *Methods Enzymol.*, 2004, **380**, 350–379.
35. I. M. Plaza del Pino, B. Ibarra-Molero and J. M. Sanchez-Ruiz, *Proteins*, 2000, **40**, 58–70.
36. M. M. Garcia-Mira, M. Sadqi, N. Fischer, J. M. Sanchez-Ruiz and V. Muñoz, *Science*, 2002, **298**, 2191–2195.
37. F. Y. Oliva and V. Muñoz, *J. Am. Chem. Soc.*, 2004, **126**, 8596–8597.
38. A. N. Naganathan, R. Perez-Jimenez, J. M. Sanchez-Ruiz and V. Muñoz, *Biochemistry*, 2005, **44**, 7435–7449.
39. A. N. Naganathan, J. M. Sanchez-Ruiz and V. Muñoz, *J. Am. Chem. Soc.*, 2005, **127**, 17970–17971.
40. J. V. Sengers and A. L. Sengers, *Chem. Eng. News*, 1968, **46**, 104–118.
41. H. B. Callen, *Thermodynamics and an Introduction to Thermostatistics*, Wiley, New York, 1985.
42. W. M. Jackson and J. F. Brandts, *Biochemistry*, 1970, **9**, 2294–2301.
43. Y. Zhou, C. K. Hall and M. Karplus, *Protein Sci.*, 1999, **8**, 1064–1074.
44. H. Kaya and H. S. Chan, *Proteins*, 2000, **40**, 637–661.
45. M. P. Irun, M. M. Garcia-Mira, J. M. Sanchez-Ruiz and J. Sancho, *J. Mol. Biol.*, 2001, **306**, 877–888.
46. A. I. Dragan and P. L. Privalov, *J. Mol. Biol.*, 2002, **321**, 891–908.
47. N. Ferguson, P. J. Schartau, T. D. Sharpe, S. Sato and A. R. Fersht, *J. Mol. Biol.*, 2004, **344**, 295–301.

48. A. Klibanov and T. J. Ahern, *Protein Engineering*, eds D. L. Oxender and C. F. Fox, Alan R. Liss, New York, 1987, 213.
49. J. M. Sanchez-Ruiz, *Biophys. J.*, 1992, **61**, 921–935.
50. S. S. Jaswal, S. M. Truhlar, K. A. Dill and D. A. Agard, *J. Mol. Biol.*, 2005, **347**, 355–366.
51. P. Hammarstrom, R. L. Wiseman, E. T. Powers and J. W. Kelly, *Science*, 2003, **299**, 713–716.
52. H. M. Petrassi, S. M. Johnson, H. E. Purkey, K. P. Chiang, T. Walkup, X. Jiang, E. T. Powers and J. W. Kelly, *J. Am. Chem. Soc.*, 2005, **127**, 6662–6671.
53. R. L. Wiseman, S. M. Johnson, M. S. Kelker, T. Foss, I. A. Wilson and J. W. Kelly, *J. Am. Chem. Soc.*, 2005, **127**, 5540–5551.

CHAPTER 6
Fast Protein Folding

MARTIN GRUEBELE

Department of Chemistry, Department of Physics, and Center of Biophysics and Computational Biology, University of Illinois, Urbana, IL 61801, USA

6.1 Introduction

Globular proteins have evolved so that a large number of backbone and side-chain coordinates arrange themselves quite accurately and reasonably quickly to fold the protein. In 1968, this prompted Cyrus Levinthal to sum up the problem: "A pathway of folding means that there exist a well-defined sequence of events which follow one another . . . If the final folded state turned out to be the one of lowest configurational energy, it would be a consequence of biological evolution not of physical chemistry."[1] We now know that proteins need not necessarily fold through a unique pathway, but the spirit of the statement remains true: a small set of important coordinates can be navigated efficiently by the folding polypeptide chain. There is a well-defined set of events sampled by protein ensembles during folding, although each individual molecule may not sample from the set in the same sequential manner.

Levinthal's remark about evolution is timely in folding kinetics, not just in folding thermodynamics. With the recent discovery of globular proteins that fold on a microsecond time scale,[2] or even at the (size-dependent) $\approx 0.5\text{--}5\,\mu s$ 'speed limit' set by backbone diffusion and chain length,[3] the question arises: Why do most natural proteins fold so slowly, on a timescale of milliseconds to seconds, when engineered proteins of the same size fold near the speed limit?[4]

Before we answer this question, it is worth considering how proteins can fold so fast that the question must be posed. Counting only an average of three coordinates per residue (two for the backbone, one per side chain), even a 30-residue mini-protein requires 90 coordinates. Diffusive movement on an unbiased

Fast Protein Folding

very-high-dimensional coordinate hypersurface would indeed occupy a lot of time to find the small region in coordinate hyperspace where the protein is compact and well structured, unless the surface has some very special properties. These properties are summarized by the principles of consistency and minimal frustration,[5,6] giving the energy landscape a funnel-like shape when energy is plotted as a function of configurational entropy (Chapter 3 by Wolynes provides a detailed description of the energy landscape approach to protein folding). Such a shape means that the number of collective coordinates needed to describe events during folding is greatly reduced from all the microscopic coordinates (such as torsion angles). The number of collective folding coordinates is not necessarily 1, as implied by the many one-dimensional free-energy surfaces in the folding literature, but at least it is manageably small.[7] Search on a low-dimensional free-energy surface can be successful in a short period of time.

The need for low-dimensional, as opposed to one-dimensional, free-energy surfaces to describe the folding of biological macromolecules makes them fairly unique. Chemical reactions of small organic molecules are usually well described by a single reaction coordinate (for example a linear combination of torsion and sp^2 to sp^3 hybridization during isomerization around a double bond). Many-particle systems undergoing phase transitions (for example water freezing) are rigorously described by a single order parameter.[8] Proteins are in neither limit, lying somewhere between a small molecule reaction and a phase transition. This added complexity is the reason folding theory is only now reaching the maturity that chemical bond theories and phase transition models attained decades earlier. Nonetheless, surprisingly simple models can describe the dynamics and thermodynamics of proteins.[9–12]

Of course, a low-dimensional surface still allows multiple pathways. Figure 6.1 shows an example, fitted for a protein that folds near the speed limit. The ideas of multiple pathways in principle, and of a single dominant pathway in practice, are not incompatible: Let us assume thermodynamic control of the folding reaction, to get a back-of-the-envelope answer. The average population ratio on two pathways with average free energies $\overline{\Delta G_1}$ and $\overline{\Delta G_2}$ then becomes

$$P_1/P_2 = e^{-(\overline{\Delta G_1} - \overline{\Delta G_2})/RT} \tag{6.1}$$

Even if the energy landscape predicts several pathways within a few RT of one another in free energy, if one lies just 3 RT below the others, it will dominate the folding kinetics with 95% of the flux. Nonetheless, the existence of other low-lying pathways confers evolutionary robustness, and occasionally heterogeneous folding dynamics can be observed even in slower folders.[4,13]

The landscape picture postulates the existence of a class of proteins useful for the direct study of heterogeneous folding dynamics: downhill folders.[6] In downhill folders, the primary free-energy barrier has been removed either by natural selection or by protein engineering, leaving fluctuations below 2–3 RT in the free energy. Thus protein populations can be large anywhere along the reaction coordinate, and normally 'hidden' features of the free-energy surface can be studied.[4]

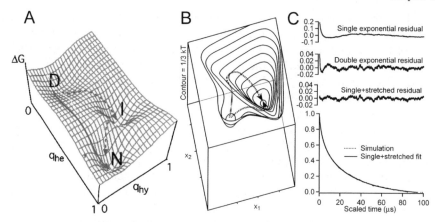

Figure 6.1 Multiple pathway scenarios. **A** Under optimal conditions, a direct pathway between D and N is preferred (solid arrow). However, a D→I→N pathway awaits only a few kT higher in free energy. If the sequence or solvent conditions are changed, adjusting the relative free energy of the pathways, both pathways may show up in the kinetics, or even just the second pathway alone (dotted arrow). **B** 2-D folding surface fitted to experimental data for the folding of λ_{6-85}.[108] Dynamics simulations on this surface show a combination of exponential (activated) and non-exponential (diffusive) kinetics characteristic of the transition from activated to downhill folding.

The existence of downhill-folding metastable intermediates,[14] natural proteins,[15] and engineered proteins[16] implies that energy barriers are not required for folding. Yet natural proteins often have significant ($>3\ RT$) barriers – at least most of those studied to date. Barriers may result from insufficient evolutionary pressure for fast folding, they may have evolved as a side effect of functional constraints, or they may have specifically evolved to reduce partial unfolding and hence aggregation and/or enzymatic protein degradation. Thus very fast folders have opened the door to new biological questions.[4]

The rest of this chapter considers experiments, theory, and simulations of fast (sub-millisecond) protein folding, and what we have learned about the physics and evolution of folding through this work.

6.2 Fast Folding: Why and How?

There are several fundamental and practical reasons for studying fast folding, some of them already alluded to earlier:

- The speed limit is correlated with protein size,[3] so many fast folders are also small, making them particularly amenable for theoretical analysis. The average rate has been correlated with contact order (measuring average loop size between contacts),[17] with chain length,[18] or with both.[19] Equally

interesting is the origin of the spread around the average, caused by "energetic frustration" of the protein (since contact order, particularly for mutants of the same protein, has already factored out "topological effects" caused by the sequence arrangement).[20] It has also been suggested that a more useful parameter than the spread is the downward deviation from the speed limit as a function of protein size.[21]
- Fast folders are amenable to atomistic simulation.[22] Simulations run at elevated temperatures have been compared with experimental unfolding rates, and distributed computing has made possible the direct comparison with experimental refolding rates,[23] with computing capacity now allowing simulations in explicit solvent. Thanks to a new generation of computers and distributed computing, molecular dynamics simulations can directly reveal heterogeneous folding dynamics and compute free-energy basins and their barriers along many reaction coordinates, and fast-folding data provide the necessary information to calibrate such simulations (see Chapter 8 by Pande for details on atomic simulations of fast-folding proteins).
- Fast folding can test important predictions of energy landscape theory that go beyond previous models of folding, such as the existence of downhill folders.[6] Even simple Go models,[15] in which energetic frustration has been removed from the picture, can be directly applied to fast folders, which presumably lack major energetic frustration and fold with a rate mainly limited by their fold topology.
- Fast folders provide the reference against which other proteins with the same fold topology can be judged. The activation barrier of a slow-folding variant of a fast folder cannot have fundamental physico-chemical causes, but must be a consequence of sequence selection by evolution (or lack thereof).[4]
- Fast folders are useful for single-molecule dynamics studies in particular. They spend a larger fraction of their time in intermediate conformations along the reaction coordinate(s), and less time in the native or unfolded wells. Thus more data about folding-unfolding transitions can be gathered (see Chapter 7 by Schuler on single-molecule studies of protein folding).[24,25]
- Fast folders are often uniquely stable, and the principles obtained from their folding kinetics and thermodynamics could be used to design unique stable protein structures to which new functions can be added.[4]

The know-how of fast folding experiments has developed rapidly since the mid 1990s, so an arsenal of techniques is now available to obtain resolution in the nanosecond-to-millisecond range.

- Equilibrium techniques based on line shape analysis have been applied to fast folding: in an NMR spectrum with two peaks assigned to the same proton in the folded and unfolded states, slow interconversion kinetics broadens these peaks, while fast kinetics eventually merges these peaks and finally yields a single narrow peak. This technique works roughly in the $50\,\mu s$ to 1 ms range, depending on the magnetic field strength used.[2]

- Equilibrium techniques based on single molecule fluorescence are getting more powerful and will soon provide the longer data samples needed to quantify rare dynamical events in the transition region of the free energy[25] (complemented by natural or engineered proteins that spend a larger fraction of their time undergoing interesting dynamics, and by new modeling approaches[26]).
- Temperature relaxation methods based on lasers span the full nanosecond-to-millisecond region, and even conventional heating systems are approaching microsecond time resolution.[27]
- Pressure relaxation methods have achieved microsecond resolution, so now the full P-T diagram can be explored, particularly useful for comparison with molecular dynamics simulation, where temperature and density are the most easily studied thermodynamic variables.[28,29]
- Optically induced relaxation techniques are beginning to find applications, from pH jumps to breakage/isomerization of constraining linkers. Such methods have been mainly tested on small peptides,[30] but should be easily generalized to proteins.
- Ultrafast mixing techniques have provided microsecond resolution for years, and are now pushing the sub-microsecond envelope.[31,32]
- Last, but not least, computer "experiments" have come of age. Free-energy surfaces can be sampled as a function of several coordinates,[22] replica exchange methods yield accurate equilibrium information,[33] and multi-microsecond trajectories of fully solvated small proteins are now possible, by studying either very large numbers of short trajectories obtained by distributed computing,[23] or single very long trajectories.[34] The experimental and simulation timescales have finally met on the microsecond timescale (see Chapter 8 by Pande).[22,23]

In the following sections, we consider briefly the dynamics of polypeptide chains as they apply to fast folding, and then discuss sub-millisecond protein folding's brief history, the instrumentation, and several case studies where modeling and experiment have gone hand in hand. Finally, downhill folding is reviewed, as the ultimate limit of fast folding, and its implications for the biological function of proteins are discussed.

6.3 Fast Dynamics of Polypeptide Chains

Proteins cannot fold any faster than their individual components (helices, sheets, loops) can form and dock against one another. Thus the dynamics of polypeptide chains forming loops contacts, secondary structure, and collapsing provides an upper limit for the folding rate of any given protein. The details of secondary structure formation are discussed elsewhere in this volume (see Chapter 1 by Doig and Chapter 2 by Doshi); the present discussion emphasizes those aspects directly connected to folding of mini-proteins and larger structures.

How fast a protein actually folds depends on both the initial and target native structure. Residual structure of the unfolded polypeptide chain can have a profound effect on folding rates. In fact, it has been proposed recently that there is a direct relationship between the height of the folding barrier and the degree of structure in the unfolded state.[35] A number of experiments and simulations have probed this residual structure, and found it to be significant even under conditions thought of as highly denaturing (pH 2, 5 M guanidine hydrochloride solutions and the like). For example, C_α chemical shifts and residual dipolar couplings have shown that apomyoglobin forms residual helical structure in 7 out of 8 helices even in the acid-unfolded state.[36] High concentrations of denaturant at high temperature yield short-range structure localized in the β-sheet (or polyproline II) basin of the Ramachandran plot.[37] Partially aligning protein molecules yields residual NMR couplings that demonstrate non-coil structure.[38] The number of such examples has increased rapidly in the last decade.

Smooth shifts of unfolded state structural content as a function of solvent conditions account for quasi-linear baselines frequently observed during temperature, pressure or denaturant titrations of proteins.[39] Even slow-folding (now meaning > 1 ms!) proteins with a significant activation barrier can rapidly rearrange unfolded structure after a jump in solvent conditions, increasing the compactness and raising secondary structure content in the unfolded basin before the main barrier is crossed (Figure 6.2). Acquisition of such structure is just as important as structure acquired after crossing the barrier. This is especially true for proteins with late transition states, where the difficult question is how all that structure manages to form before the barrier is crossed.[40] Large and rapid increases in unfolded state structure, observed for some proteins when conditions are switched to favor the native state, can be seen as an early type of partial downhill folding, discussed in detail later.

6.3.1 Loop Formation

In Kramers' analytical theory for activated condensed phase reactions, the rate of barrier crossing along a single collective reaction coordinate is given by[41]

$$k_{forward} = \nu^\dagger e^{-\Delta G^\dagger/RT} \quad (6.2)$$

Without a barrier, reaction occurs at the attempt frequency, or prefactor, $\nu^\dagger$. Kramers' model provides formulas for the prefactor based on well-defined assumptions about the free-energy surface. Using the diffusion coefficient for a Zimm chain (a freely jointed chain of N beads of mass m connected by links of length b subject to viscosity η),[42] and assuming a harmonic well and activation barrier with characteristic frequencies ω_{well} and ω_{TS}, the prefactor becomes

$$\nu^\dagger_{Zimm} = \frac{4\omega_{well}\omega_{TS}m}{3\pi^2\eta b\sqrt{6\pi N}} \quad (6.3)$$

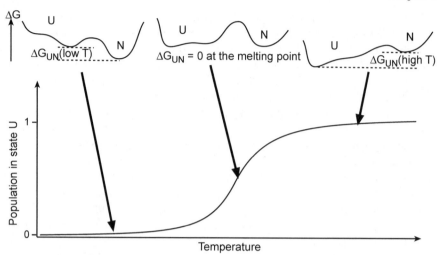

Figure 6.2 Structure can form in three ways during folding: downhill under conditions strongly favoring the native state (*e.g.* within the U well on the left side); climbing up the barrier (in all three examples), and *en route* from the barrier to the native state (particularly in the middle example). If a protein is suddenly switched from destabilizing (right) to stabilizing (left) conditions, the local minimum in the "U" well can move substantially towards the native state. The resulting additional structure formed downhill is no less important for folding than that structure formed climbing to or dropping from the activation barrier.

The characteristic frequencies are the inverses of the isomerization times for the links of the Zimm chain. In real proteins, this is the timescale for rotation about the Ramachandran angles or for individual residue basin hopping, which itself is controlled by a "micro-barrier." Such chemical barriers have been investigated by methods ranging from full quantum mechanics in the gas phase to implicit solvent or solvated molecular dynamics simulations.[43] For conjugated systems, that timescale is $\approx 10-100$ ps. We will use the lower value to obtain more conservative (higher) estimates for folding barriers. Assuming the bulk viscosity of 0.001 Pa s holds for the microscopic chain motion, $m \approx 100$ Da, $b \approx 4$ Å, and $N = 100$ for a 50-residue chain connecting two regions of secondary structure, we obtain $v^\dagger \approx (100 \text{ ns})^{-1}$.

This is indeed close to the timescales measured experimentally and computed from more sophisticated models for chain contacts for this length scale.[44,45] Three types of experiments have been conducted: quenching of tryptophan triplet states by cysteines,[46] energy transfer between two labels connected to the polypeptide chain,[47] and electron transfer-induced loop formation in proteins.[48] The fastest loop formers are glycines (*ca.* 5–10 ns for a 5-residue loop), while trans-prolines greatly increase the activation barrier for loop formation. For chains of length 100, the time rises to 100 ns, in agreement with Equation (6.3). The electron transfer experiments carried out in the environment of a whole protein also yield a slightly longer timescale, 250 ns for a 15-residue loop.

Likewise, experiments on interior (as opposed to end-to-end) loop formation also show a slower reaction rate.[49]

It is worth noting that Equation (6.3) does not imply that there are no barriers associated with chain contact formation: the 10 ps isomerization time clearly involves a micro-barrier for torsion about the Ramachandran angles, and the 100 ns attempt frequency arises from an effective entropic barrier in the Zimm model. However, these micro-barriers are not the ones crossed by motion along the collective folding coordinate (we stick to one collective coordinate for simplicity here). Instead, the micro-barriers reduce the diffusion constant along the collective reaction coordinate, and make the diffusion constant coordinate-dependent (e.g. ω_{well} and ω_{TS} need not be identical in Equation (6.3)). The activation energy along the collective folding coordinate is included explicitly in Equation (6.2), whereas the micro-barriers are included in the temperature-dependent prefactor. Hence the term micro-barrier is used here for local coordinates of which there are many (e.g. amide torsions), and "activation barrier" is reserved for the barrier encountered by collective folding coordinates, of which there are only a few (e.g. helix content coordinates and docking coordinates, as postulated by the diffusion-collision model, to give just one example).[50] Partitioning between prefactor and activation free energy thus requires a rigorously defined reaction coordinate.[51]

Interestingly, the main difficulty in comparing computed and experimental folding barriers lies not in any ambiguity of the coordinates, but in their compatibility. For example, an infrared spectroscopy measurement provides an order parameter based on an amide I band absorption signal averaged over all peptide bonds in the molecule.[52] This order parameter may not be a rigorous reaction coordinate according to the dynamical definition of saddle-point crossings.[53] Most likely, principal component analysis of a larger number of spectroscopic probes would reveal that the IR signal is not one of the principal components, and that more than one principal component is needed; nonetheless the IR order parameter is well defined. Likewise, counting i to $i + 4$ hydrogen bonds in a well-defined angular range precisely defines a computable helical hydrogen bonding order parameter for a molecular dynamics simulation. Here too, principal component analysis[54] may reveal that other computable order parameters are needed to describe the reaction, and that the H-bond connectivity coordinate is not a rigorous reaction coordinate.[55] Nonetheless, the experimental (IR) and computed (H-bond count) order parameters are closely related for a helix bundle protein, and provide similar description of secondary structure formation. Thus one expects similar barriers for both. An important goal for fast-folding experiments and theory in the next several years is to make this approximate agreement completely quantitative by comparing multiple order parameters and extracting consistent approximations to reaction coordinate(s). Examples of such experiment-modeling compatibility already exist: the radius of gyration R_g can be measured accurately by experiment, and computed accurately by molecular dynamics simulation. Other order parameters come close: FRET experiments make some assumptions about orientational averaging, but can be fairly directly compared to model calculations. Yet other computable order parameters cannot be measured currently with sub-ms time resolution: the fraction of native contacts Q is an example. And, *vice versa*, some experimental measures of folding have so far defied accurate *ab-initio* theoretical analysis: CD spectra of proteins are an example (with the exception of tryptophan couplets[56]).

Equation (6.3) should also not be taken to imply that the actual speed limit for folding of a 100-residue protein will be 100 ns. Rather, 100 ns is still a lower limit: several tertiary contact formation events are required to define the topology of a folded chain (on the order of 5–10 contacts for 100 residues depending on how complex the fold is). Even folding without an enthalpic barrier (*i.e.* a perfectly funneled landscape) therefore involves several diffusive searches, which manifest themselves as an entropic barrier along the reaction coordinate. Indeed, experimentally measured barriers tend to be more entropic under conditions optimized for folding, as seen for example in WW domain.[40] Ultimately, such effects will give rise to small barriers along any set of order parameters, no matter how closely they correspond to true reaction coordinates. If nothing else, the impossibility of perfect packing and avoiding all non-native contact energies with an alphabet of only 20 amino acids will make sure that at least micro-barriers exist. This observation can also be cast in terms of internal friction of a protein (sometimes accounted for by exponents $d < 1$ in the η^{-d} prefactor or by expressions of the type $(\eta_{Protein} + \eta_{Solvent})^{-1}$ replacing the bulk viscosity dependence in Equation (6.3)).[57]

Different N-dependences for the prefactor have been suggested. Based on homopolymer models, an N^{-1} scaling is possible.[3] Exponential scaling with contact order has also been suggested, since native topology should play an important role when folding occurs downhill to the native state.[21] Currently, the experimental data pool for peptides and fast-folding proteins is not sufficiently large and does not have sufficiently good size coverage to favor one of these models (or other powers of N such as in Equation (3)) decisively. Several theoretical models exist for the length dependence of contact formation.[58,59] They differ from Equation (6.3) in important details, such as the exponent d of the η^{-d} viscosity dependence and in their N-dependence. These models posit a turnover in the rate below a certain N (which can be as low as 1 in some models), caused by the persistence length of the polypeptide chain. Experiments indicate that no complete turnover occurs even for small (<10-residue) chains.[46,47,60]

6.3.2 Protein Collapse

Protein collapse is idealized by the transition from an extended to a compact random coil. In that limit, Equation (6.3) can also be employed, and yields a similar lower limit on collapse. Recent experiments have shown that collapse can occur on the 0.1 µs timescale,[61] and thermodynamic studies of "intermediates" can be interpreted *via* barrierless collapse instead of two-state folding,[39] but the absence of barriers is not obligatory in coil–coil transitions. Rapid-flow mixing experiments on cytochrome c show that collapse occurs on a 45-µs timescale,[62] corresponding to a 6 RT barrier even with the relatively large

(0.1 μs)$^{-1}$ prefactor deduced from Equation (6.3).[63] Direct comparison of ubiquitin secondary structure formation (*via* circular dichroism) with collapse (*via* small-angle X-ray scattering measurement of the radius of gyration) demonstrates that secondary structure formation can be much faster than collapse, as proposed by the classic "framework models."[64] For collapse, as for folding, there is no universal "no barrier" or "barrier" mechanism.

6.3.3 Secondary Structure Formation

Formation of helical secondary structures, another important event during folding, is discussed in great detail elsewhere in this book (see Chapter 2 by Doshi). Here we note only that measured helix propagation times of about 1 ns, and initiation times of 10 ns, are in league with the time required for the formation of small loops.[65,66] Beta sheet peptides have proved somewhat slower, but still quite fast. The original observation of beta hairpin formation kinetics yielded a few μs for a hairpin from GB-1.[67] The more stable trpzip 2 peptide can form native-like structure in several hundred nanoseconds, the fastest timescale observed for hairpin folding thus far.[68] Even somewhat more complex beta-strand structures can approach this limit.[69,70]

6.3.4 Timescales

Thus loop formation, collapse, and secondary structure formation have ultimate timescales that are quite comparable, in the 5–100 ns range (corresponding to a 3 *RT* energy range). From a structural point of view, this is an important reason why protein folding has proved so difficult to study: secondary and tertiary structure formation timescales are comparable, and cannot be separated easily. In this, proteins differ from RNA, where counter ion concentration can be used to examine secondary and tertiary structure separately. On the other hand, having the energy and timescales for secondary structure comparable allows for a richer interplay between the two, and hence a greater variety of structures. The greater variety could have contributed to evolutionary selection of proteins over RNA as the main carrier of phenotype.

6.4 Microsecond Protein Folding

When is protein folding fast? Here is what theory and experiment have to say.

6.4.1 History

In 1995, it was proposed that some proteins might fold without a significant activation barrier along the collective reaction coordinate.[6] This was labeled the "type 0" scenario in contrast with the "type 1" scenario of two-state folding over an activation free-energy barrier. The free-energy surface of such a protein cannot be completely smooth: simulations and landscape theory show that free energy surfaces projected along one dimension fluctuate on the order of 1–2 RT along the order parameter in regions other than the highest barrier.[71–73] Models based on energy landscape theory can be used to compute what fraction of these barriers is "longitudinal" (*i.e.* contributes to the reaction coordinate), and what fraction is "transverse" (*i.e.* should be incorporated in the prefactor, together with other frictional effects).[55,74,75] However, without a complete set of order parameters to describe the reaction, this partitioning must remain somewhat ambiguous.[76]

Experiments on peptides are in agreement with such fluctuations of the free energy. According to Equation (6.2) and the approximate Zimm chain parameters, this corresponds to a folding time of no faster than 1 μs for a small (<100-residue) protein. This number is also supported by the notion of a "speed limit" for folding, first derived from chain contact formation in unfolded cytochrome c.[77] On the other hand, proteins folding in about 1 ms or longer should have a fairly robust activation barrier of ≈ 7 RT or more.

The activation barrier range from 1–7 RT is where experiment and modeling finally meet. The first molecular dynamics simulation of a mini-protein that could be called "sub-millisecond" from the computational point of view (where longer is better) was reported in 2002: an implicit solvent simulation to 0.7 ms.[78] Experiments on mini-proteins that could be called microsecond folders (here shorter is better), with folding times below 10 μs, were also reported during the last few years.[3] Full atom simulations are also reducing the need for extrapolation, beginning with studies at elevated temperature where the dynamics is faster, and progressing towards physiological temperature.

The history of sub-millisecond protein folding experiments is brief, and begins with a laser photolysis study of cytochrome c by Roder, Eaton, and co-workers in 1993.[79] Relaxation methods (where a sudden switch in conditions, such as removal of a bound CO from the heme group of cytochrome c) are one way of studying such fast kinetics. Equilibrium exchange studies are another. In 1995, Oas and co-workers reported the ≈ 200 μs equilibrium between the unfolded and folded states of a fragment of lambda repressor by poising the protein at the middle of the denaturation transition using urea, and monitoring broadening of the two histidine resonances from the folded and unfolded proteins.[2] Shortly thereafter, nanosecond unzipping,[80] helix–coil equilibria of short helices,[66] sub-millisecond, and microsecond events during protein refolding were reported.[81,82] In the late 1990s and beyond, additional techniques such as continuous flow mixers,[31,83] sub-millisecond pressure jumps,[28] pH jumps,[84] and photochemical triggering of restraining groups (*e.g.* disulfides or azo compounds)[30,85] were added to the arsenal. Since then, there has been a rapid growth of sub-millisecond folding data, on systems ranging in size from 12-mer hairpins to multi-domain proteins with over 400 residues.

On the computational side stands the landmark study by Duan and Kollman, who followed the mini-protein VHP-36 for one microsecond, observing the formation of a compact globular state (although not the native state).[34] In 2000, Caflisch and co-workers obtained µs trajectories at multiple temperatures, opening the door to thermodynamic sampling of fast folders.[86] Since then, simulations have steadily been pushing towards the millisecond timescale, both for single trajectories and for replica-exchange sampling to provide accurate thermodynamic information from simulations. Explicit solvent simulations in the tens of µs range are now the state-of-the-art (see Chapter 8 by Pande).

6.4.2 Sub-millisecond Instrumentation

A plethora of instrumentation has been applied to sub-ms folding studies, some of it newly designed, and some of it adaptations of pre-existing techniques to this new field. Broadly speaking, current fast kinetics experiments can be subdivided into equilibrium and relaxation techniques. In addition, the subdivision into bulk and single-molecule experiments is becoming more relevant, as single molecule experiments are just now beginning to provide sub-ms dynamical information (see Chapter 7 by Schuler). Figure 6.3 summarizes the timescales accessed by the various techniques currently in use.

The equivalence of relaxation and equilibrium techniques relies in principle on Onsager's fluctuation–dissipation theorem for linear response processes.[8]

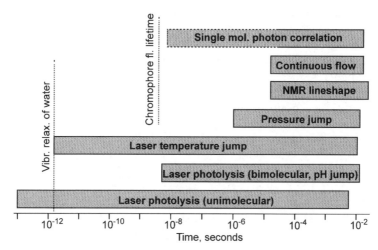

Figure 6.3 Timescales of sub-ms folding techniques currently reached or likely to be reached in the near future. Bimolecular photolysis is limited by diffusion, temperature jumps by water vibrational relaxation, lineshape analysis by motional narrowing and continuous flow by channel dimension. Photon correlation measurements are limited by chromophore lifetime; the dotted line indicates the ultimate limit, although current generation experiments have > 50 µs time resolution. All timescales indicated refer to kinetic time resolution, not probe technique resolution (*e.g.* fluorescence lifetime can be measured to ≈ 50 ps by correlated photon counting, to ps by up-conversion techniques).

According to this theorem, for a small enough perturbation, the relaxation following a jump in external conditions from a set $\{c_i\}$ to a set $\{c_f\}$ has the same time dependence as the time correlation function at the condition $\{c_f\}$. Under certain circumstances, such as two-state exponential relaxation over a substantial barrier, the requirement for a small perturbation can even be lifted. For fast-folding experiments in particular, where barriers are small and the initial and final population distributions may overlap, this criterion cannot be relaxed. This should be kept in mind when comparing different types of experiments, especially experiments with different probes. While two-state folding guarantees probe and jump-size independence, downhill or multi-state kinetics can be identified by probe and jump-size dependence.

The most common bulk equilibrium technique applied to fast folding is proton NMR lineshape analysis.[2] Interconversion between unfolded and folded protein acts as an additional relaxation mechanism, broadening NMR peaks. Thus lineshape analysis does not require a fast initiation step, but rather is used under equilibrium conditions. The advantage of this technique is simplicity of the experiment. The main disadvantage is that the protein has to be biased near the middle of its unfolding transition ($K = k_f/k_u \approx 0.1\text{--}10$), so peaks corresponding to both the unfolded and folded states can be observed. The reaction cannot therefore be studied under native conditions. Rates can be directly measured over a $< 100\,\mu s$ to $> 50\,ms$ range by proton NMR. The most extensively studied protein by this technique remains lambda repressor fragment.[2,87–93] Transverse relaxation dispersion experiments on ^{15}N isotopically labeled proteins, which were pioneered by the Palmer and Lewis groups, allow measuring relaxations involving very small population shifts (e.g. $<1\%$), but are still limited to processes slower than $100\,\mu s$.[94] This approach has been recently applied to a small downhill folding protein.[95]

The most common relaxation techniques currently in use are either mixing- or laser-based. We begin the discussion with continuous flow mixing. Unlike stopped-flow, which is limited to timescales near 1 ms in current generation designs, continuous flow mixers have reached $<50\,\mu s$ time resolution in folding experiments.[83,96] In one mixing approach, two solutions are mixed in a turbulent region (e.g. two capillaries merging at a small ball). The mixed jet then flows through a capillary at constant speed. Thus the distance along the capillary can be mapped into time, and a detector sliding along the capillary can be used to detect kinetics. The disadvantage is that reliable time resolution into the ns region is still lacking, and that continuous flow requires large amounts of sample in current-generation setups. The great advantage of this technique is its versatility as far as the type of perturbation is concerned: denaturant, temperature, pH, and concentration jumps are all easily feasible.[32]

The most common laser-relaxation technique is the temperature jump, either by resistive heating[97] or induced by a laser pulse.[98] The latter method has been most widely adapted, as it is capable of reaching the 10 ns timescale where the elementary processes discussed earlier occur. In the laser temperature-jump technique, the solvent is heated by a pulsed near-infrared laser source.[99] Relaxation of vibrational energy during T-jumps in aqueous solvents is complete in under 10 ps,[100] and vibrational relaxation of high frequency modes ($>200\,\text{cm}^{-1}$) of proteins is equally fast or faster. Thus T-jumps can reach the picosecond timescale, and in fact peptides have been studied on that timescale.[80] The upper end of the time window is set by thermal diffusion, as the sample recools. For typical sample thicknesses in the 0.1–1 mm range, this sets an upper limit in the millisecond range. T-jumps can be used to study refolding (from cold denatured states)[81,82] and unfolding (to heat denatured states);[66,101] this difference becomes particularly important when the protein is not a two state folder (when $k_{observed} \neq k_{fold} + k_{unfold}$). The disadvantage is that T-jumps address only one of many interesting thermodynamic variables. The great advantage of T-jumps is their generality: all proteins with a hydrophobic core cold- and heat-denature.

Laser-induced electron transfer processes have also been extensively used to study folding processes.[102] This approach makes use of the fact that different oxidation states of redox proteins have different folding equilibrium constants. A protein can thus be poised unfolded in one redox state, and then rapidly fold when the redox state is switched, or *vice versa*. Initial experiments were conducted in the millisecond timescale, but current-generation studies are capable of microsecond resolution and have studied even fast loop contact formation processes.[48] The main disadvantage is that the technique is limited to redox proteins, usually including prosthetic groups. The advantage is that irreversible switching is possible, so processes can be studied over many timescales.

A number of additional bulk relaxation techniques have been applied to μs events. Piezo-induced pressure jumps up to 200 bar and as fast as 50 μs have been applied to study folding of a cold shock protein.[28] A pressure jump capable of switching up to 2500 atm in $<2\,\mu\text{s}$ has recently been developed for protein folding applications.[103] Laser photolysis was the first technique applied to fast-folding experiments (CO dissociation from the heme group of cytochrome c).[104] Laser photolysis (*e.g.* by cleavage of disulfide bridges, or isomerization of azobenzene linkers)[30,85] has seen rapidly growing application to folding studies of simple model systems. Developments in cleavable amino acids will further extend this technique to larger proteins. An optical proton switch (*o*-nitrobenzaldehyde) has been used to study pH jumps as a function of pH,[84] isolating contributions from different side chains based on their pK_a values. All of these relaxation and equilibrium fluctuation techniques together sample the full range of thermodynamic parameters.

6.4.3 Spectroscopic Signatures Used in Fast Folding

Bulk relaxation studies have been coupled to a large number of probes.[27] This is particularly important for fast-folding kinetics, as different probes provide information about different reaction pathways and can ultimately be combined to yield a complete picture of the important reaction coordinates for fast-folding proteins. Infrared, circular dichroism,[105] and Raman spectroscopies[31,106] probe secondary structure. The amide I band in particular (1600–1700 cm^{-1}) has been studied extensively for changes in signatures corresponding to formation or loss of α-helix, β-sheet, and coil structure. Current experiments utilize either scanning high-resolution lasers,[107] or dispersed femtosecond IR pulses[108] to follow the time evolution of the protein infrared spectrum. Step-scan FTIR has not been utilized, although it is capable of μs time resolution.[109] Thanks to isotopically labeled or modified amino acids, IR spectroscopy can also yield localized information.[110] Very recently, two-dimensional IR techniques have been applied to protein folding. These measurements allow multiple amide chromophores to be orientationally correlated, and enhance the resolution of the IR experiment, an important factor for the rather broad amide bands.[111,112]

Fluorescence spectroscopy has been applied to probe several different aspects of folding. Tryptophan fluorescence wavelength provides information on solvent exposure (a 325–345 nm fluorescence maximum corresponds to total or partial burial, a 350–360 nm maximum to complete solvent exposure), while tryptophan intensity and lifetime (usually in the 2–6 ns range) provide information about electron- or proton-transfer quenching, which is short range (1–3 Å) and usually stronger in the folded protein.[113–116] Fluorescence quenching or wavelength shift upon resonant energy transfer between dye labels (Förster resonant energy transfer, or FRET) has also been extensively used.[96,117] Electron paramagnetic resonance of cysteine mutants with spin labels also provides accurate distance information.[118] Other techniques include ultraviolet absorption spectroscopy,[60] photoacoustic spectroscopy[84] (to monitor volume changes during folding/unfolding), and small angle X-ray scattering (applied by slowing down sub-ms events through the use of highly viscous solvents).[119] Amide exchange, an NMR technique generally used in the ms regime, is described in detail elsewhere in this volume (see Chapter 4 by Bai).

In contrast to equilibrium or relaxation bulk studies stand single molecule experiments. During the last few years, single molecule experiments have begun to make important contributions to fast protein folding. The advantage compared to bulk studies is obvious: individual proteins may differ significantly in their behavior from the ensemble average, and these differences provide important clues about the structure of the free-energy surface. Although higher moments of distributions can be measured in principle by bulk experiments, and multi-step experiments can be used to dissect heterogeneous populations, this is most easily done by studying single molecules or very small ensembles of molecules.

Single molecule fast-folding studies can be force-resolved or time/position-resolved (see Chapter 7 by Schuler).[25] The latter are currently of main interest in the context of fast-folding kinetics.[24,120,121] In these experiments, a protein molecule is generally labeled with a donor and acceptor dye, and allowed to enter the detection region by diffusion. The emitted photons are sorted by color, arrival time spacing, or by their time delay from the pulsed excitation laser. This provides distance distribution or lifetime histograms correlated with the state of the single protein molecule. Recent studies have revealed that unfolded states under native-like conditions contribute large baselines,[24] and multiple-correlation studies have revealed separate ensembles during folding.[122] Studies where the protein is confined and immobilized are also promising.[123]

The main difficulty with single molecule experiments thus far has been the limited sampling. A typical single molecule can be observed for milliseconds, too little time for slow-folding proteins; only a few molecules per second or so are observed in diffusion limited experiments based on confocal microscopy; in addition, relatively few photons are extracted in this short time span, even when extrinsic dye labels of much higher longevity and quantum yield than natural labels such as tryptophan are used. Two developments are changing this. The first is the discovery of microsecond folders, which allow a single molecule observation to span multiple folding–unfolding events. Just as importantly, microsecond folders spend a larger fraction of time in the interesting region of the free-energy surface where the transition between states occurs. Assuming conservatively that the transition time is on the order of $v^\dagger \approx (100\,\text{ns})^{-1}$ in Equation (6.2), and the folding time is $10\,\mu\text{s}$, a fast folder will spend at least 1% of its time in the dynamically most interesting region of the free energy surface, as opposed to sitting almost exclusively in the folded or unfolded well. The second development is highly automated data collection with pulsed lasers, which can provide lifetime and photon spacing information for large numbers of molecules, allowing a meaningful statistical analysis.[122,124] This is important when only a few percent of the observed photons correspond to the transition of the protein through the region where heterogeneous kinetics can be observed.

Current single molecule studies are generally carried out as equilibrium measurements. However, as the sampling problem is resolved, relaxation studies will also become possible. In the next section, comparison between fast-folding results, theory, and model calculations will be made. For a discussion of the developments in computer instrumentation that have accessed the sub-ms regime from the low end *in silico*, the reader is referred to Chapter 8 by Pande.

6.4.4 Case Studies

Equation (6.1) makes fast-folding experiments extraordinarily diverse, yet it also provides a unifying interpretation. Fast-folding proteins have been found to collapse to compact globules over small barriers,[63,82] or essentially downhill;[61,125] to fold with detectable pre-barrier intermediates,[126] with post-barrier intermediates,[127] on-barrier intermediates,[128] two-state-like,[129] or with a diffusive downhill component;[16] to show dispersion of different thermodynamic probes,[15] or not;[108] to fold/unfold *via* multiple pathways,[129] or mainly a single pathway;[130] to form secondary structure first,[131] or after collapse.[132]

All of these observations taken together indicate a free-energy surface for the fastest folders whose roughness is on a par with the principal activation free energy. Altering such a landscape by mutation or environment will most effectively shift populations to different local minima, and also shift the most important dynamical interconversion processes (Figure 6.4).[133] This roughness presumably exists equally for slower folders or multi-domain folders,[72] but the

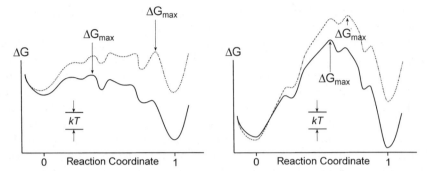

Figure 6.4 Low barrier (left) and high barrier (right) free-energy surfaces under strong (solid line) and weak (dotted line) native bias. The locations of the highest activation barrier and of populated minima are much more sensitive to environmental conditions if the overall barrier is small (left) than if the overall barrier is large (right). Nonetheless, the free-energy surface of the slow folder also has roughness on the kT energy scale, and such "high-energy" intermediates have been observed.[171]

unique advantage of fast folders is that it can be studied directly, unobscured by large barriers. From a physico-chemical point of view, small free-energy fluctuations of engineered proteins are the ultimate rate-limiting factor to be incorporated either in the prefactor (transverse roughness) or activation energy (longitudinal roughness). From a biological point of view, they provide the reference against which to judge the larger folding barriers of most natural proteins, and their possible evolutionary origins.[4] This subsection considers a number of fast folders, organized by what has been learned from them. For additional information, the reader is referred to several recent reviews.[3,4]

For many single domain proteins, collapse to a compact globule without a desolvated core is an important early step towards the folded state. A wide range of timescales exists for this process, from tens of nanoseconds to milliseconds and beyond. The collapse process thus extends from nearly barrier-free processes to highly activated processes. For example, the collapse time for the 40-residue mini-protein BBL at pH 3, where it cannot fold to the native state, has been measured at 60 ns, and it increases above 308 K.[61] The temperature trend is in agreement with a hydrophobic driving force for the collapse process and, indeed, BBL has a small hydrophobic core. The timescale is in agreement with barrier-free collapse, estimated at ≈ 50 ns from the Zimm model in Equation (6.3). Electron transfer-induced fast-folding experiments on a zinc-substituted cytochrome c yield a loop contact formation time of 250 ns within the context of the protein.[48] This is slightly slower than the BBL result even when scaled for chain length. In contrast, the ABGH core of apomyoglobin (153 residues) collapses and forms some tertiary structure in ≈ 10 μs at room temperature from the cold denatured state in the absence of denaturant, although it still slightly slows down at higher temperature.[82] Based on Equation (6.3) this process, although fast, still corresponds to a barrier of about 4–5 RT. Local minima of apomyoglobin stabilized by varying solvent conditions also show relaxations on this timescale.

The collapse time for cytochrome c, measured under denaturing conditions in 1.5 M guanidine hydrochloride buffer, is 40 μs at room temperature.[77] In the absence of denaturant, the intact protein collapses in ≈ 60 μs, whereas a 1–65 fragment collapses in 25 μs, and much faster at higher temperature.[62] Electron paramagnetic resonance studies of cytochrome c mutants with spin label-derivatized cysteines also yield timescales between 20 and 60 μs.[118] The speedup with temperature and the 500-fold slower timescale (in contrast to BBL) are compatible with a substantial activation barrier of 6 RT or more. Of course, cytochrome c differs from simple small proteins in that very specific contacts to a bulky heme group have to be made. Other proteins collapse even more slowly, so collapse is clearly not sequence-optimized in all natural proteins. It is currently not known whether systematic removal of hydrophobic residues can turn a rapid collapse into a slow one, although this is strongly suggested by the data, as some proteins or protein domains of topological complexity comparable to slowly collapsing proteins are known to collapse rather rapidly. Another effect needs to be potentially accounted for when assigning barrier heights using an equation like Equation (6.2): the possibility of anomalous diffusion. Stretched kinetics have been invoked in a variety of folding scenarios,[14,30,134] including end-to-end recombination of peptide chains whose disulfide bridges were photolyzed.[135] In one case, the recombination dynamics spanned timescales from ps to ms, and could not be described by a rate constant.[85] This suggests a very rough free-energy surface, where both fast and slow mechanisms co-exist depending on the region of configuration space accessed by the chain.[136]

The size of the unfolded and native states of proteins that rapidly form compact states are also subject to significant variation with environment. Single molecule experiments, where the unfolded state can be examined separately from the folded state under native conditions, show that the unfolded state becomes very compact at low temperature.[24] This high degree of structure is reached *via* migration of the unfolded free-energy minimum along the reaction coordinate, not by a barrier crossing. This is important because it shows that acquisition of structure during folding does not have to be coupled with an increase in free energy up to a barrier. The native state can be similarly sensitive: fast pH jump experiments on native cytochrome c at pH 7 show that protonation of its histidines reduces the molar volume by -82 ml mole^{-1}, a rather large amount.[84]

Although many small proteins fold as two-state folders, several fast folders form metastable intermediates. Apomyoglobin was already mentioned above; its compact globular state forms about 100 000 times faster than the native fold,[82,137] even though the heme-binding CDEF domain (whose folding is rate limiting) is topologically no more complex than the ABGH core. The main difference between the two is that the ABGH core contains highly packed hydrophobic residues, whereas the CDEF core relies on a heme group to provide its packing. Here is a case where hydrophobicity and packing must cooperate to allow fast folding and, indeed, enlarging hydrophobic side chains in the CDEF core increases the overall folding rate of apomyoglobin.[138] As another example, the structure of the intermediate of the small protein barstar formed from the cold denatured state was analysed by site-directed mutagenesis (Phi-value analysis).[81] Phi-value analysis identifies residues that are structured in the intermediate by determining whether replacing them has a substantial effect on the rate of intermediate formation. Similarly to apomyoglobin, barstar forms a compact intermediate within 200 μs, yet the native state forms much later, within 300 ms. An even smaller protein, engrailed homeodomain, still shows biphasic kinetics, which have been analysed in terms of a compact intermediate state;[126] the overall fast-folding rate of about 20 μs now limits the barrier to $<5\ RT$ even with the conservative speed limit of Equation (6.3). The intermediate ensemble observed in this case has been stabilized by a point mutation,[139] also indicative of a folding barrier comparable to the free energy roughness: in such cases it should be particularly easy to stabilize different free energy local minima by small perturbations. In that interpretation of fast biphasic kinetics, the phases arise from a marginal barrier, which allows both barrier-crossing and diffusive (downhill) populations to be monitored.[16] Multiple minima lead to multiple timescales, even when the barriers separating them are too shallow to qualify as well-separated "thermodynamic intermediates." Even a simple hairpin peptide, trpzip 2, has shown thermodynamic and kinetic evidence of small free-energy minima; the sub-μs rates observed under some conditions indicate that the overall activation barrier is $\leq 3\ RT$.[68,73,140]

Apomyoglobin and barstar are cases where a fast-folding intermediate occurs before the rate-limiting step. Evidence has equally been uncovered for intermediates following the rate-limiting step. Proton exchange NMR experiments hint that cytochrome b562 forms some of its helices *en route* to the transition state, but others after the barrier has been passed.[127] In yet other cases, the location of intermediates has been controversial. For example, a re-analysis of ubiquitin folding showed that fluorescence monitors single-phase kinetics, contrary to earlier reports.[141,142] A further analysis showed that other probes, such as SAXS and CD, differ after all in their timescale from fluorescence, in a way that can be explained by an early intermediate.[64] The solvent conditions in that experiment were significantly different from room temperature aqueous buffer studies, but this serves only to make the point of Figure 6.4: that free energy local minima can be found anywhere along the reaction coordinate, and that the sensitivity of experiments to various local minima depends on sequence and environment, most sensitively so for fast folders. These local minima vary in depth, and when the depth exceeds $3\ RT$, we begin to call them "intermediates."

However, even in large proteins these intermediates need not be separated from the unfolded state by large barriers: both the C- and N-terminal domains of phosphoglycerate kinase, a hinged two-domain protein of 415 residues, form compact states with secondary structure within tens of μs after folding is initiated from the cold denatured state.[14]

Fast folders have been used to analyse the thermodynamic and kinetic equivalence of cold and heat denatured states of proteins.[87] A study of the β-hairpin MrH3a in 8% hexafluoropropanol yields a much faster refolding rate to the native state from the cold denatured state than from the heat denatured state, assuming two separate two state equilibria of the form CN and NH.[143] In contrast, analysis of a "slow" (100 μs) folding variant of the 80-residue λ-repressor fragment shows that the folding rates from the cold and heat denatured states are equivalent, if the solvent viscosity is taken into account.[144] Indeed, in P-T phase diagrams the two states can interconvert without going through a folding transition. The difference between these two cases lies in the presence or absence of a hydrophobic core. MrH3a does not have one, and residual hairpin structure controls the dynamics; this structure differs substantially between the cold and heat denatured states. λ_{6-85} has a large hydrophobic core, which drives folding from both the heat and cold denatured states, mitigating the effect of loops on the folding kinetics.

A comparison of fast-folding kinetics with structural information from molecular dynamics simulations has revealed different mechanisms by which fast folders reach the native state, even when they are apparently two-state (see Chapter 8 by Pande for a more detailed discussion). A mutant of the miniprotein BBA5, a zinc finger-like structure (β-turn-α), originally designed by Imperiali and co-workers,[145] had its folding rate constants and equilibrium constant directly compared with 700 μs of distributed computing implicit solvent molecular dynamics trajectories.[78] The molecule is only weakly cooperative, with a room temperature folding rate of about 8 μs. Several of the short simulations folded to native structures, by a variety of paths that included helix or turn formation as the first step. Although there is some bias towards a helix-first mechanism, the bias is not strong, and BBA5 exemplifies folding on a funnel-like energy surface down several paths. Good agreement was obtained between computed and measured kinetics and thermodynamics. Similar experiments and calculations were also carried out for the 20-residue miniprotein tryptophan cage, and for the 35 residue villin headpiece (a three-helix bundle).[146–149] In both cases experimental and computed rates also agreed within uncertainty, and the VHP mechanism allows different helix-pairing combinations. More recently, VHP and BBA5 have also been simulated in explicit solvent (TIP3P water), in one case up to 0.5 ms total.[23] The explicit solvent simulations largely support the earlier results obtained in implicit solvent. At least for small systems without large hydrophobic cores and very small (if any) folding barriers, computational simulation with current-generation force fields thus agrees closely with experiment. Also for small systems, where desolvation of extensive hydrophobic surfaces is not required to reach the native state, implicit solvent models perform well. The next generation of studies tackling mini-proteins with larger hydrophobic cores (*e.g.* the lambda repressor fragment) will show how well this translates when hydrophobic surface desolvation plays a key role.

Beta sheet-only proteins, often thought of as slower folders, have also been successfully engineered towards folding rates $>(10\,\mu s)^{-1}$. With a folding time of 40 μs (depending on mutant and truncation at the termini), the FBP28 WW domain, a triple-stranded β-sheet based on the FBP wild type, is already a fast folder. Inserting its hairpin 1 or similar short hairpins into the hPin1 WW domain structure produces very fast folders, some with folding time constants below 10 μs.[70] Smaller peptides fold even faster: the first hairpin studied in detail folds in a few μs, trpzip 2 in under 1 μs under some conditions.[67,68] Nonetheless, β-sheet structures have not reached the speed of comparably sized helical mini-proteins and peptides. This indicates that turn formation alone, which should occur in 100 ns or less for optimized turns,[150] is not the factor controlling kinetics. Of course turns can be rate limiting in some cases: Turn 1 of hPin WW domain is formed during the rate-limiting step of folding,[40] and its removal speeds up folding as mentioned above. This turn is an unusually long six-residue sequence, evolved to support the binding function of the hPin WW domain. Thus evolution for function and ease of folding seem to be at odds in this case. Molecular dynamics simulations have contributed much to interpreting fast-folding experiments on β-sheets (see Chapter 8 by Pande). Both 1-D and 2-D free-energy surfaces generated for hairpins as well as larger triple-stranded sheets reveal local minima in the free-energy surface, which can be deepened or removed by mutation.[151,152] Although the simulations may not produce quantitative agreement (*i.e.* the details of local minima may differ from experiment), they underscore the role that free-energy roughness plays during folding.

Some of the earliest direct comparisons of computation and experiment are between unfolding rates of fast folders and fully solvated molecular dynamics simulations.[153] The fastest-folding example is the engrailed homeodomain, a three-helix bundle with a topology similar to VHP.[154] As computing power has increased, the extrapolated experimental unfolding rate (about 50 ns at 75 °C) and the computed unfolding time to a transition ensemble from which large structural changes occur are in good agreement.[126] One result from these studies (in contradiction to the general scheme in Figure 6.4) is that the transition state structure of engrailed homeodomain seems almost temperature independent even though the barrier must be relatively small for this microsecond folder. It is possible that the molecule has a late transition state, even at low temperatures favoring the native state, and that this transition state cannot become much more native-like when the temperature is raised. Possible anti-Hammond effects have also been discussed in other folding contexts.[133]

Larger proteins and multi-domain proteins are also capable of fast dynamics – up to a point. Based on Equation (6.3), on homopolymer collapse models, or on the inverse correlation between folding rate and contact order or chain length, larger proteins ultimately fold more slowly. However, the size of a single domain has an upper limit near 200 ± 50 residues in most proteins. Beyond that, proteins usually form multiple domains connected by tethers or hinges. In the case of PGK, the individual domains collapse to a compact state independently on the sub-ms timescale, as shown by temperature jump studies of the whole protein and of protein fragments.[134] If this holds true in general, there will be an upper limit to the complexity of fold topologies (the most complex structure possible with ≈ 200 residues), yielding an upper limit on the folding time of optimized sequences (naturally, unoptimized sequences can be arbitrarily slower). A protein near this upper limit may have been discovered already: the large protein cyclophilin a folds on the sub-ms timescale,[155] whereas proteins of a similar size fold in ≈ 1 second on average.

6.5 Downhill Folding

Downhill folding is the most extreme manifestation of fast folding. If a protein has no barriers > 3 RT along the folding coordinate(s), barrier top populations become measurable even in experiments whose typical signal-to-noise ratio does not exceed 100 : 1. In essence, a kinetics experiment gains direct access to quantities such as $v^{\dagger}$ and ω_{barrier} in Equations (6.2) and (6.3) by directly observing excited protein states. The kinetics measurement thus becomes a dynamics measurement. At the same time the folding rate approaches the speed limit for diffusion along the collective reaction coordinate(s). As discussed earlier, for a small protein the speed limit should be five times slower than the diffusive prefactor of $(100 \text{ ns})^{-1}$ of Section 6.3 because several diffusive events are required to define the fold topology. This number is very close to the 1-μs limit originally estimated by Hagen and Eaton.[77] The current status of downhill folding has been reviewed in references 3, 4, 156, and 157; here follows a brief presentation of the salient points.

Two general downhill scenarios have been proposed. The original scenario emerged from statistical mechanical analysis of the landscape model, as summarized by Bryngelson *et al.* in 1995.[6] In this scenario, proteins make a downhill to two-state transition when they are stressed by mutation or by an environment unfavorable for the native state. Figure 6.5 shows how a protein under optimal conditions is a downhill folder, and then becomes a two-state folder when the native state free energy is raised. At even higher stress, downhill unfolding results. Kinetics with an exponential–non-exponential–exponential transition as a function of temperature provided the first support for this scenario,[14] which is now also supported by data for folding to the native state.[16,21,108,125] As discussed by Hagen,[158] and shown experimentally,[125,129] pure downhill folding is not necessarily non-exponential. The proof of kinetic downhill folding in the Bryngelson scenario relies on a transition from exponential folding (two-state) to non-exponential folding (low barrier), and back towards exponential folding (pure downhill with simple diffusion process) with strongly correlated kinetic amplitudes and phases along the way.[14,21,125,159] As a final note, experiments, simulations, and theory find that downhill folders in the Bryngelson scenario have sigmoidal denaturation transitions when stress is applied because they revert to two-state folding.[108,159] This rules out the usual approaches for identifying such proteins as two-state folders by titration experiments.

The second scenario is also shown in Figure 6.5, and is founded on the observation by Muñoz and co-workers that some proteins with broad thermodynamic folding–unfolding transitions cannot be characterized by a single transition temperature when multiple probes are used.[15,160] In this type of

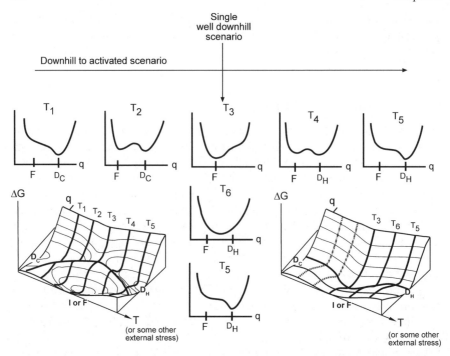

Figure 6.5 Two downhill folding scenarios. Bottom left and arrow across, adapted from Sabelko et al.:[14] stress induces a downhill to two-state transition. D_H is the heat denatured state, to which downhill *un*folding occurs at a very high temperature T_5. As the temperature is decreased to T_4, an activation barrier forms because the free energy in the transition region increases together with the unfolded free energy. At the optimal folding temperature T_3, the free energy has tilted towards the native state, and downhill folding now occurs. The whole process reverses at the lower T, eventually populating the cold denatured state D_C at T_1. Bottom right and down arrow, based on Muñoz:[160] stress induces a shift of the free-energy minimum, but a single minimum remains throughout. The free-energy minimum shifts smoothly from the heat denatured state to the native state. The cold denaturation side has not yet been observed.

downhill folding, a barrier does not appear when stress is applied to the protein. Instead, the free-energy minimum gradually shifts towards the unfolded state when stress is applied, taking the protein population with it in a "one-state" manner. The folding is downhill under all conditions. It is worth noting that such one-state scenarios can demonstrably produce sigmoidal denaturation curves, albeit with large baselines.[160,161] Baselines are often observed in proteins labeled as two-state, fitted, subtracted from the data, and then ignored. This time-honored approach is clearly not viable when investigating potential downhill folders. Baselines must be explained along with the rest of the data.

Experimental observations of the two downhill scenarios differ in one important aspect: mini-proteins with small hydrophobic cores (*e.g.* BBL[15]) may be able to follow the "one-state" scenario, while fast-folding proteins with larger hydrophobic cores (*e.g.* phosphoglycerate kinase[14] or lambda repressor fragment[16]) may have to switch from "one-state" to "two-state" under stress. Large hydrophobic surfaces need to be desolvated before burial in the core, and this makes the process more prone to having a barrier under stress. Recent experiments and models studying volume changes in the transition state upon hydrophobic–hydrophilic substitution support the idea that desolvation is a major contribution to the folding barrier of larger proteins.[162,163]

So far, no large natural protein folds downhill to the native state. For example, BBL is a substrate-moving domain from a larger protein,[15] and the fragment from DNA-binding λ-repressor had to be engineered to increase its folding rate by at least another factor of 20 to fold downhill.[16,125] BBL clearly shows the multiple thermodynamic melts characteristic of a single-well downhill folder, while λ-repressor has shown titrations characteristic of the downhill-to-two-state transition. However, recently a version of λ-repressor has been engineered that shows large deviations at the melting midpoint, approaching the single-well scenario also.[125] Thus it appears that both natural and engineered small proteins can fold in a single well even at temperatures approaching their melting temperatures. Among large proteins, cyclophilin a may be a candidate for a large natural downhill folder, but only if $\ln k_{folding}$ scales linearly or faster with sequence length or contact order.[21]

In summary, the experimental evidence for downhill folding stems from: observations of probe dependence and large baselines in thermodynamic titration curves;[15,35,95,156,160,164] observation of probe-dependent kinetics with specific transitions between exponential and non-exponential kinetics as the native state is stabilized;[14,16,108,125] dependence of these kinetic transitions only on overall folding rate, not on specific mutants or solvent conditions;[21,125,157] different viscosity dependence of the diffusive phase compared to the activated (two-state) phase;[68] and molecular dynamics simulations revealing free-energy surfaces with barriers $<2\ RT$.[16,71,73]

Several reasons connected with protein evolution have been proposed for the scarcity of wild-type proteins as downhill folders:[4] Even if they are much more stable than the unfolded state, barrierless native states may be more prone to aggregation because they can partially unfold with ease. Folding kinetics can be compromised by evolution for protein function, which often favors solvated binding pockets, flexibility-enhancing glycines, long functional loops,[165] or prosthetic groups, to the detriment of hydrophobic stabilization. And, lastly, evolutionary pressure for fast folding has its limits because cytoplasmic crowding, ribosomal expression, and chaperoning influence cellular folding processes. Nonetheless, downhill folding, if universal for all fold topologies, provides a baseline against which to judge such evolutionary pressures. The difference between the wild-type protein folding rate and the downhill rate is accounted for by evolutionary pressures from aggregation, function, and protein environment.

Conversely, evolution could also select *for* downhill folding of proteins in several ways:[15] Proteins with a mechanical function, such as spring-like extension during substrate transport, benefit from a single well that avoids unproductive trapping of the module in the extended or contracted local minimum. Downhill folding can also play a role in folding–binding interactions, for instance when unbinding of one substrate induces a downhill conformational switch that rapidly prepares the protein for uptake of another substrate.[156] It could turn out that many natural downhill folders exist because of evolutionary pressure for function. Such proteins may have been overlooked because they do not produce nice sigmoidal denaturation curves, like natively disordered proteins for instance.

There has been a lively discussion in the literature as to whether downhill folders of either type (one-/two-state or purely one-state) could be explained by folding intermediates.[159,164,166,167] Basically, models involving a succession of low-barrier intermediates and/or traps can be invoked to explain multi-probe thermodynamics and kinetics instead of downhill folding. Countering this, downhill models have added logarithmic oscillations to explain residual structure in non-exponential kinetics.[14,168] It is worth pointing out that no protein is likely to fold downhill monotonically on a completely smooth free-energy surface: a 20-amino-acid code cannot optimize all native interactions and eliminate all non-native ones, leaving residual roughness on the free-energy surface. Experiment, modeling and theory have pegged this roughness at 1–3 RT.[16,20,68,71,72,169] The key is this: folding does not occur at zero temperature, so folding should be considered downhill when thermal excitations RT are comparable to the residual folding barriers, leading to a breakdown of transition state theory. The current generation of experiments and simulations certainly demonstrates downhill or near-downhill folding on the 1–3 RT scale. In light of Equation (6.3), the many fast folders observed at any size between 12 and 500 residues[21,70,129,147,148,155,170] will always beg the question of the "reverse-Levinthal paradox:" how come many proteins fold so slowly?

6.6 Outlook

Fast-folding experiments are testing some of the most critical predictions of energy landscape theory: the sensitivity of the folding mechanism to protein sequence and environment caused by roughness of the free-energy surface (*i.e.* Equation (6.1)), and the existence of nearly barrier-free folding, limited only by residual roughness of the free-energy surface on the order of 1–3 RT (Equation (6.3)). Full atom simulation in conjunction with experiment is providing a rich picture of folding, not limited to a fixed sequence of structures, but revealing multiple free-energy minima whose populations are exponentially sensitive to free energy. For this reason, truly parallel folding paths are rare, but a number of low-energy paths exist to confer robustness in the context of sequence evolution. The number of such paths is limited by the need to avoid too high a density of states of partially folded states, which would unduly stabilize non-native globules. Proteins that sample long-lived intermediates are skirting one edge of the compromise between too few and too many low-energy folding paths; proteins that fold two-state but, very slowly, the other edge.

Comparison of experiment with theory and computational modeling points to many important issues that must be resolved before fast folding is fully understood: Multi-probe and single molecule experiments must become routine to probe free-energy surface complexity. Observed and theoretical collective coordinates must be related quantitatively, and optimal folding coordinates must be deduced from such a consistent set; only then can folding barriers be defined precisely, distinguishing them quantitatively from "micro-barriers" caused by single bond isomerization or local solvent rearrangement. This problem is analogous to the old conundrum of transition state theory when some degrees of freedom are averaged out: should we write $k = kT/h\,\sigma\,e^{-E^\dagger/kT}$, or $k = kT/h\,e^{-(E^\dagger - kT\ln\sigma)/kT} = kT/h\,e^{-G^\dagger/kT}$? Does the steric factor σ belong in the prefactor, or in the exponent? If we include all the microscopic coordinates, it belongs in the exponent. If we include only collective reaction coordinates, it belongs in the prefactor. Protein folding is fairly efficient and can be described using a small number of coordinates, so the latter picture will bring us closer to truly understanding the folding process.

Acknowledgement

This work was supported by grant MCB 0316925 from the National Science Foundation and by grant 37333-AC7 of the American Chemical Society Petroleum Research Fund.

References

1. C. Levinthal, *J. Chim. Phys.*, 1968, **65**, 44–45.
2. G. S. Huang and T. G. Oas, *Proc. Natl. Acad. Sci. USA*, 1995, **92**, 6878–6882.
3. J. Kubelka, J. Hofrichter and W. A. Eaton, *Curr. Opinion. Struct. Biol.*, 2004, **14**, 76–88.
4. M. Gruebele, *Comptes Rendue Biologies*, 2005, **328**, 701–712.
5. N. Go, *Annu. Rev. Biophys. Bioeng.*, 1983, **12**, 183–210.
6. J. D. Bryngelson, J. N. Onuchic, N. D. Socci and P. G. Wolynes, *Proteins Struct. Funct. Genet.*, 1995, **21**, 167–195.
7. O. M. Becker, *Theor. Chem.*, 1997, **398–399**, 507–516.
8. D. Chandler, *Modern Statistical Mechanics*, Oxford University Press, Oxford, 1989.
9. D. Thirumalai and C. Hyeon, *Biochemistry*, 2005, **44**, 4957–4970.
10. Y. Levy, S. S. Cho, T. Shen, J. N. Onuchic and P. G. Wolynes, *Proc. Natl. Acad. Sci. USA*, 2005, **102**, 2373–2378.
11. H. Kaya and H. S. Chan, *J. Mol. Biol.*, 2002, **315**, 899–909.
12. S. B. Ozkan, K. A. Dill and I. Bahar, *Biopolymers*, 2003, **68**, 35–46.
13. O. Bieri, G. Wildegger, A. Bachmann, C. Wagner and T. Kiefhaber, *Biochemistry*, 1999, **38**, 12460–12470.

14. J. Sabelko, J. Ervin and M. Gruebele, *Proc. Natl. Acad. Sci. USA*, 1999, **96**, 6031–6036.
15. M. Garcia-Mira, M. Sadqi, N. Fischer, J. M. Sanchez-Ruiz and V. Muñoz, *Science*, 2002, **298**, 2191–2195.
16. W. Y. Yang and M. Gruebele, *Nature*, 2003, **423**, 193–197.
17. K. W. Plaxco, K. T. Simons and D. Baker, *J. Mol. Biol.*, 1998, **277**, 985–994.
18. A. N. Naganathan and V. Munoz, *J. Am. Chem. Soc.*, 2005, **127**, 480–481.
19. D. N. Ivankov, S. O. Garbuzynskiy, E. Alm, K. W. Plaxco, D. Baker and A. V. Finkelstein, *Protein Sci.*, 2003, **12**, 2057–2062.
20. J. E. Shea, J. N. Onuchic and I. C. L. Brooks, *J. Chem. Phys.*, 2000, **113**, 7663–7671.
21. W. Yang and M. Gruebele, *Biophys. J.*, 2004, **87**, 596–608.
22. J. Shea and C. L. Brooks, *Annu. Rev. Phys. Chem.*, 2001, **52**, 499–535.
23. C. D. Snow, E. J. Sorin, Y. M. Rhee and V. S. Pande, *Annu. Rev. Biophys. Biomol. Struct.*, 2005, **34**, 43–69.
24. B. Schuler, E. A. Lipman and W. A. Eaton, *Nature*, 2002, **419**, 743–747.
25. B. Schuler, Single-molecule fluorescence spectroscopy of protein folding, *Chem. Phys. Chem.*, 2005, **6**, 1206–1220.
26. V. B. P. Leite, J. Onuchic, G. Stell and J. Wang, *Biophys. J.*, 2004, **87**, 3633–3641.
27. M. Gruebele, *Annu. Rev. Phys. Chem.*, 1999, **50**, 485–516.
28. M. Jacob, G. Holtermann, D. D. Perl, J. J. Reinstein, T. Schindler, M. A. Geeves and F. X. Schmid, *Biochemistry*, 1999, **38**, 2882–2891.
29. C. A. Royer, *Biochim. Biophys. Acta.*, 2002, **1595**, 201–209.
30. J. Bredenbeck, J. Helbing, J. R. Kumita, G. A. Woolley and P. Hamm, *Proc. Natl. Acad. Sci. USA*, 2005, **102**, 2379–2384.
31. S. Takahashi, S. Yeh, T. K. Das, C. Chan, D. S. Gottfried and D. L. Rousseau, *Nat. Struct. Biol.*, 1997, **4**, 44–50.
32. H. Roder, K. Maki, H. Cheng and M. C. R. Shastry, *Methods*, 2004, **34**, 15–27.
33. Y. Sugita and Y. Okamoto, *Chem. Phys. Lett.*, 1999, **314**, 141–151.
34. Y. Duan and P. A. Kollman, *Science*, 1998, **282**, 740–744.
35. A. N. Naganathan, U. Doshi and V. Munoz, *J. Am. Chem. Soc.*, 2007, **129**, 5673–5682.
36. H. J. Dyson and P. E. Wright, *Chem. Rev.*, 2004, **104**, 3607–3622.
37. W. Y. Yang, E. Larios and M. Gruebele, submitted.
38. K. W. Plaxco and M. Gross, *Nat. Struc. Biol.*, 2001, **8**, 659–660.
39. M. J. Parker and S. Marqusee, *J. Mol. Biol.*, 1999, **293**, 1195–1210.
40. M. Jäger, H. Nguyen, J. Crane, J. Kelly and M. Gruebele, *J. Mol. Biol.*, 2001, **311**, 373–393.
41. H. A. Kramers, *Physica*, 1940, **7**, 284.
42. B. H. Zimm, *J. Chem. Phys.*, 1956, **24**, 269–278.
43. M. H. Zaman, M.-Y. Shen, R. S. Berry, K. F. Freed and T. R. Sosnick, *J. Mol. Biol.*, 2003, **331**, 693–711.

44. A. Szabo, K. Schulten and Z. Schulten, *J. Chem. Phys.*, 1980, **72**, 4350–4357.
45. D. Thirumalai, *J. Phys. Chem. B*, 1999, **103**, 608–610.
46. L. J. Lapidus, W. A. Eaton and J. Hofrichter, *Proc. Natl. Acad. Sci. USA*, 2000, **97**, 7220–7225.
47. O. Bieri, J. Wirz, B. Hellrung, M. Schutkowski, M. Drewello and T. Kiefhaber, *Proc. Natl. Acad. Sci. USA*, 1999, **96**, 9597–9601.
48. I. J. Chang, J. C. Lee, J. R. Winkler and H. B. Gray, *Proc. Natl. Acad. Sci. USA*, 2003, **100**, 3838–3840.
49. B. Fierz and T. Kiefhaber, *J. Am. Chem. Soc.*, 2007, **129**, 672–679.
50. D. Bashford, D. L. Weaver and M. Karplus, *J. Biomol. Struct. Dyn.*, 1984, **1**, 1243–1255.
51. D. C. Clary, *Science*, 1998, **279**, 1879–1882.
52. H. Susi and D. M. Byler, *Methods Enzymol.*, 1986, **130**, 290–311.
53. B. J. Berne, in *Activated Barrier Crossing: Applications in Physics, Chemistry and Biology*, ed., P. Hänggi, G. R. Fleming, World Scientific, Singapore, 1993, pp. 82–119.
54. O. M. Becker, *J. Comput. Chem.*, 1998, **19**, 1255–1267.
55. R. B. Best and G. Hummer, *Proc. Natl. Acad. Sci. USA*, 2005, **102**, 6732–6737.
56. I. B. Grishina and R. W. Woody, *Faraday Discuss.*, 1994, **99**, 245–262.
57. L. L. Qiu and S. J. Hagen, *Chem. Phys.*, 2005, **312**, 325.
58. C. J. Camacho and D. Thirumalai, *J. de Physique I*, 1995, **5**.
59. K. Schulten, Z. Schulten and A. Szabo, *J. Chem. Phys.*, 1981, **74**, 4426–4457.
60. J. Faraone-Mennella, H. B. Gray and J. R. Winkler, Early events in the folding of four-helix-bundle heme proteins, *Proc. Natl. Acad. Sci. USA*, 2005, **102**, 6315–6319.
61. M. Sadqi, L. J. Lapidus and V. Muñoz, *Proc. Natl. Acad. Sci. USA*, 2003, **100**, 12117–12122.
62. L. L. Qiu, C. Zachariah and S. J. Hagen, *Phys. Rev. Lett.*, 2003, **90**, 168103.
63. S. J. Hagen and W. A. Eaton, *J. Mol. Biol.*, 2000, **297**, 781–789.
64. E. Larios, J. S. Li, K. Schulten, H. Kihara and M. Gruebele, *J. Mol. Biol.*, 2004, **340**, 115–125.
65. W. A. Eaton, V. Muñoz, P. A. Thompson, E. R. Henry and J. Hofrichter, *Acc. Chem. Res.*, 1998, **31**, 745–753.
66. S. Williams, T. P. Causgrove, R. Gilmanshin, K. S. Fang, R. H. Callender, W. H. Woodruff and R. B. Dyer, *Biochemistry*, 1996, **35**, 691–697.
67. V. Muñoz, P. A. Thompson, J. Hofrichter and W. A. Eaton, *Nature*, 1997, **390**, 196–199.
68. W. Y. Yang and M. Gruebele, *J. Am. Chem. Soc.*, 2004, **126**, 7758–7759.
69. Y. Xu, P. Purkayashta and F. Gai, *J. Am. Chem. Soc.*, 2006, **128**, 15836–15842.
70. H. Nguyen, M. Jaeger, J. Kelly and M. Gruebele, *J. Phys. Chem. B*, 2005, **109**, 15182–15186.

71. T. V. Pogorelov and Z. Luthey-Schulten, *Biophys. J.*, 2004, **87**, 207–214.
72. J. J. Portman, S. Takada and P. G. Wolynes, *J. Chem. Phys.*, 2001, **114**, 5069–5081.
73. W. Y. Yang, J. Pitera, W. Swopes and M. Gruebele, *J. Mol. Biol.*, 2004, **336**, 241–251.
74. S. Takada, J. J. Portman and P. G. Wolynes, *Proc. Natl. Acad. Sci. USA.*, 1997, **94**, 2318–2321.
75. J. J. Portman, S. Takada and P. G. Wolynes, *J. Chem. Phys.*, 2001, **114**, 5082–5096.
76. A. Szabo, private communication.
77. S. J. Hagen, J. Hofrichter, A. Szabo and W. A. Eaton, *Proc. Natl. Acad. Sci. USA*, 1996, **93**, 11615–11617.
78. C. Snow, H. Nguyen, V. Pande and M. Gruebele, *Nature*, 2002, **420**, 102–106.
79. C. M. Jones, E. R. Henry, Y. Hu, C. K. Chan, S. D. Luck, A. B. a, H. Roder, J. Hofrichter and W. A. Eaton, *Proc. Natl. Acad. Sci. USA*, 1993, **90**, 11860–11864.
80. C. M. Phillips, Y. Mizutani and R. M. Hochstrasser, *Proc. Natl. Acad. Sci. USA*, 1995, **92**, 7292–7296.
81. B. Nölting, R. Golbik and A. R. Fersht, *Proc. Natl. Acad. Sci. USA*, 1995, **92**, 10668–10672.
82. R. M. Ballew, J. Sabelko and M. Gruebele, *Proc. Natl. Acad. Sci. USA*, 1996, **93**, 5759–5764.
83. M. C. Ramachandra Shastry, S. D. Luck and H. Roder, *Biophys. J.*, 1998, **74**, 2714–2721.
84. S. Abbruzzetti, E. Crema, L. Masino, A. Vecli, C. Viappiani, J. R. Small, L. J. Libertini and E. W. Small, *Biophys. J.*, 2000, **78**, 405–415.
85. M. Volk, Y. Kholodenko, H. S. M. Lu, E. A. Gooding, W. F. DeGrado and R. M. Hochstrasser, *J. Phys. Chem.*, 1997, **101**, 8607–8616.
86. P. Ferrara, J. Apostolakis and A. Caflisch, *J. Phys. Chem. B*, 2000, **104**, 5000–5010.
87. G. S. Huang and T. G. Oas, *Biochemistry*, 1996, **35**, 6173–6180.
88. R. E. Burton, G. S. Huang, M. A. Daugherty, P. W. Fullbright and T. G. Oas, *J. Mol. Biol.*, 1996, **263**, 311–322.
89. R. E. Burton, J. K. Myers and T. G. Oas, *Biochemistry*, 1998, **37**, 5337–5343.
90. R. E. Burton, G. S. Huang, M. A. Daugherty, T. L. Calderone and T. G. Oas, *Nat. Struct. Biol.*, 1997, **4**, 305–310.
91. J. K. Myers and T. G. Oas, *Biochemistry*, 1999, **38**, 6761–6768.
92. S. Ghaemmaghami, J. M. Word, R. E. Burton, J. S. Richardson and T. G. Oas, *Biochemistry*, 1998, **37**, 9179–9185.
93. G. S. Huang and T. G. Oas, *Biochemistry*, 1995, **34**, 3884–3892.
94. A. G. Palmer, C. D. Kroenke and J. P. Loria, *Methods Enzymol.*, 2001, **339**, 204–238.
95. M. Sadqi, D. Fushman and V. Munoz, *Nature*, 2006, **442**, 317–321.

96. C. Chan, Y. Hu, S. Takahashi, D. L. Rousseau, W. A. Eaton and J. Hofrichter, *Proc. Natl. Acad. Sci. USA*, 1997, **94**, 1779–1784.
97. H. Hoffmann, E. Yeager and J. Stuehr, *Rev. Sci. Instrum.*, 1968, **39**, 649–653.
98. D. H. Turner, G. W. Flynn, S. K. Lundberg, L. D. Faller and N. Sutin, *Nature*, 1972, **239**, 215–217.
99. R. M. Ballew, J. Sabelko, C. Reiner and M. Gruebele, *Rev. Sci. Instrum.*, 1996, **67**, 3694–3699.
100. H. R. Ma, C. Z. Wan and A. H. Zewail, *J. Am. Chem. Soc.*, 2006, **128**, 6338–6340.
101. P. A. Thompson, W. A. Eaton and J. Hofrichter, *Biochemistry*, 1997, **36**, 9200–9210.
102. G. A. Mines, T. Pascher, S. C. Lee, J. R. Winkler and H. B. Gray, *Chem. Biol.*, 1996, **3**, 491–497.
103. C. Dumont and M. Gruebele, *Rev. Sci. Instrum.*, 2008, in press.
104. C. M. Jones, E. R. Henry, Y. Hu, C. Chan, S. D. Luck, A. Bhuyan, H. Roder, J. Hofrichter and W. A. Eaton, *Proc. Natl. Acad. Sci. USA*, 1993, **90**, 11860–11864.
105. E. Chen, M. J. Wood, A. L. Fink and D. S. Kliger, *Biochemistry*, 1998, **37**, 5589–5598.
106. I. K. Lednev, A. S. Karnoup, M. C. Sparrow and S. A. Asher, *J. Am. Chem. Soc.*, 1999, **121**, 4076–7077.
107. R. B. Dyer, F. Gai, W. H. Woodruff, R. Gilmanshin and R. H. Callender, *Acc. Chem. Res.*, 1998, **31**, 709–716.
108. H. Ma and M. Gruebele, *Proc. Natl. Acad. Sci. USA*, 2005, **102**, 2283–2287.
109. J. A. Bailey, F. L. Tomson, S. L. Mecklenburg, G. M. MacDonald, A. Katsonouri, A. Puustinen, R. B. Gennis, W. H. Woodruff and R. B. Dyer, *Biochemistry*, 2002, **41**, 2673–2683.
110. Z. Getahun, C. Y. Huang, T. Wang, B. D. Leon, W. F. DeGrado and F. Gai, *J. Am. Chem. Soc.*, 2003, **125**.
111. P. Mukherjee, I. Kass, I. Arkin and M. T. Zanni, *Proc. Natl. Acad. Sci. USA*, 2006, **103**, 3528–3533.
112. H. S. Chung, M. Khalil, A. W. Smith, Z. Ganim and A. Tokmakoff, *Proc. Natl. Acad. Sci. USA*, 2005, **102**, 612–617.
113. J. R. Lakowicz, *Methods Enzymol.*, 1986, **131**, 518–567.
114. P. Callis and T. James, *Biophys. J.*, 2001, **80**, 2093–2109.
115. Y. Chen and M. D. Barkley, *Biochemistry*, 1998, **37**, 9976–9982.
116. J. Ervin, J. Sabelko and M. Gruebele, *J. Photochem. Photobiol., B*, 2000, **54**, 1–15.
117. P. Lillo, M. T. Mas and J. M. Beechem, *Biophys. J.*, 1998, **74**, Tu-Pos-130.
118. V. M. Grigoryants, K. A. DeWeerd and C. P. Scholes, *J. Phys. Chem. B*, 2004, **108**, 9463–9468.
119. Z. Qin, J. Ervin, E. Larios, M. Gruebele and H. Kihara, *J. Phys. Chem. B*, 2002, **106**, 13040–13046.

120. X. Zhuang, T. Ha, H. D. Kim, T. Centner, S. Labeit and S. Chu, *Proc. Natl. Acad. Sci. USA*, 2000, **97**, 14241–14244.
121. S. Weiss, *Nat. Struct. Biol.*, 2000, **7**, 724–729.
122. T. A. Laurence, X. X. Kong, M. Jaeger and S. Weiss, *Proc. Natl. Acad. Sci. USA*, 2005, **102**, 17348–17353.
123. E. Rhoades, M. Cohen, B. Schuler and G. Haran, *J. Am. Chem. Soc.*, 2004, **126**, 14686–14687.
124. P. Tinnefeld, D. Herten and M. Sauer, *J. Phys. Chem. A*, 2001, **105**, 7989–8003.
125. F. Liu and M. Gruebele, *J. Mol. Biol.*, submitted, 2007.
126. U. Mayor, N. R. Guydosh, C. M. Johnson, J. G. Grossmann, S. Sato, G. S. Jas, S. M. Freund, D. O. Alonso, V. Daggett and A. R. Fersht, *Nature*, 2003, **421**, 863–867.
127. H. Q. Feng, J. Takei, R. Lipsitz, N. Tjandra and Y. W. Bai, *Biochemistry*, 2003, **42**, 12461–12465.
128. A. Bachmann and T. Kiefhaber, *J. Mol. Biol.*, 2001, **306**, 375–386.
129. Y. Zhu, D. O. V. Alonso, K. Maki, C.-Y. Huang, S. J. Lahr, V. Daggett, H. Roder, W. F. DeGrado and F. Gai, *Proc. Natl. Acad. Sci. USA*, 2003, **100**, 15486–15491.
130. D. T. Leeson, F. Gai, H. M. Rodriguez, L. M. Gregoret and R. B. Dyer, *Proc. Natl. Acad. Sci. USA*, 2000, **97**, 2527–2532.
131. R. M. Ballew, J. Sabelko and M. Gruebele, *Nat. Struct. Biol.*, 1996, **3**, 923–926.
132. T. R. Sosnick, L. Mayne and S. W. Englander, *Proteins: Struct. Funct. Genet.*, 1996, **24**, 413–426.
133. I. E. Sánchez and T. Kiefhaber, *Biophys. Chem.*, 2003, **100**, 397–407.
134. S. Osváth, J. Sabelko and M. Gruebele, *J. Mol. Biol.*, 2003, **333**, 187–199.
135. R. Metzler, J. Klafter, J. Jortner and M. Volk, *Chem. Phys. Lett.*, 1998, **293**, 477–484.
136. N. D. Socci, J. N. Onuchic and P. G. Wolynes, *Proc. Natl. Acad. Sci. USA*, 1999, **96**, 2031–2035.
137. P. Jennings and P. Wright, *Science*, 1993, **262**, 892–895.
138. C. Garcia, C. Nishimura, S. Cavagnero, H. J. Dyson and P. E. Wright, *Biochemistry*, 2000, **39**, 11227–11237.
139. T. L. Religa, J. S. Markson, U. Mayor, S. M. V. Freund and A. R. Fersht, *Nature*, 2005, **437**, 1053–1056.
140. C. D. Snow, L. L. Qiu, D. G. Du, F. Gai, S. J. Hagen and V. S. Pande, *Proc. Natl. Acad. Sci. USA*, 2004, **101**, 4077–4082.
141. S. Khorasanizadeh, I. Peters and H. Roder, *Nat. Struct. Biol.*, 1996, **3**, 193–205.
142. B. A. Krantz and T. R. Sosnick, *Biochemistry*, 2000, **39**, 11696–11701.
143. R. B. Dyer, S. J. Maness, S. Franzen, R. M. Fesinmeyer, K. A. Olsen and N. H. Andersen, *Biochemistry*, 2005, **44**, 10406–10413.
144. W. Y. Yang and M. Gruebele, *Philos. Trans. R. Soc. London, Ser. B*, 2005, **43**, 13018–13025.

145. M. D. Struthers, R. C. Cheng and B. Imperiali, *J. Am. Chem. Soc.*, 1996, **118**, 3073–3081.
146. C. D. Snow, B. Zagrovic and V. S. Pande, *J. Am. Chem. Soc.*, 2002, **124**, 14548–14549.
147. L. L. Qiu, S. A. Pabit, A. E. Roitberg and S. J. Hagen, *J. Am. Chem. Soc.*, 2002, **124**, 12952–12953.
148. J. Kubelka, W. A. Eaton and J. Hofrichter, *J. Mol. Biol.*, 2003, **329**, 625–630.
149. B. Zagrovic, C. D. Snow, M. R. Shirts and V. S. Pande, *J. Mol. Biol.*, 2002, **323**, 927–937.
150. F. Krieger, A. Moglich and T. Kiefhaber, *J. Am. Chem. Soc.*, 2005, **127**, 3346–3353.
151. J. Karanicolas and C. L. Brooks, III, *Proc. Natl. Acad. Sci. USA*, 2003, **100**, 3954–3959.
152. W. C. Swope, J. W. Pitera, F. Suits, M. Pitman, M. Eleftheriou, B. G. Fitch, R. S. Germain, A. Rayshubski, T. J. C. Ward, Y. Zhestkov and R. Zhou, *J. Phys. Chem. B*, 2004, **108**, 6582–6594.
153. A. G. Ladurner, L. S. Itzhaki, V. Daggett and A. R. Fersht, *Proc. Natl. Acad. Sci. USA*, 1998, **95**, 8473–8478.
154. U. Mayor, C. M. Johnson, V. Daggett and A. R. Fersht, *Proc. Natl. Acad. Sci. USA*, 2000, **97**, 13518–13522.
155. T. Ikura, T. Hayano, N. Takahashi and K. Kuwajima, *J. Mol. Biol.*, 2000, **297**, 791–802.
156. V. Munoz, *Annu. Rev. Biophys. Biomol. Struct.*, 2007, **36**, 395–412.
157. M. Gruebele, *Proteins: Struct. Func. Bioinfo.*, in press, 2008.
158. S. J. Hagen, *Proteins: Struct. Funct. Genet.*, 2003, **50**, 1–4.
159. H. Ma and M. Gruebele, *J. Comput. Chem.*, 2005, **27**, 125–134.
160. V. Muñoz, *Int. J. Quantum Chem.*, 2002, **90**, 1522–1528.
161. V. Munoz and J. M. Sanchez-Ruiz, *Proc. Natl. Acad. Sci. USA*, 2004, **101**, 17646–17651.
162. L. Brun, D. G. Isom, P. Velu, B. Garcia-Moreno and C. Royer, *Biochemistry*, in press.
163. M. Collins, G. Hummer, M. L. Quillin, B. W. Matthews and S. M. Gruner, *Proc. Natl. Acad. Sci. USA*, 2005, **102**, 16668–16671.
164. A. N. Naganathan, R. Perez-Jimenez, J. M. Sanchez-Ruiz and V. Munoz, *Biochemistry*, 2005, **44**, 7435–7449.
165. M. Jäger, J. Zhang, J. Bieschke, H. Nguyen, G. Dendle, M. Bowman, J. Noel, M. Gruebele and J. Kelly, *Proc. Natl. Acad. Sci. USA*, 2006, **108**, 10648–10653.
166. N. Ferguson, P. J. Schartau, T. D. Sharpe, S. Sato and A. R. Fersht, *J. Mol. Biol.*, 2004, **344**, 295–301.
167. S. F. Chekmarev, S. V. Krivov and M. Karplus, *J. Phys. Chem.*, in press.
168. R. Metzler, J. Klafter and J. Jortner, *Proc. Natl. Acad. Sci. USA*, 1999, **96**, 11085–11089.

169. U. H. E. Hansmann and J. N. Onuchic, *J. Chem. Phys.*, 2001, **115**, 1601–1606.
170. R. B. Dyer, S. J. Maness, E. S. Peterson, S. Franzen, R. M. Fesinmeyer and N. H. Andersen, *Biochemistry*, 2004, **43**.
171. G. Pappenberger, C. Saudan, M. Becker, A. E. Merbach and T. Kiefhaber, *Proc. Natl. Acad. Sci. USA*, 2000, **97**, 17–22.

CHAPTER 7
Single Molecule Spectroscopy in Protein Folding: From Ensembles to Single Molecules

BENJAMIN SCHULER

Biochemisches Institut, Universität Zürich, Winterthurerstr. 190, 8057 Zürich, Switzerland

7.1 Introduction

We are used to depicting chemical reactions and biochemical processes in terms of individual molecules. We do so on various levels of complexity, ranging from simple textbook cartoons to the atomic detail of molecular dynamics simulations (see Chapter 8 by Pande). But, remarkably, the vast majority of our knowledge about these systems has been derived from experiments on large ensembles of molecules, which yield only average values of observable properties. The sequence of molecular events describing the underlying reactions is typically inferred from testing a model by systematic variation of parameters. For kinetic ensemble studies, the reactions need to be synchronized, which is often difficult. The concept of observing single molecules is particularly appealing for processes with a large degree of conformational and dynamic heterogeneity, protein folding being a case in point.

Only recently has it become feasible to investigate the folding of single protein molecules. These techniques offer a fundamental advantage beyond our mere fascination for the direct observation of molecular processes: they can resolve and quantify the properties of individual molecules or subpopulations inaccessible in ensemble experiments. Fluorescence spectroscopy is a particularly appealing technique, owing to its extreme sensitivity and versatility. With

Förster resonance energy transfer (FRET), we can investigate intramolecular distance distributions and conformational dynamics of single proteins. In combination with fluctuation methods such as fluorescence correlation spectroscopy (FCS), it should be possible to obtain a detailed picture of the dynamic processes of protein folding only limited by the timescales of fluorescence photophysics (*i.e.* down to the sub-nanosecond range). In the following, some of the history and the basic underlying concepts of single molecule detection and analysis will be presented, illustrated with examples from single molecule protein folding or conceptually related experiments.

7.2 History and Principles of Single Molecule Detection

Remarkably, the very first single molecules detected appear to have been proteins. In 1970, step-like changes in ion conductance were observed for artificial lipid films containing small numbers of gramicidin A molecules.[1] Since then, the recording of such signals from single channels has matured to a method that now dominates the investigation of ion channels.[2] Due to this long tradition, the single channel field provides a wealth of examples and analysis methods for the stochastic chemical kinetics observed in single molecules.

A very different type of methods allowed the first observation of single atoms and molecules on surfaces in the early 1980s: scanning tunnelling microscopy and atomic force microscopy[3] (AFM), where a very fine tip is used to probe the surface of a sample with atomic resolution. The simplicity of the instrumentation quickly made AFM a standard method, and soon it allowed the imaging of samples in solution,[4] including proteins. In 1997, the mechanical unfolding of individual domains of titin molecules was reported, with both AFM[5] and laser tweezers.[6,7] The investigation of proteins under mechanical force has made accessible completely new aspects of protein folding and stability that can clearly not be studied in ensemble experiments.[8–11] This approach is of particular interest for proteins involved in mechanical functions, but the mechanical stability of other proteins is also of fundamental interest.

The optical detection of single molecules dates back to the 1970s, when it became possible to measure fluorescence from single atoms in dilute atomic beams, *i.e.* in the gas phase, where the background problem is minimal. Observing single molecules in the condensed phase is much more difficult, because Rayleigh and Raman scattering produce a huge background, not to mention contaminants, making great demands on the purity of the matrix. Moreover, an atom in vacuum is a chemically very stable system, even in its excited state, whereas fluorescent molecules in the condensed phase survive only a limited number of excitation–emission cycles before they are irreversibly destroyed – a process termed photobleaching. Single molecule detection in a solid or liquid matrix therefore requires additional measures. First of all, an optical method must be used that provides as strong a signal from an individual molecule as possible relative to the background from the large number of matrix molecules. The most popular option is fluorescence, where a dye molecule resonantly interacts with the excitation light. Due to the Stokes shift, the emitted light can be selected spectrally, typically with interference filters. To aid the signal from an individual molecule to be observed against the background signal from matrix molecules, it is helpful to reduce the size of the detection volume as much as possible, because the background level is proportional to the number of matrix molecules in the observed volume. Such spatial selection can be achieved, for instance, by very tightly focusing a laser beam into the sample, combined with confocal detection (Figure 7.1), or by total internal reflection fluorescence (TIRF) microscopy.

The first detection of individual chromophores in solid matrices was achieved at liquid helium temperature with modulated absorption spectroscopy in 1989.[12] Soon after, fluorescence spectroscopy was shown to yield a superior signal-to-noise ratio.[13] Within a couple of years, successful fluorescence experiments at room temperature followed,[14–18] leading to an explosion of the field, and opening the way for single molecule detection of suitably labeled biomolecules.[19] The introduction of FCS with a confocal detection scheme[20] and single molecule sensitivity was an important step for single molecule studies in solution. The demonstration of FRET in individual, labeled DNA molecules[21] in 1996 triggered a revival of the "spectroscopic ruler,"[22,23] enabling distance measurements in single biomolecules. The strongly distance-dependent radiationless energy transfer between an acceptor and a donor chromophore[24] had been used to study protein folding and dynamics for decades[25] and, indeed, the first experiments applying single molecule FRET to protein folding followed soon.[26–29] Especially the ability of the method to separate subpopulations, e.g. folded and unfolded protein molecules (Figure 7.1), and its potential to investigate intramolecular dynamics has made it into a central component of the growing field of single molecule protein folding.

Instead of reviewing the progress in the field chronologically and comprehensively, which has been done elsewhere,[10,30–34] I will present some of the main ideas of single molecule experiments and analysis to illustrate how they can be used to study protein folding, and conclude with a perspective of what we might expect from this new approach in the future. As the focus of this article is on conceptual issues, the practical aspects of single molecule experiments will also not be treated in detail. They can be found in a number of recent reviews.[31,35–39]

7.3 Kinetics: From Ensembles to Single Molecules

Assuming an infinitely large homogeneous system (a good approximation for a standard protein-folding experiment in a test tube[i]) consisting of a set of n subpopulations ("states") separated by sufficiently large free energy barriers, the

---

[i] 100 μl of a 1 μM sample still contain $6 \cdot 10^{13}$ molecules.

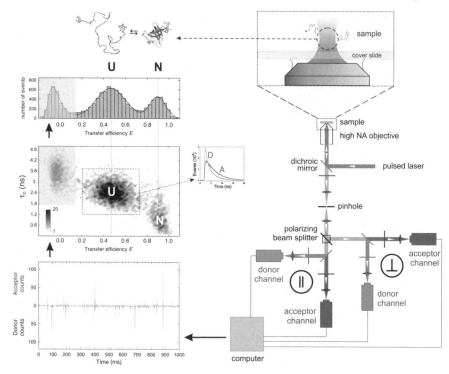

Figure 7.1 Schematic of a confocal single molecule experiment on freely diffusing molecules (right side). In this example, the signal is first separated by polarization and then by wavelength into two detection channels each, corresponding to emission from donor and acceptor chromophores. The left side shows an example of data from an experiment on Csp*Tm* molecules (labeled with Alexa 488 and Alexa 594 fluorophores) freely diffusing in a solution containing 1.5 M guanidinium chloride, conditions close to the unfolding midpoint. One second of a fluorescence intensity measurement (total acquisition time is 60 min) is shown on the lower left, with large bursts of photons originating from individual molecules diffusing through the confocal volume. From each of the bursts (discriminated from background by a combination of thresholds), a transfer efficiency E and a fluorescence lifetime τ (for both donor and acceptor, the donor lifetime is shown in the figure) can be calculated and plotted in a two-dimensional density graph (middle, left side). Subpopulations can be selected based on this graph (dashed box) to calculate subpopulation-specific properties, such as fluorescence lifetimes or correlation functions. A one-dimensional histogram of transfer efficiencies is depicted on top. The plots show populations with transfer efficiencies of $E \approx 0.9$ for folded and $E \approx 0.45$ for unfolded molecules. This separation of subpopulations allows changes in transfer efficiencies and other properties to be analysed individually for each state. The third peak at a transfer efficiency close to zero (shaded) is due to molecules without an acceptor chromophore. The measurement was done with a MicroTime 200 time-resolved fluorescence microscope (PicoQuant) with an excitation wavelength of 470 nm.

concentrations of molecules assigned to these states will evolve in time according to a set of coupled, first order, ordinary differential equations: the reaction rate equations. The overall reaction is thus treated as a continuous, deterministic process. The time course of the reaction will be described by a sum of $n-1$ exponential terms. Their individual time constants, however, will be complex algebraic expressions of *all* the rate constants in the mechanism and therefore have, in general, no simple physical significance.

But, in fact, the system is neither continuous nor deterministic. It is discrete, because molecules come in whole numbers, and molecular populations only change by integer amounts. Moreover, the system is stochastic, because the transitions of individual molecules from one state to another are ultimately initiated by thermally driven diffusive fluctuations and molecular collisions in the system. If we observe the signal from a sufficiently small number of molecules, these fluctuations about the average behavior become large enough to be measured, and can be used to extract dynamic information about the system, an insight conceived in the context of light scattering experiments[40-43] and membrane channel recordings,[44] and now used routinely, for instance in FCS[20,43,45] (*cf.* Section 7.4). Suppose, for example, we observe $N = 1000$ protein molecules under conditions corresponding to the unfolding midpoint. The probability of an individual molecule being folded is $p = 0.5$, so the variation in the number of folded molecules is given by the standard deviation of the binomial distribution as $[Np(1-p)]^{1/2} \approx 16$. The number of protein molecules that are folded at equilibrium is therefore not constant, but is 500 ± 16, where the standard deviation reflects the random fluctuations in the number of folded molecules from instant to instant.

The smaller the number of molecules we observe, the larger the relative fluctuations of the signal will be. Ultimately, if only a single molecule is observed, we expect step-like transitions between the states the molecule can populate. In the simplest example of a two-state folding reaction

$$U \underset{k_u}{\overset{k_f}{\rightleftharpoons}} N \qquad (7.1)$$

where equilibration within each state is fast relative to the transition rates between the states, the molecule will reside in either the folded (N) or the unfolded state (U) for extended periods of time, with intermittent rapid jumps between them. The average dwell times in N and U are given by the inverse rate coefficients of the unfolding reaction k_u and the folding reaction k_f, respectively. Such simple stochastic chemical kinetics have been observed for single immobilized molecules of the small cold shock protein Csp*Tm* (Figure 7.2), a two-state folder also by all criteria accessible in ensemble experiments.[46-48]

7.3.1 Rate Constants and Probabilities

If we consider the temporal behavior of an individual molecule with transitions between a folded and an unfolded state, the term "fraction of molecules folded

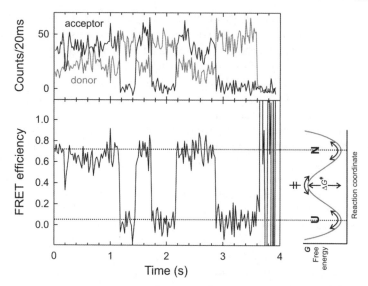

Figure 7.2 Example of data from an experiment on fluorescently labeled, vesicle-encapsulated Csp*Tm* molecules immobilized on a surface.[91] The experiment was performed at 2.0 M guanidinium chloride, the denaturation midpoint of Csp*Tm*. Under these conditions, the protein would be expected to remain in the folded and unfolded states for extended periods of time, with rapid, intermittent jumps across the barrier between the two corresponding free-energy minima. The top panel shows the fluorescence intensity trajectories recorded from the donor and acceptor chromophores of an individual protein. The anticorrelated changes in their emission intensities result in clear jumps of the transfer efficiency (bottom panel), reflecting the expected behavior of a two-state protein.

at equilibrium" must be replaced by the "fraction of time for which the molecule is folded," a quantity that can be measured accurately from a single protein molecule only with observation times much longer than the effective relaxation time τ of the two-state system, *i.e.* the inverse sum of the folding and unfolding rate constants k_f and k_u.

$$\tau = (k_f + k_u)^{-1} \quad (7.2)$$

While observation times in single molecule fluorescence experiments are still very much limited by the photostability of current organic chromophores, paradigmatic examples of analogous behavior are available from single channel recordings, where the ion current across an individual channel can sometimes be measured for hours, allowing a thorough kinetic and thermodynamic analysis of the channel opening and closing mechanism from an individual molecule.

In thinking about the folding kinetics of a single molecule, the question is: given the protein molecule is, say, unfolded at time t, what is the unfolded state lifetime τ_u of the molecule? More generally, we would like to know the probability density function (pdf) of the times the protein molecule stays folded or unfolded. To this end, we first define $U(t)$ as the probability that the molecule remains unfolded throughout the time from 0 to t. If we now assume that, for an unfolded molecule, the probability P of folding during a short[ii] time interval Δt is just proportional to the length of this time interval, we get the conditional probability

$$P(t + \Delta t | \text{unfolded at } t) = \alpha \Delta t \tag{7.3}$$

where α is a constant, independent of time. This latter assumption means that the probability of folding is independent of what has happened prior to t. This is a fundamental characteristic of this type of random process (a Markov process). As the molecule can either fold or not fold during Δt, and the probabilities of these two alternatives must add up to 1, we also know that the probability of remaining unfolded during Δt is $1 - \alpha \Delta t$. We can thus express the probability that the molecule remains unfolded during the entire interval from 0 to $t + \Delta t$ as

$$U(t + \Delta t) = U(t)(1 - \alpha \Delta t) \tag{7.4}$$

taking into account the Markov assumption. After slightly reorganizing the equation, we get in the limit $\Delta t \to 0$

$$\lim_{\Delta t \to 0} \frac{U(t + \Delta t) - U(t)}{\Delta t} = \frac{dU(t)}{dt} = -\alpha U(t) \tag{7.5}$$

by the definition of the derivative. As $U(0) = 1$, we obtain as a solution

$$U(t) = e^{-\alpha t} \tag{7.6}$$

Here, and from the corresponding probability density function of the unfolded state lifetime

$$p(t) = \alpha e^{-\alpha t} \tag{7.7}$$

we recognize immediately that our constant α is the inverse time constant τ_u, and thus the rate constant k_f. It is this connection with the single molecule world that illustrates why exponential decays are so abundant in macroscopic chemical kinetics.

Another illustrative way of getting to the exponential distribution[2] pictures the unfolded protein as randomly diffusing within its unfolded state free-energy minimum, driven by the thermal energy of the system. If this motion takes place on a timescale much faster than the inverse folding rate (the separation of timescales, a prerequisite for a two-state process), we can interpret each thermal

---

[ii] Δt must be small relative to $1/\alpha$, so we can neglect processes such as multiple transitions during Δt.

kick as a binomial trial with some probability, p, of successfully crossing the barrier to the folded state. The chance of success must therefore be very small in each step, so a large number of trials, N, will be necessary on average before the protein folds. Briefly, in this case the binomial distribution approaches the Poisson distribution, which gives the probability of not folding in time t essentially as $e^{-\alpha t}$. This is the reason for the close connection between Poisson statistics and exponential probability density functions, both highly abundant in single molecule statistics. If, as in the above example, the distribution of time intervals between events is exponential, then the mean number of events during a given time interval will follow a Poisson distribution. Another remarkable property of the exponential distribution is that it is the only random function with no memory.

The lack of memory that is assumed in the above derivations is an interesting issue in protein dynamics. It has been suggested from single molecule experiments on enzymatic reactions[49] that there may be a "molecular memory" on unexpectedly long timescales, rendering successive enzymatic turnovers dependent on each other. A similar phenomenon is conceivable in protein folding, if there are additional dynamics coupled to folding that approach the timescales of the folding and unfolding rate coefficients. This may be expected for slow processes such as peptidyl-prolyl *cis-trans* isomerization. Whereas fast transitions in agreement with a simple two-state process were observed for individual Csp*Tm* molecules (Figure 7.2), slow dynamics have been postulated based on single molecule FRET experiments on immobilized adenylate kinase[50] and ribonuclease H[51] (*cf.* Section 7.4).

7.4 Correlation Analysis

One of the most powerful methods to investigate such "molecular memory" effects is correlation analysis. Currently, the most common approach for investigating the relaxation kinetics of a reaction is to perturb the entire ensemble, *e.g.* by a rapid change in denaturant concentration in a stopped-flow instrument or by a laser-induced temperature jump, and then to observe the system return to equilibrium under the new set of conditions. According to the fluctuation-dissipation theorem,[52] the rates of relaxation of a system to equilibrium after a small macroscopic perturbation, and the time correlation of spontaneous fluctuations of the undisturbed system at equilibrium, are described by the same rate coefficients. Correlation spectroscopy can therefore provide kinetic information about chemical reactions or molecular dynamics even at equilibrium.

If we think of the system under investigation, *e.g.* our folding protein molecule, as possessing some sort of internal coherence, this coherence is lost, or dissipated, through the interaction with the randomly fluctuating solvent molecules. For instance, if we observe a small volume of a solution, the concentrations of reactants fluctuate about their equilibrium mean values as a result of random variations both in the number of molecules formed or eliminated by the chemical reaction and in the number of molecules which enter or leave the observation region by diffusion. Hence, both the rate coefficients for diffusion and the reaction may be determined from observations of the rates of decay of spontaneous concentration fluctuations without disturbing the equilibrium of the reaction system, *i.e.* in an entirely non-invasive fashion.

The central quantity for analysing such fluctuations and their loss of coherence are correlation functions.[53] The autocorrelation function of the property A, for instance, is defined as[iii]

$$\langle A(t)A(t+\tau)\rangle = \lim_{T\to\infty} \frac{1}{T} \int_0^T A(t)A(t+\tau)\,dt \qquad (7.8)$$

(In any experimental measurement the averaging is of course done over finite time.) Cross-correlation functions between different properties or signals, *e.g.* fluorescence emission from a donor and an acceptor chromophore undergoing FRET, can be defined analogously. The autocorrelation function of a non-conserved, non-periodic property decays from its initial value $\langle A^2 \rangle$ to the final value $\langle A \rangle^2$ with a time constant characteristic of the fluctuation of A, where $A(t)$ and $A(t + \tau)$ are expected to become totally uncorrelated at long times (Figure 7.3A). In many cases, the autocorrelation function decays like a single exponential with a characteristic *relaxation time* or *correlation time* of the property, but often it takes a more complicated functional form. The most common example is a fluorescence correlation experiment where fluorescently labeled molecules diffuse through a confocal volume with a three-dimensional Gaussian shape. The simplest setup (Figure 7.1) involves focusing a laser beam into the

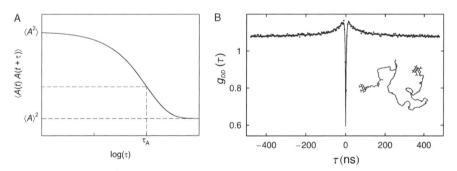

Figure 7.3 (A) A general, unnormalized correlation function (see text). (B) Donor intensity autocorrelation function from Hanbury Brown and Twiss experiments on Csp*Tm* labeled with a donor and acceptor dye, showing the rapid global unfolded chain dynamics in the tens of nanoseconds range.[85] The fast component in the range of a few nanoseconds is caused by photon antibunching.[110]

[iii] Strictly speaking, this definition is only true for an ergodic system, where the averaging is independent of the starting time t.

sample with a high numerical aperture objective, thus forming a diffraction-limited focal volume. The fluorescence of molecules in this region of the sample is collected through the same objective, out-of-focus light is removed with a pinhole in the image plane of the microscope, and the signal is then typically detected with avalanche photodiodes. The resulting intensity autocorrelation function normalized by the mean intensity squared is

$$G(\tau) = 1 + \frac{1}{\langle N \rangle}\left(1 + \frac{\tau}{\tau_D}\right)^{-1}\left(1 + \frac{\tau}{\omega^2 \tau_D}\right)^{-1/2}\left(1 + K e^{-\tau/\tau_r}\right) \quad (7.9)$$

where $\langle N \rangle$ is the average number of molecules in the observed volume, τ_D is the characteristic time it takes a molecule to diffuse through the observation volume, ω is the aspect ratio of the volume, and K is the equilibrium constant of a reaction with a relaxation time τ_r resulting in fluctuations of the emission intensity. In this case there are thus two mechanisms contributing to the observed intensity fluctuations: diffusion of molecules in and out of the confocal volume, and fluctuations in the concentrations of molecular species caused by the reaction, which could, for example, be a protein folding reaction (Equation (7.2)). From Equation (7.9), it is obvious that the observation of reaction dynamics is limited to timescales smaller than the diffusion timescale, because the diffusive terms will decay to zero for $t \gg \tau_D$.

Correlation analysis can be performed on molecules freely diffusing in solution, as in a typical FCS experiment, but it can equally be applied to the signal recorded from single immobilized molecules. Both approaches have been used for the investigation of protein folding.[54] For example, Chattopadhyay et al.[55] investigated conformational fluctuations in unfolded intestinal fatty acid binding protein doubly labeled with the fluorophore TMR, such that intramolecular self-quenching of the dyes can lead to fluctuations in the fluorescence emission rate. The corresponding signal autocorrelation exhibits a component with a relaxation time of 1.6 μs and amplitude that changes in response to denaturant concentration in parallel with the population of the unfolded state, suggesting that it corresponds to the unfolded state dynamics of the protein. The authors suggest additional intramolecular interactions or increased chain stiffness as a reason for these dynamics to be significantly slower than expected from experiments on unstructured peptides[56–58] (see Chapter 6 by Gruebele for more details on fast dynamics in proteins). With similar methods, Neuweiler et al.[59] investigated Trp-cage folding via the quenching of the fluorophore MR121 by tryptophan. Their results are in good agreement with the μs-folding kinetics observed in laser-induced temperature jump experiments[60] and revealed an additional kinetic component on the 100 ns timescale. The interpretation was complicated by complex formation between Trp and the dye, but it illustrates the potential of correlation spectroscopy for sub-microsecond dynamics, which have turned out to be particularly useful for clarifying the diffusive behavior of unfolded polypeptide chains (see Section 7.5) and its implications for the protein folding reaction (see Chapter 6 by Gruebele).

7.5 FRET Efficiency Distributions and Distance Dynamics

In FCS experiments, the sample concentration is optimized for the best signal-to-noise ratio of the autocorrelation function. As the rate of photon emission increases with $\langle N \rangle$ (the mean number of molecules in the observation volume) the photon statistics will be better, and the statistical noise will be smaller for larger $\langle N \rangle$. On the other hand, the amplitude of the correlation function is inversely proportional to $\langle N \rangle$ (cf. Equation (7.9)), and the signal itself is thus larger for small concentrations. It turns out that the optimal signal-to-noise ratio is achieved in a concentration regime where on average 10 or more molecules reside in the observation region.[61] In practice, this usually corresponds to concentrations in the nanomolar range. Increasing the sample concentration further will eventually lead to complete averaging of uncorrelated fluctuations, and the single molecule aspect of the experiment will be lost. In the other extreme, for a single molecule experiment in its pure form, it will be necessary to dilute the solution such that $\langle N \rangle \ll 1$. In confocal experiments on freely diffusing molecules, this will result in bursts of photons originating from single molecules diffusing through the confocal volume, separated by intervals with background signal (Figure 7.1). The fluorescence signal from each molecule can thus be identified individually using suitable threshold criteria.

7.5.1 Single Molecule FRET Experiments

In single molecule FRET experiments (Figure 7.1), the distance dependence of the transfer efficiency E is related to the donor and acceptor fluorescence according to

$$E = \frac{n_A}{n_A + n_D} = \frac{1}{1 + (r/R_0)^6} \quad (7.10)$$

which can be used to obtain distance information from individual molecules. Here, n_A and n_D are the number of donor and acceptor photons in an individual burst (corrected for fluorescence quantum yields, detection efficiencies, crosstalk between the detection channels, and direct excitation of the acceptor[62]), r is the inter-dye distance, and R_0 is the Förster radius of the donor/acceptor pair, which is calculated as

$$R_0^6 = \frac{9000 \, (\ln 10) \, \kappa^2 Q_D J}{128 \, \pi^5 n^4 N_A} \quad (7.11)$$

where J is the overlap integral between the donor emission and the acceptor absorption spectra, Q_D is the donor's fluorescence quantum yield, κ^2 is a factor depending on the relative orientation of the chromophores' transition dipoles,

n is the refractive index of the medium between the dyes, and N_A is Avogadro's number.[24,63]

Performing a single molecule FRET experiment on protein folding requires several steps (for details, see ref. 62). First, protein samples have to be prepared for labeling, either by chemical synthesis or by recombinant expression in combination with site-directed mutagenesis. After identifying a suitable dye pair with the Förster radius in the right range, the fluorophores need to be attached to the protein as specifically as possible to avoid chemical heterogeneity.[64] The equilibrium and kinetic properties of the labeled protein should then be measured in ensemble FRET experiments and compared directly to unlabeled protein to ensure that the folding mechanism is not altered. It is helpful to prepare control molecules, such as polyproline peptides[29,65,66] or double-stranded DNA,[67] with the same chromophores as the protein. After customizing the instrument for the sample, data can be taken directly either on freely diffusing molecules or on immobilized molecules, if observation times greater than a few milliseconds are desired. Finally, the data need to be processed[62] to correct for background contributions and other effects, to identify fluorescence bursts in diffusion experiments, to calculate transfer efficiencies, fluorescence lifetimes, and fluorescence intensity correlation functions.

In single molecule FRET experiments on chymotrypsin inhibitor 2 (CI2) it was first demonstrated that this approach is able to separate the signal from folded and unfolded subpopulations by histogramming the transfer efficiencies calculated from a large number of fluorescence bursts.[27] Such FRET efficiency histograms (Figure 7.1) have become a common way of analysing this type of experiment, allowing the subpopulations to be investigated individually and thus avoiding the signal averaging between folded and unfolded molecules typical of ensemble experiments. This has led to the discovery of the equilibrium collapse of the unfolded state even in two-state folders such as the small cold shock protein CspTm[29] and many other small proteins.[27,68–75] It may be tempting to convert the efficiency histograms directly into distance distributions[26–28,76] and further into potentials of mean force. However, this is rarely justified, and requires detailed knowledge about the timescales of the dynamic processes involved.

7.5.2 Timescales and Distance Distributions

The relative magnitude of the timescales of at least four different processes will have an influence on the position and the width of the FRET efficiency histogram: (a) the rotational correlation time of the chromophores, (b) the fluorescence lifetime of the donor, (c) the intramolecular dynamics of the chain connecting the fluorophores, and (d) the observation timescale.

The rotational correlation time of the chromophores influences the value of the orientation factor κ^2 (Equation (7.11)): if dye reorientation is sufficiently fast such that the relative orientation of the donor and acceptor dipoles average out while the donor is in the excited state, κ^2 can be assumed to equal 2/3. If, in the other extreme, the donor fluorescence decay is much faster than dye reorientation, a static distribution of relative dye orientations can be assumed. Intermediate cases are difficult to treat analytically,[63] and simulations become the method of choice. $\kappa^2 = 2/3$ is often a good approximation, because the rotational correlation times of typical dyes are in the range of a few hundred picoseconds, while their fluorescence lifetimes are in the nanosecond range (although they may be shortened significantly by the transfer process).

The characteristic times of the fluorescence decay and of inter-dye distance changes are often less clearly separated. If they are in a similar range, the distribution of transfer rates resulting from the distance distribution will give rise to highly non-exponential fluorescence decays, which in favorable cases can even be used to obtain information about the shape of the distance distribution.[77]

Finally, the relative magnitudes of the inter-dye distance distribution and the observation timescale (more accurately, the inter-photon times[78]) will affect the width of the measured transfer efficiency distributions. As shown by Gopich and Szabo,[78,79] the observation time must be approximately an order of magnitude smaller than the relaxation time of the donor–acceptor distance to obtain physically meaningful distance distributions or corresponding potentials of mean force. Otherwise, only the mean value of the transfer efficiency of the respective subpopulation can be used to extract information about the distance distribution, and an independent model for the shape of the distance distribution is needed. In practice, this means that distance distributions can be determined from free diffusion experiments on proteins if the underlying dynamics are on a timescale greater than about 1 ms, assuming photon count rates of $\sim 10^5\,\mathrm{s}^{-1}$ typically achieved during fluorescence bursts.[78] A noticeable influence of dynamics on the width, however, is already expected for fluctuations in the 10 to 100 μs timescale,[78] which has been used to set bounds on the pre-exponential factor for protein folding.[29] Recently, methods to obtain reaction dynamics from the shape of transfer efficiency histograms have been developed,[80] but to date no experimental applications have been reported.

The quantitative influence of these different timescales is illustrated by a recent analysis of the transfer efficiencies obtained in single molecule experiments on polyproline peptides of different lengths labeled with a FRET pair[66] in combination with Langevin molecular dynamics simulations used to find the shape of the end-to-end distance distribution $P(r)$ of these stiff peptides. The three physically plausible limits for the possible averaging regimes and the resulting mean transfer efficiencies $\langle E \rangle$ are:

1. If the rotational correlation time τ_c of the chromophores is small relative to the fluorescence lifetime τ_f of the donor (i.e. $\kappa^2 = 2/3$), and the dynamics of the peptide chain (with relaxation time τ_p) are slow relative to τ_f,

$$\langle E \rangle = \int_a^{l_c} E(r) P(r) dr \quad \text{with} \quad E(r) = (1 + (r/R_0)^6)^{-1} \qquad (7.12)$$

where $P(r)$ is the normalized inter-dye distance distribution, a is the distance of closest approach of the dyes, and l_c is the contour length of the peptide.

2. If $\tau_c \ll \tau_f$ and $\tau_p \ll \tau_f$,[iv]

$$\langle E \rangle = \frac{\int_a^{l_c} (R_0/r)^6 P(r) dr}{1 + \int_a^{l_c} (R_0/r)^6 P(r) dr} \qquad (7.13)$$

3. If $\tau_c \gg \tau_f$ and $\tau_p \gg \tau_f$,

$$\langle E \rangle = \int_0^4 \int_a^{l_c} E(r, \kappa^2) P(r) p(\kappa^2) dr\, d\kappa^2 \quad \text{with} \quad E(r, \kappa^2) = \left(1 + \frac{2}{3\kappa^2}(r/R_0)^6\right)^{-1} \qquad (7.14)$$

The theoretical isotropic probability density $p(\kappa^2)$ for the case in which all orientations of the donor and acceptor transition dipoles are equally probable[63,81] is

$$p(\kappa^2) = \begin{cases} \left[\dfrac{1}{2\sqrt{3\kappa^2}} \ln(2 + \sqrt{3})\right] & 0 \leq \kappa^2 \leq 1 \\ \left[\dfrac{1}{2\sqrt{3\kappa^2}} \ln\left(\dfrac{2 + \sqrt{3}}{\sqrt{\kappa^2} + \sqrt{\kappa^2 - 1}}\right)\right] & 1 \leq \kappa^2 \leq 4 \end{cases} \qquad (7.15)$$

with $\kappa^2 = (\cos\theta_T - 3\cos\theta_D \cos\theta_A)^2$, where θ_T is the angle between the donor and acceptor transition dipoles, and θ_D and θ_A are the angles between the transition moments and the line connecting the centers of the donor and acceptor, respectively.

It is important to recognize that even for a molecule with a single fixed distance or very rapid conformational averaging, the resulting FRET efficiency histogram is relatively broad. A fundamental source of broadening is shot noise, the variation in count rates about fixed means due to the discrete nature of the signals (only small numbers of photons is observed from an individual molecule!), but in practice broader histograms than expected from shot noise alone are usually observed. The origin of this excess width is currently unclear,[29,72] but there are factors apart from slow distance fluctuations that could potentially contribute, such as fluctuating fluorescence quantum efficiencies, or confocal volumes for donor and acceptor channels that are either of different size or misaligned. Consequently, without a suitable reference it is difficult to interpret a width in excess of shot noise in terms of slow conformational dynamics.

The separation of the signals from various subpopulations, such as the thermodynamic states in two-state protein folding reactions, can be used to measure a variety of parameters[82] that are difficult to determine otherwise. In a confocal experiment with pulsed excitation and four detection channels (Figure 7.1), for example, the emission wavelength range (*i.e.* whether it is a donor or an acceptor photon), the polarization, and the time of emission relative to the excitation

---

[iv] Note that the averaging has to be done over the transfer rate k_t, i.e. $\langle E \rangle = 1/(1 + k_0/\int_a^{l_c} k_t(r) P(r) dr)$, where $k_t(r) = k_0(R_0/r)^6$, and k_0 is the fluorescence decay rate of the donor in the absence of the acceptor.

pulse become available for every detected photon. Consequently, we can determine for each burst of photons from a single molecule the transfer efficiency (Equation (7.10)), the donor and acceptor fluorescence lifetimes, the fluorescence anisotropy, and a number of auxiliary parameters that aid the interpretation of the results. In a second step, the bursts from subpopulations can be grouped, *e.g.* to obtain fluorescence decays of an individual subpopulation, devoid of signal contributions from other molecules. Whereas, for example, the fluorescence lifetime from an individual burst can only be estimated with relatively large uncertainty, the combination of all photons from a subpopulation can result in decays that are suitable for more detailed analyses.

Hoffmann *et al.*,[83] for instance, investigated variants of the small two-state protein Csp*Tm* with the dye labels positioned such that different segments of the chain could be probed. A combined analysis of FRET efficiency histograms and subpopulation-specific fluorescence lifetimes (Figure 7.1) gave good agreement with intramolecular distance distributions of a Gaussian chain for all variants, even at low denaturant concentrations, where the chain is compact. This indicates that any residual structure can affect only short segments and is probably highly dynamic. Kinetic synchrotron radiation circular dichroism experiments in fact provided evidence for the presence of some β-structure in the compact unfolded state,[83] and kinetic ensemble FRET experiments probing a short segment that forms a β-strand in the folded state indicate the local formation of extended structure in the compact unfolded state of a closely related cold shock protein, Csp*Bc*.[84] However, there are also examples for deviations from Gaussian chain behavior. Laurence *et al.*[68] used subpopulation-specific fluorescence lifetime analysis to investigate structural distributions upon collapse of CI2 and ACBP and inferred the presence of transient residual structure in the unfolded state.

7.5.3 Dynamics from Transfer Efficiency Fluctuations

A versatile way of probing distance dynamics over a broad range of timescales is to monitor the resulting *fluctuations* of the transfer rate between donor and acceptor.[28] In combination with the separation of subpopulations, the full potential of such correlation analyses (*cf.* Section 7.4) can be used to selectively measure intramolecular dynamics on timescales from nanoseconds[85] to seconds and longer.[70]

An example for the investigation of very rapid dynamics with such methods has recently been reported by Nettels *et al.*,[85] who analysed the single molecule photon statistics of the fluctuations in intensity of donor and acceptor fluorophores that result from distance fluctuations in the unfolded subpopulation of the small two-state protein Csp*Tm*. The basic idea of their experiment is the following: if, for example, a donor photon is emitted at time $\tau = 0$, the chain ends are likely to be far apart at that instant, corresponding to a low rate of energy transfer. A very short time later, the ends will still be far apart, and the likelihood of emitting another donor photon will still be high. However, at times much greater than the reconfiguration time of the chain, the molecule will have lost the "memory" of its initial configuration, and the probability of donor emission will be determined by the average transfer efficiency. Thus the autocorrelation of the emission intensity is expected to decay approximately on the timescale of chain reconfiguration (Figure 7.3B). With information on the unfolded state dimensions available from previous experiments,[83] and using a model that describes chain dynamics as a diffusive process on a one-dimensional free energy surface,[85–87] the very rapid reconfiguration time could be extracted. This time decreases from about 60 ns to 20 ns between 0 and 6 M guanidinium chloride, after correcting for solvent viscosity. The addition of denaturant thus not only expands the chain, but the reduced transient intra-molecular interactions decrease the contribution of internal friction to chain diffusion. Correlation spectroscopy indicates that at times less than ~100 μs there are no additional long-range dynamics in unfolded CspTm.[88] Moreover, since the dye-labeled protein folds in ~10 ms in the absence of denaturant with an exponential time course,[29] the requirement that the unfolded state dynamics be fast compared to the folding times indicates that there are no slower unfolded state dynamics in this protein.

An example for dynamics from correlation analysis of FRET experiments on longer timescales is the work of Kuzmenkina *et al.*,[51] who investigated surface-immobilized RNase H from microseconds to minutes. In this case, a component of the correlation function on the 20 μs timescale was assigned to polypeptide reconfiguration; the actual folding reaction was observed on a 100 s timescale. Additionally, step-like transfer efficiency changes with a wide range of amplitudes were detected even *between* efficiency values assigned to the unfolded state. A similar behavior had previously been reported for adenylate kinase A.[50] Cross-correlation analysis of the experiments on RNase H yielded a relaxation time of about 2 s, which was interpreted as slow conformational transitions with correspondingly large barriers in the unfolded state. While a clear structural interpretation of these observations is not yet available, they illustrate that unfolded state dynamics may occur over a wide range of timescales.

7.6 Pleasure, Pain, and Promise of Single Molecule Experiments

As illustrated above, single molecule analyses can reveal the underlying distributions of distances, timescales, or forces averaged out in ensemble measurements, and will thus contribute substantially to our understanding of dynamically heterogeneous systems such as proteins. Even though the method is relatively new, integrated commercial instrumentation is already available,[89] which allows these methods to be applied even in laboratories that do not develop their own instrumentation. However, due to the complexity of sample preparation, instrumentation, and data analysis, we are also faced with the risk of jumping to conclusions or introducing a subjective bias by selecting individual molecules that agree with our expectations while ignoring others. It will therefore remain crucial to complement single molecule experiments with ensemble data for validation and, possibly even more importantly, to include control molecules to separate effects originating from photophysics and other complications from the molecular property under study.

Single molecule methods should be able to advance our understanding of many of the elementary properties of the energy surfaces of protein folding that have been put forward by theory (see Chapter 3 by Wolynes), but have been difficult to test experimentally. Some of the topics currently being pursued intensively include the structure and dynamics of the collapsed unfolded state as observed under conditions favoring the native structure,[29,90] and a mechanistic analysis of folding reactions from trajectories of individual protein molecules,[50,91–93] including their potential kinetic and thermodynamic heterogeneity. An ultimate goal will be to time-resolve the transitions between folded and unfolded states (as opposed to the mere redistribution of populations observed in the classical chemical kinetics of protein folding). In this context, particular promise lies in the analysis of the different scenarios of "downhill folding" (see Chapter 6 by Gruebele). As originally proposed by Bryngelson et al.,[94] downhill folding or unfolding becomes accessible in kinetic experiments when the bias for the folded or the unfolded state, respectively, becomes sufficiently large.[95,96] By significantly populating all structures along the reaction coordinate, this could allow the observation of the actual folding process in single protein molecules and the distribution of microscopic pathways taken by a folding protein. In the other scenario, only one thermodynamic state exists under a whole range of conditions, resulting in a gradual shifting of a single free-energy minimum from unfolded to folded structures.[97,98] Single molecule methods should be extremely helpful in quantifying the structural distributions involved in such "one-state" folding, even though this will require improvements in photon emission rates and collection efficiencies to cope with the microsecond folding times of these molecules. It might be problematic to identify barrierless folding if the barrier exists at high denaturant concentration and disappears as the protein is stabilized by decreasing the denaturant concentration.[73,98]

Key goals in method development will be to overcome the current limitations of the available techniques. In single molecule fluorescence, the photophysics and the photochemistry of the fluorophores are probably the most limiting factor. Even very stable dyes survive only a certain number of excitation cycles, seriously limiting the number of photons that can be observed from a single molecule. Intermittent excitation can be used to extend the acquisition time,[51] but only at the expense of time resolution. The photon emission rate and thus the time resolution of the experiment are ultimately limited by the fluorescence lifetime of the chromophores, typically in the nanosecond range. Assuming an optimistic photon collection efficiency of 10%, and one excitation every 10 ns under optical saturation, this yields maximum average count rates in the 10 MHz range. In practice, however, the rates are significantly lower

($\sim$100 kHz), mostly due to the population of long-lived dark states. The field would thus gain enormously from fluorophores without populated triplet or radical dark states, higher emission rates, and greater photostability. At the same time, instrumentation development will continue to be crucial for pushing the limits of single molecule spectroscopy.

Finally, one of the great expectations for single molecule protein folding is its developing convergence with biophysical theory, especially molecular dynamics simulations, which have already been very instructive for the analysis of mechanical unfolding experiments[99,100] and single molecule FRET[66,101] (see Chapter 8 by Pande). Moreover, theory will continue to play a crucial role for data analysis (*e.g.* the optimal use of photon statistics for the extraction of dynamic parameters[79,80,102–104]), for the quantitative treatment of FRET results even when the timescales of the underlying processes overlap,[66,78] for the extraction of equilibrium parameters from non-equilibrium experiments,[105,106] and for a variety of other fundamental aspects.[107–109] This multifaceted combination of methods certainly contributes to both the challenge and the fascination in single molecule studies. And maybe individual molecules are indeed more prone to give away the secrets of protein folding than large crowds.

Acknowledgements

I thank Daniel Nettels for helpful comments on the manuscript. This work has been supported by the Schweizerische Nationalfonds, the Deutsche Forschungsgemeinschaft, the VolkswagenStiftung, and the Human Frontier Science Program.

References

1. S. B. Hladky and D. A. Haydon, *Nature*, 1970, **225**, 451.
2. B. Sakmann and E. Neher, *Single Channel Recording*, Plenum Press, 1995.
3. G. Binnig, C. F. Quate and C. Gerber, *Phys. Rev. Lett.*, 1986, **56**, 930.
4. B. Drake, C. B. Prater, A. L. Weisenhorn, S. A. Gould, T. R. Albrecht, C. F. Quate, D. S. Cannell, H. G. Hansma and P. K. Hansma, *Science*, 1989, **243**, 1586.
5. M. Rief, M. Gautel, F. Oesterhelt, J. M. Fernandez and H. E. Gaub, *Science*, 1997, **276**, 1109.
6. M. S. Kellermayer, S. B. Smith, H. L. Granzier and C. Bustamante, *Science*, 1997, **276**, 1112.
7. L. Tskhovrebova, J. Trinick, J. A. Sleep and R. M. Simmons, *Nature*, 1997, **387**, 308.
8. M. Carrion-Vazquez, A. F. Oberhauser, T. E. Fisher, P. E. Marszalek, H. Li and J. M. Fernandez, *Prog. Biophys. Mol. Biol.*, 2000, **74**, 63.
9. R. B. Best and J. Clarke, *Chem. Commun.*, 2002, 183.
10. X. Zhuang and M. Rief, *Curr. Opin. Struct. Biol.*, 2003, **13**, 88.

11. C. Bustamante, Y. R. Chemla, N. R. Forde and D. Izhaky, *Annu. Rev. Biochem.*, 2004, **73**, 705.
12. W. E. Moerner and L. Kador, *Phys. Rev. Lett.*, 1989, **62**, 2535.
13. M. Orrit and J. Bernard, *Phys. Rev. Lett.*, 1990, **65**, 2716.
14. J. J. Macklin, J. K. Trautman, T. D. Harris and L. E. Brus, *Science*, 1996, **272**, 255.
15. S. M. Nie, D. T. Chiu and R. N. Zare, *Science*, 1994, **266**, 1018.
16. X. S. Xie and R. C. Dunn, *Science*, 1994, **265**, 361.
17. W. P. Ambrose, P. M. Goodwin, J. C. Martin and R. A. Keller, *Phys. Rev. Lett.*, 1994, **72**, 160.
18. E. Betzig and R. J. Chichester, *Science*, 1993, **262**, 1422.
19. *"Single Molecules" Special Issue, Science*, **283**.
20. R. Rigler, U. Mets, J. Widengren and P. Kask, *Eur. Biophys. J.*, 1993, **22**, 169.
21. T. Ha, T. Enderle, D. F. Ogletree, D. S. Chemla, P. R. Selvin and S. Weiss, *Proc. Natl. Acad. Sci. USA*, 1996, **93**, 6264.
22. L. Stryer and R. P. Haugland, *Proc. Natl. Acad. Sci. USA*, 1967, **58**, 719.
23. L. Stryer, *Annu. Rev. Biochem.*, 1978, **47**, 819.
24. T. Förster, *Annalen der Physik*, 1948, **6**, 55.
25. E. Haas, *Chem. Phys. Chem.*, 2005, **6**, 858.
26. Y. W. Jia, D. S. Talaga, W. L. Lau, H. S. M. Lu, W. F. DeGrado and R. M. Hochstrasser, *Chem. Phys.*, 1999, **247**, 69.
27. A. A. Deniz, T. A. Laurence, G. S. Beligere, M. Dahan, A. B. Martin, D. S. Chemla, P. E. Dawson, P. G. Schultz and S. Weiss, *Proc. Natl. Acad. Sci. USA*, 2000, **97**, 5179.
28. D. S. Talaga, W. L. Lau, H. Roder, J. Tang, Y. Jia, W. F. DeGrado and R. M. Hochstrasser, *Proc. Natl. Acad. Sci. USA*, 2000, **97**, 13021.
29. B. Schuler, E. A. Lipman and W. A. Eaton, *Nature*, 2002, **419**, 743.
30. G. Haran, *J. Physics-Condensed Matter*, 2003, **15**, R1291.
31. B. Schuler, *Chem. Phys. Chem.*, 2005, **6**, 1206.
32. X. Michalet, S. Weiss and M. Jäger, *Chem. Rev.*, 2006, **106**, 1785.
33. G. U. Nienhaus, *Macromol. Biosci.*, 2006, **6**, 907.
34. B. Schuler and W. A. Eaton, *Curr. Opin. Struct. Biol.*, in press.
35. W. E. Moerner, *J. Phys. Chem. B*, 2002, **106**, 910.
36. C. Eggeling, S. Berger, L. Brand, J. R. Fries, J. Schaffer, A. Volkmer and C. A. Seidel, *J. Biotechnol.*, 2001, **86**, 163.
37. T. Ha, *Methods*, 2001, **25**, 78.
38. X. Michalet, A. N. Kapanidis, T. Laurence, F. Pinaud, S. Doose, M. Pflughoefft and S. Weiss, *Annu. Rev. Biophys. Biomol. Struct.*, 2003, **32**, 161.
39. B. Schuler, in *Protein Folding and Stability*, ed. Y. Bai and R. Nussinov, Plenum Press, Totowa, NJ, USA, 2007, vol. **366**.
40. B. J. Berne and H. L. Frisch, *J. Chem. Phys.*, 1967, **47**, 3675.
41. L. Blum and Z. W. Salsburg, *J. Chem. Phys.*, 1968, **48**, 2292.
42. Y. Yeh and R. N. Keeler, *J. Chem. Phys.*, 1969, **51**, 1120.
43. D. Magde, W. W. Webb and E. Elson, *Phys. Rev. Lett.*, 1972, **29**, 705.

44. B. Katz and R. Miledi, *Nature*, 1970, **226**, 962.
45. R. Rigler and E. L. Elson, *Fluorescence Correlation Spectroscopy: Theory and Applications*, Springer, Berlin, 2001.
46. D. Perl, C. Welker, T. Schindler, K. Schröder, M. A. Marahiel, R. Jaenicke and F. X. Schmid, *Nat. Struct. Biol.*, 1998, **5**, 229.
47. B. Schuler, W. Kremer, H. R. Kalbitzer and R. Jaenicke, *Biochemistry*, 2002, **41**, 11670.
48. D. Wassenberg, C. Welker and R. Jaenicke, *J. Mol. Biol.*, 1999, **289**, 187.
49. H. P. Lu, L. Xun and X. S. Xie, *Science*, 1998, **282**, 1877.
50. E. Rhoades, E. Gussakovsky and G. Haran, *Proc. Natl. Acad. Sci. USA*, 2003, **100**, 3197.
51. E. V. Kuzmenkina, C. D. Heyes and G. U. Nienhaus, *Proc. Natl. Acad. Sci. USA*, 2005, **102**, 15471.
52. H. B. Callen and T. A. Welton, *Phys. Rev.*, 1951, **83**, 34.
53. B. J. Berne and R. Pecora, *Dynamic Light Scattering*, Dover, Mineola, NY, 2000.
54. C. Frieden, K. Chattopadhyay and E. L. Elson, *Adv. Protein Chem.*, 2002, **62**, 91.
55. K. Chattopadhyay, E. L. Elson and C. Frieden, *Proc. Natl. Acad. Sci. USA*, 2005, **102**, 2385.
56. O. Bieri, J. Wirz, B. Hellrung, M. Schutkowski, M. Drewello and T. Kiefhaber, *Proc. Natl. Acad. Sci. USA*, 1999, **96**, 9597.
57. L. J. Lapidus, W. A. Eaton and J. Hofrichter, *Proc. Natl. Acad. Sci. USA*, 2000, **97**, 7220.
58. L. J. Lapidus, P. J. Steinbach, W. A. Eaton, A. Szabo and J. Hofrichter, *J. Phys. Chem. B*, 2002, **106**, 11628.
59. H. Neuweiler, S. Doose and M. Sauer, *Proc. Natl. Acad. Sci. USA*, 2005, **102**, 16650.
60. L. L. Qiu, S. A. Pabit, A. E. Roitberg and S. J. Hagen, *J. Am. Chem. Soc.*, 2002, **124**, 12952.
61. O. Krichevsky and G. Bonnet, *Rep. Prog. Phys.*, 2002, **65**, 251.
62. B. Schuler, in *Protein Folding Protocols*, ed. Y. Bai and R. Nussinov, Humana, Totowa, NJ, 2007, vol. **366**.
63. B. W. Van Der Meer, G. Coker III and S. Y. S. Chen, *Resonance Energy Transfer: Theory and Data*, VCH Publishers, Inc., New York, Weinheim, Cambridge, 1994.
64. A. N. Kapanidis and S. Weiss, *J. Chem. Phys.*, 2002, **117**, 10953.
65. R. Best, K. Merchant, I. V. Gopich, B. Schuler, A. Bax, and W. A. Eaton, *Proc. Natl. Acad. Sci. USA*, 2007, **104**, 18964.
66. B. Schuler, E. A. Lipman, P. J. Steinbach, M. Kumke and W. A. Eaton, *Proc. Natl. Acad. Sci. USA*, 2005, **102**, 2754.
67. A. N. Kapanidis, N. K. Lee, T. A. Laurence, S. Doose, E. Margeat and S. Weiss, *Proc. Natl. Acad. Sci. USA*, 2004, **101**, 8936.
68. T. A. Laurence, X. X. Kong, M. Jager and S. Weiss, *Proc. Natl. Acad. Sci. USA*, 2005, **102**, 17348.

69. E. V. Kuzmenkina, C. D. Heyes and G. U. Nienhaus, *J. Mol. Biol.*, 2006, **357**, 313.
70. E. V. Kuzmenkina, C. D. Heyes and G. U. Nienhaus, *Proc. Natl. Acad. Sci. USA*, 2005, **102**, 15471.
71. E. Sherman and G. Haran, *Proc. Natl. Acad. Sci. USA*, 2006, **103**, 11539.
72. K. A. Merchant, R. B. Best, J. M. Louis, I. V. Gopich and W. A. Eaton, *Proc. Natl. Acad. Sci. USA*, 2007, **104**, 1528.
73. F. Huang, S. Sato, T. D. Sharpe, L. M. Ying and A. R. Fersht, *Proc. Natl. Acad. Sci. USA*, 2007, **104**, 123.
74. T. Tezuka-Kawakami, C. Gell, D. J. Brockwell, S. E. Radford and D. A. Smith, *Biophys. J.*, 2006, **91**, L42.
75. S. Mukhopadhyay, R. Krishnan, E. A. Lemke, S. Lindquist and A. A. Deniz, *Proc. Natl. Acad. Sci. USA*, 2007, **104**, 2649.
76. B. D. Slaughter, J. R. Unruh, E. S. Price, J. L. Huynh, R. J. B. Urbauer and C. K. Johnson, *J. Am. Chem. Soc.*, 2005, **127**, 12107.
77. E. Haas, E. Katchalskikatzir and I. Z. Steinberg, *Biopolymers*, 1978, **17**, 11.
78. I. V. Gopich and A. Szabo, *J. Phys. Chem. B*, 2003, **107**, 5058.
79. I. V. Gopich and A. Szabo, *J. Chem. Phys.*, 2005, **122**, 1.
80. I. V. Gopich and A. Szabo, *J. Phys. Chem. B*, 2007.
81. R. E. Dale, J. Eisinger and W. E. Blumberg, *Biophys. J.*, 1979, **26**, 161.
82. J. Widengren, V. Kudryavtsev, M. Antonik, S. Berger, M. Gerken and C. A. M. Seidel, *Anal. Chem.*, 2006, **78**, 2039.
83. A. Hoffmann, A. Kane, D. Nettels, D. E. Hertzog, P. Baumgartel, J. Lengefeld, G. Reichardt, D. A. Horsley, R. Seckler, O. Bakajin and B. Schuler, *Proc. Natl. Acad. Sci. USA*, 2007, **104**, 105.
84. C. Magg, J. Kubelka, G. Holtermann, E. Haas and F. X. Schmid, *J. Mol. Biol.*, 2006, **360**, 1067.
85. D. Nettels, I. V. Gopich, A. Hoffmann and B. Schuler, *Proc. Natl. Acad. Sci. USA*, 2007, **104**, 2655.
86. I. V. Gopich and A. Szabo, *J. Chem. Phys.*, 2006, **124**, 154712.
87. Z. S. Wang and D. E. Makarov, *J. Phys. Chem. B*, 2003, **107**, 5617.
88. D. Nettels, A. Hoffmann and B. Schuler, *J. Phys. Chem. B*, in press.
89. M. Wahl, F. Koberling, M. Patting, H. Rahn and R. Erdmann, *Curr. Pharm. Biotechnol.*, 2004, **5**, 299.
90. E. V. Kuzmenkina, C. D. Heyes, and G. U. Nienhaus, *J. Mol. Biol.*, 2006, in press.
91. E. Rhoades, M. Cohen, B. Schuler and G. Haran, *J. Am. Chem. Soc.*, 2004, **126**, 14686.
92. J. M. Fernandez and H. B. Li, *Science*, 2004, **303**, 1674.
93. C. Cecconi, E. A. Shank, C. Bustamante and S. Marqusee, *Science*, 2005, **309**, 2057.
94. J. D. Bryngelson, J. N. Onuchic, N. D. Socci and P. G. Wolynes, *Proteins*, 1995, **21**, 167.
95. J. Sabelko, J. Ervin and M. Gruebele, *Proc. Natl. Acad. Sci. USA*, 1999, **96**, 6031.

96. W. Y. Yang and M. Gruebele, *Nature*, 2003, **423**, 193.
97. M. M. Garcia-Mira, M. Sadqi, N. Fischer, J. M. Sanchez-Ruiz and V. Muñoz, *Science*, 2002, **298**, 2191.
98. V. Munoz, *Annu. Rev. Biophys. Biomol. Struct.*, 2007, **36**, 395.
99. M. Rief and H. Grubmüller, *Chem. Phys. Chem.*, 2002, **3**, 255.
100. R. B. Best, S. B. Fowler, J. L. T. Herrera, A. Steward, E. Paci and J. Clarke, *J. Mol. Biol.*, 2003, **330**, 867.
101. M. Margittai, E. Schweinberger, J. Widengren, D. Fasshauer, R. Jahn, G. Schröder, H. Grübmuller, E. Haustein, S. Felekyan, M. König and C. A. M. Seidel, *Biophys. J.*, 2002, **82**, 305a.
102. H. Yang and X. S. Xie, *J. Chem. Phys.*, 2002, **117**, 10965.
103. G. F. Schröder and H. Grubmüller, *J. Chem. Phys.*, 2003, **119**, 9920.
104. G. Haran, *Chem. Phys.*, 2004, **307**, 137.
105. G. Hummer and A. Szabo, *Proc. Natl. Acad. Sci. USA*, 2001, **98**, 3658.
106. J. Liphardt, S. Dumont, S. B. Smith, I. Tinoco Jr. and C. Bustamante, *Science*, 2002, **296**, 1832.
107. J. Wang and P. Wolynes, *Phys. Rev. Lett.*, 1995, **74**, 4317.
108. C. Bai, C. Wang, X. S. Xie and P. G. Wolynes, *Proc. Natl. Acad. Sci. USA*, 1999, **96**, 11075.
109. J. Wang, *J. Chem. Phys.*, 2003, **118**, 952.
110. L. Fleury, J. M. Segura, G. Zumofen, B. Hecht and U. P. Wild, *Phys. Rev. Lett.*, 2000, **84**, 1148.

CHAPTER 8

Computer Simulations of Protein Folding

VIJAY S. PANDE, ERIC J. SORIN, CHRISTOPHER D. SNOW AND YOUNG MIN RHEE

Department of Chemistry and Biophysics Program, Stanford University, Stanford, CA 94305, USA

8.1 Introduction: Goals and Challenges of Simulating Protein Folding

Computer simulation holds great promise to significantly complement experiment as a tool for biological and biophysical characterization. Simulations offer the promise of atomic spatial detail with femtosecond temporal resolution. However, the application of computational methodology has been greatly limited due to fundamental computational challenges: put simply, for much of what one would want to examine, atomistic simulations would require decades to millennia to complete. Below, we detail current methods to tackle these challenges as well as recent applications of this methodology.

8.1.1 Simulating Protein Folding

Proteins play a fundamental role in biology. With their ability to perform numerous biological functions, including acting as catalysts, antibodies, and molecular signals, proteins today realize many of the goals to which modern nanotechnology aspires. However, before proteins can carry out these remarkable molecular functions, they must perform another amazing feat – they must assemble themselves. This process of protein self-assembly into a particular

---

RSC Biomolecular Sciences
Protein Folding, Misfolding and Aggregation: Classical Themes and Novel Approaches
Edited by Victor Muñoz
© Royal Society of Chemistry 2008

shape, or "fold", is called protein folding. Due to the importance of the folded state in the biological activity of proteins, recent interest in misfolding related diseases[1] (see Chapter 10 by Esteras-Chopo *et al.*), and a fascination with how this process occurs,[2-4] there has been much work to unravel the mechanism of protein folding[5] (see Chapter 3 by Wolynes).

While there are several questions relating to the "protein folding problem", including structure prediction[6,7] and protein design (see Chapter 9 by Lehmann and co-workers), here we will concentrate on another aspect of folding: *how* do proteins fold into their final folded structure? Experimentally characterizing the detailed nature of the protein folding mechanism is considerably more difficult than characterizing the static structure. We therefore turn to the combination of experiment and atomistic models (that can readily yield the desired spatial and temporal detail), but we must in turn ask "how quantitatively predictive are these simulations?" The true test is statistical significance. The very act of *statistically* comparing with experiment is critical, and leads to either model validation or an indication that further model refinement is necessary.

There are two approaches one can take in molecular simulation. One direction is to perform coarse-grained simulations using simplified, or "minimalist", models. These models typically either make simplifying assumptions (such as Go models, which use simplified Hamiltonians[8]), or employ coarse-grained representations (such as using alpha-carbon only models to represent the protein[9]) or potentially both. While these methods are often first considered due to their computational efficiency, perhaps an even greater benefit of simplified models is their ability to potentially yield insight into general properties involved in protein folding. However, with any model there are limitations and the cost for such potential insight into general properties of folding is the limitation of restricted applicability to any *particular* protein system.

Alternatively, one can examine more detailed models. These models typically have full atomic detail, often for both the protein and solvent alike. Detailed models have the obvious benefit of potentially greater fidelity to experiment. However, this comes at two great costs. First, the computational demands for performing the simulation become enormous. Second, the added degrees of freedom lead to an explosion of extra detail and simulation-generated data; the act of gleaning insight from this sea of data is no simple task and is often underestimated, especially in light of the more straightforward (although still often difficult) task of simply performing the simulations. We emphasize that the relevant question is not whether a given method is "correct" in some absolute sense (as all models have limitations), but whether the model is predictive to some degree of accuracy.

Why are detailed models worth this enormous effort in both simulation and analysis? First, quantitative comparison between theory and experiment is critical for validating simulation as well as lending interpretation to experimental results. While it is generally held that experiments will not be able to yield the detail and precision available in simulations (and that simulations may likely be the only way one can fully understand the folding mechanism[10]), without quantitative validation of simulations there is no way to know whether the simulation model or methodology are sufficiently accurate to yield a faithful reproduction of reality. Indeed, without a quantitative comparison to experiment, there is no way to decisively arbitrate the relative predictive merits of one model over another.

Second, detailed models potentially have a greater predictive power. In principle, a detailed model should allow one to start purely from the protein sequence and, by simulating the physical dynamics of protein folding, yield everything that one can measure experimentally, including folding and unfolding rates, free energies, and the detailed geometry of the folded state. In practice, the ability of detailed models to achieve these lofty goals rests both on the ability to carry out the computationally demanding kinetics simulations as well as the ability of current models (force fields) to yield sufficiently accurate representations of inter-atomic interactions.

8.1.2 What are the Challenges for Atomistic Simulation?

First, one must consider the source of the great computational demands of molecular simulation at atomic detail. To simulate dynamics, typically one numerically integrates Newton's equations for all of the atoms in the system. By choosing models with atomic degrees of freedom, one must simulate the dynamics at the timescales of atomic motion (femtoseconds). Indeed, if the timestep involved in numerical integration is pushed too high (without constraining degrees of freedom), the numerical integration becomes unstable. This leads to the trivial problem that if one wants to reach the millisecond timescale by taking femtosecond steps, many (10^{12}) steps must be taken. While modern molecular dynamics codes are extremely well optimized and perform typically millions of steps per CPU day, this clearly falls short of what is needed (see Figure 8.1).

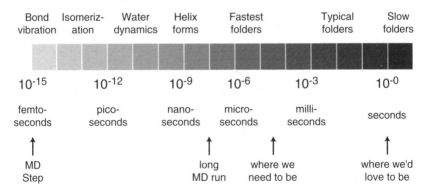

Figure 8.1 Relevant timescales for protein folding. While detailed simulations must start with femtosecond timesteps, the timescales one would like to reach are much longer, requiring billions (microseconds) to trillions (milliseconds) of iterations. Typical fast, modern CPUs can do approximately a million iterations in a day, posing a major challenge for detailed simulation.

However, even if one could reach the relevant timescales, the next question is whether our models would be sufficiently accurate. In particular, would we reach the folded state, would the folded state be stable (with free energy of stability comparable to experiment), and would we reach the folded state with a rate comparable to experiment? Indeed, if one could quantitatively predict protein folding rates, free energies of stability, and structure of the relevant states at equilibrium, one would be able to predict essentially *everything* that can be measured experimentally. While rates and free energies themselves can only indirectly detail the nature of how proteins fold, clearly the ability to quantitatively predict all experimental observables is a necessary prerequisite for any successful theory or simulation of protein folding.

However, a quantitative prediction of all experimental observables is necessary but not sufficient. If a simulation could only reproduce experiments, the simulation would not yield any new insight, which is the goal of simulations in the first place. This leads to a third important challenge for simulation: gaining new insight. Indeed, as one adds detail to simulations, the burden of analysis becomes greater and greater. Atomistic simulations can easily generate gigabytes of data to be processed, but the volume of data does not reduce the inherent complexity of the physical process. A vast number of degrees of freedom from time-resolved protein and water coordinates can obscure any simple, direct analysis of the folding mechanism.

Additionally, analysis of such simulations may reflect the seemingly arbitrary state definitions used by the one performing the analysis, and great care must therefore be taken in defining the relevant states prior to data analysis. This, of course, often presents the most notable issue in interpreting simulation data, due to the sheer difficulty in collecting adequate data to define the states, and microstates, that the model would predict. As detailed below, this issue is most often overcome by employing simplified models. These models are generally built around the known or desired states prior to simulation, but suffer the obvious lack of predicting metastable, misfolded, or intermediate states that may be observable when using atomistic simulation models.

8.2 Protein Folding Models: from Atomistic to Simplified Representations

8.2.1 Atomic Force Fields

Atomistic models for protein folding typically utilize a classical force field, which attempts to reproduce the physical interaction between the atoms in the protein and solvent. The energy of the system is defined as the sum of interatomic potentials, which consist of several terms:

$$E = E_{LJ} + E_{Coulomb} + E_{bonded} \tag{8.1}$$

The van der Waals interaction between atoms is most commonly modeled by a Lennard–Jones energy (E_{LJ})

$$E_{LJ} = \Sigma_{ij} \varepsilon_{ij} [(\sigma_{ij}/r_{ij})^{12} - (\sigma_{ij}/r_{ij})^{6}] \qquad (8.2)$$

where σ_{ij} is related to the size of the atoms i and j and ε_{ij} is related to the strength of their interaction. While van der Waals attraction is relatively weak, the LJ potential also serves an important role in providing hard core repulsion between atoms. The bonded interactions modeled in E_{bonded} handle the specific stereochemistry of the molecule – in particular, the nature of the covalent bonds and steric constraints in the angles and dihedral angles of the molecule. These interactions are clearly local, but they play a very important role in determining the conformational space of the molecule; changes to the backbone dihedral potentials in such a model can lead to greatly diverging simulation results.[11] $E_{Coulomb}$ corresponds to the familiar Coulomb's law:

$$E_{Coulomb} = \Sigma_{ij} q_i q_j / r_{ij} \qquad (8.3)$$

where q_i is the charge on atom i and r_{ij} is the distance between atoms i and j. To best parameterize atomic force fields, such as accounting for quantum mechanical effects between nearby atoms, some force fields also include scaling coefficients for the pairwise E_{LJ} and $E_{Coulomb}$ terms between atoms separated by three covalent bonds (so-called "1-4 scaling"), and it has recently been demonstrated that modifying these scaling terms can significantly alter simulation results.[11]

It is perhaps most natural to handle the pairwise interactions explicitly as in Equation (8.1). However, this leads to simulation codes whose performance scales as N^2, where N is the number of atoms being simulated. Clearly, this is very computationally demanding. To reduce this demand, the calculation can ideally be made to scale linearly with N. For inherently short range interactions, it is natural to do this with cutoffs and long range corrections, *i.e.* to set the potential to zero smoothly once the distance is beyond some cutoff, such as 12 Å. Such cutoff procedures have been shown to lead to qualitatively incorrect results for Coulomb interactions[12] and reaction field or Ewald-based methods have been suggested as alternatives that can obtain significantly better results.[13]

Clearly there are many parameters in the above formulas. Indeed, these numbers grow further when one considers the fact that the chemical environment of atoms causes even the same type of chemical element (*e.g.* carbon) to act very differently. For example, carbon in a hydrocarbon chain will behave fundamentally differently from carbon in an aromatic ring. In order to handle such purely quantum mechanical effects in a classical model, one creates multiple atom types (corresponding to the different relevant environments) for each physical atomic element. In this example, one would define different carbon atom types. Thus, while there are only a handful of relevant physical atoms involved (primarily carbon, hydrogen, oxygen, and nitrogen), there can be tens to hundreds of different atom types.

Although this is clearly the natural way to handle the role of chemical environment in a classical model, this leads to an explosion of parameters needed in the model, leading to a modelling challenge in the determination of these parameters. Several groups have risen to this challenge and have developed parameterizations for the force field functionals similar to the form above. Typically, these parameterizations are divided into terms for proteins (such as AMBER,[14] CHARMM,[15] and OPLS[16]) and for the solvent (such as TIP or SPC models). Additionally, these force fields are typically parameterized using a specific water model, and may also be associated with specific molecular dynamics packages. One should thus be careful in combining protein and solvent models and also not confuse atomic force fields with the molecular dynamics software for which they were derived.

8.2.2 Implicit Solvation Models

With the parameterization described above for the physical forces between atoms, one can simulate all relevant interactions: protein–protein, protein–solvent, and solvent–solvent. However, in typical simulations with solvent represented explicitly (*i.e.* directly simulating the solvent atom by atom), the number of solvent atoms is much larger than the number of protein atoms and thus the majority of the computational time (*e.g.* 90%) goes into simulating the solvent. Clearly the solvent plays an important role since the hydrophobic and dielectric properties of water are essential to protein stability.[17,18] However, an alternative to explicit simulation of water is to include these properties *implicitly* by using a continuum model of solvent properties.

Typically, these models account for hydrophobicity in terms of some free-energy price for solvent exposed area on the protein. These surface area (SA) based methods vary somewhat in terms of how the surface area is calculated as well as the energetic dependence on this exposed surface area. We stress that one should not *a priori* expect that a simpler (and perhaps less accurate) calculation of the surface area yields worse results than a more geometrically accurate SA calculation. Indeed, since SA is itself an approximation, what is important for the fidelity of the model is not the geometric accuracy of the surface area but rather whether the SA term faithfully reproduces the physical effect as judged by comparison to experiment.

The dielectric contribution of water to the free energy is in some ways a more difficult contribution for which to account. The canonical method follows the Poisson–Boltzmann (PB) equation. To demonstrate the philosophy of implementing PB calculations, consider a protein immersed in solvent where the protein and solvent are modeled as dielectric media with dielectric constants of ε_{in} and ε_{out} respectively (thus making the dielectric a function of spatial position, $\varepsilon(x, y, z)$). Also, consider that the protein will likely have charges with a spatial density $\rho_{protein}(x, y, z)$ and that there will be counter-ions in the solvent with a charge density $\rho_{countert}(x, y, z)$. In this case, we can describe the resulting electrostatic potential and charge density as

$$\nabla[\varepsilon(x, y, z)\nabla\phi] = -4\pi\rho(x, y, z)$$
$$= -4\pi[\rho_{\text{protein}}(x, y, z) + \rho_{\text{countert}}(x, y, z)] \quad (8.4)$$

where the total charge density $\rho(x, y, z)$ is comprised of both the protein and counter-ion charges. If one assumes that the counter-ion density is driven thermodynamically to its free energy minimum, we can make the "mean field"-like approximation that

$$\rho_{\text{countert}}(x, y, z) = \Sigma_I n_i q_i \exp[-q_i \phi(x, y, z)/kT] \quad (8.5)$$

where n_i is the bulk number density of counter-ion species i and q_i is its charge. Thus, this method handles counter-ions implicitly as well as aqueous solvent. Including this term leads to the so-called non-linear Poisson–Boltzmann equation. If the Boltzmann term is Taylor expanded for small $\phi(x, y, z)/kT$ (*i.e.* high temperature, low counter-ion concentration, or low potential strength), one gets the so-called linearized Poisson–Boltzmann equation.

In general, the Poisson–Boltzmann equation is considered by many to be the "gold standard" for implicit solvation calculations. It can be used for both energy and force calculation[19] and is thus suitable for molecular dynamics. However, PB calculation is also typically very computationally demanding and there has been much effort to develop more computationally tractable, empirical approximations to the PB equation. For example, Still and co-workers developed an empirical approximation to PB.[20] Based on a generalization of the Born equation for the potential of atoms, Still's Generalized Born (GB) model (and its subsequent variants from Still's group and other groups) have been shown to be both computationally tractable and quantitatively accurate for some problems, including the solvation free energy of small molecules[20] and protein folding kinetics.[21]

8.2.3 Minimalist Models

To further simplify the model, the protein force field can be generated from the experimental structure. Using the information of the native conformation, attractive parts of the LJ potentials for all non-native contact pairs can be reduced or turned off altogether. Such a potential may lead to minimized frustration for folding (*i.e.* smoothing the energy landscape by removing small energetic barriers and metastable microstates, as shown in Figure 8.2), enabling much faster folding simulations. In many cases, the model is built by considering each amino acid residue as one particle (coarse-graining) to maximize the simplification. Using explicit or implicit solvent models mentioned in the above paragraphs is technically possible, though such an approach will lose the benefit of using the minimalist model itself. Therefore, solvent effects are usually considered using Langevin dynamics (random forces imparted on each

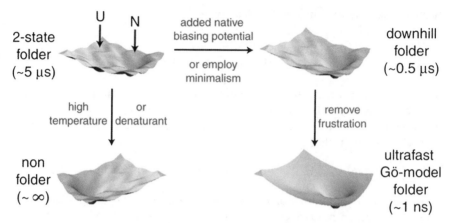

Figure 8.2 Example free-energy surface for a simple two-state folder and related surfaces derived by adding external forces or simplifications to the simulation model, demonstrating the variation in necessary simulation timescales for sampling of various models. Some sampling methods, such as REMD and umbrella sampling, make use of several landscapes by adding biasing potentials or including a large range of temperatures, while minimalist models remove landscape frustration and/or the presence of a non-native free-energy basin.

simulated body to represent solvent viscosity) or can be incorporated explicitly in the pairwise protein non-bonded interaction potential.[22]

8.2.4 How Accurate are the Models?

Any question of accuracy must consider the desired experimental observable. One natural quantity to examine is the solvation free energy of small molecules, such as amino acid side chains.[23] With recent advances in high-precision free-energy methods,[23,24] one can directly compare the models to experiment within experimental error.

For explicit solvent models,[24] the solvation free energies of small molecule analogs to amino acid side chains show a systematic shift (towards being less soluble). This would lead to an artificial stabilization of proteins (since the unfolded state would be less stable) and could have a significant impact on predicted protein–protein and protein–ligand free energies. These results suggest natural force-field improvements; recent work in this direction removes this systematic shift, leading to models with zero mean error with solvation free energy experiments and a surprisingly low RMSD (~ 0.4 kcal mol^{-1}).[24]

How accurate are implicit solvent models? While the GB models are somewhat empirical, they have been shown to agree reasonably well with PB calculations. More importantly, GB models have been able to accurately predict experimental results, such as the solvation free energy of small molecules.[20,25] In the end, experiment must of course be the final arbiter of any theoretical method. Moreover, while PB is on a much firmer mathematical footing (*i.e.* one can derive it directly from the Poisson equation), one must consider that PB itself is empirical in nature in some respects. The concept of a dielectric is macroscopic; it is an approximation to apply this macroscopic concept to the microscopic world of small molecules and proteins (hundreds to thousands of atoms). However, the success of PB as a predictive tool demonstrates the validity (or, at the very least, predictive power) of such methods and approximations.

8.3 Sampling: Methods to Tackle the Long Timescales Involved in Folding

Simulating the mechanism of protein folding is a great computational challenge due to the long timescales involved. Below, we briefly summarize some methods that have been used to address this challenge. As in any computational method, each has its own limitations and it is natural to consider the regime of applicability of each method (Figure 8.2).

8.3.1 Tightly Coupled Molecular Dynamics (TCMD)

To simulate molecular dynamics (MD) one typically integrates Newton's equations numerically for the atoms in the system with femtosecond timesteps to include the fast timescales of atomic motion. Thus, to reach the millisecond timescale, many (10^{12}) steps must be taken. While modern molecular dynamics codes are extremely well optimized and perform typically 10^6 steps per CPU day, this clearly falls short of what is needed. Using multiple CPUs in a tightly coupled fashion to speed a single trajectory is appealing, but is an inefficient use of CPU power (*i.e.* one does not get a 100× speed increase with 100 CPUs) and thus has not been widely used to get beyond the nanosecond timescale, with the notable exception of Duan and Kollman's single 1 μs trajectory of the villin headpiece.[26]

8.3.2 Replica Exchange Molecular Dynamics (REMD)

Replica Exchange Molecular Dynamics[27–31] has become a powerful technique to explore the free-energy landscapes of proteins, with speed increases[32] of roughly 10× over traditional MD. Moreover, REMD efficiently parallelizes with only slightly coupled networking required. However, REMD achieves its speed increase by using a non-physical form of kinetics (in temperature replica space). This method yields a Boltzmann-weighted ensemble after sufficient convergence,[32] but the trajectories cannot themselves be used to predict any direct kinetic properties, although aspects related to the kinetics (such as possibly kinetically relevant intermediates) can be inferred from the resulting free-energy landscapes.[29]

8.3.3 High-temperature Unfolding

While folding times are very long from a simulation point of view, unfolding (especially under high denaturation conditions) can be very fast – on the nanosecond timescale.[33–35] Under extreme denaturing conditions (*e.g.* ~400 K temperature), one would expect the folded state to become only metastable, with a low barrier to unfolding. Daggett and Levitt[33] first took advantage of this scenario, and Daggett's group has subsequently pioneered this method to examine a variety of proteins and compare their results to experiment, especially with a comparison of ϕ values calculated at high-temperature folding *vs.* experimental measurements.[10,36] One note of caution is that the transition state character is dependent on temperature. For example, the Gruebele lab has found temperature-sensitive ϕ-values.[37] Of particular significance of the impact of this approach has been the ability to closely connect simulation predictions to experiment. However, applying extreme temperatures to models developed under ambient/biological temperatures (*i.e.* 300 ± 10 K) must be done with caution: it has recently been shown that even force fields that appear to be extremely accurate for the system studied fail to reproduce experimentally observed temperature-dependent trends at high and low temperatures.[11] While it is possible to study protein unfolding under conditions that approximate experiment, simulations to date trade authentic recapitulation of the experimental kinetics in favor of computational tractability.

8.3.4 Low-viscosity Simulation Coupled with Implicit Solvation Models

This is another common means to try to tackle long timescales.[38–41] In regular simulations with an implicit solvent model, one typically uses the Langevin equation for dynamics and employs a damping term consistent with water-like viscosity. However, water is relatively viscous and such simulations can be very costly. Instead, many groups have proposed the use of viscosities only 1/100 to 1/1000 that of water (or even no viscosity at all). While lowering the viscosity greatly speeds the kinetics,[38] the effect of such non-physical modeling inherently assumes a potential risk of altering not only the rate but also the nature of the overall kinetics of the system.[42] Assuming simulation convergence, the correct thermodynamics should be obtained, but it must also be understood that the thermodynamics will be based on the model, and therefore may also miss microstates that are coupled with properties of the solvent (such as, in this case, viscosity).

8.3.5 Coarse-grained and Minimalist Models

These kinds of models have played a large role in recent simulations of protein folding.[6,22,43] The idea is to largely trade chemical complexity for computational tractability. Coarse-grained models allow one to directly address a range of hypotheses relating to general properties of folding. However, in their generality, by construction they may lack the ability to access more detailed questions of folding (depending on the nature of the question of interest). Whereas detailed models cannot, in general, be used to collect ensemble statistics for large biomolecular systems, this is not true for minimalist models, and a recent study used such a model to make a direct connection between individual folding pathways and the bulk observed folding mechanism for a system consisting of ~ 5000 atoms.[44]

8.3.6 Path Sampling

Given an initial trajectory between the unfolded and folded regions, which can be generated *via* high-temperature unfolding or similar means, this method generates an ensemble of different pathways that join the unfolded and folded regions. For example, Bolhuis and co-workers determined the formation order of hydrogen bonds and the hydrophobic core in a β-hairpin.[45] Using the fluctuation-dissipation theorem,[46] it is possible to calculate folding rates from these ensembles. More recently, a new method called transition interface sampling[47] introduced an alternate method to calculate transition rates. Since path-sampling methods are very computationally demanding, it is interesting to consider whether one can construct an algorithm that can more efficiently utilize simulation data (*e.g.* folding trajectories) in order to predict folding rates and mechanisms.

8.3.7 Graph-based Methods

Graph-based methods sample configuration space and connect nearby points with weights according to their transition probabilities. From these graphs, it is possible to calculate such properties as most probable path, p_{fold} values[48] as well as to analyse the order in which secondary structures form.[49] However, the graph representation of protein-folding pathways does not solve the sampling problem, but recasts it, and sampling any continuous, high-dimensional space is still a difficult challenge. Previous graph-based methods have sampled configuration space uniformly (*i.e.* choosing conformations at random) or used sampling methods biased towards the native state. Clearly, as the protein size increases, it becomes very difficult to sample the biologically important conformations with random sampling.

8.3.8 Markovian State Model Methods[50–53]

These methods have recently shown promise to allow for an atomically detailed model with quantitative prediction of kinetics. They can take advantage of the benefits of many of the methods above, such as in the generation of initial nodes, as well as build upon the methods of path sampling and graph-based methods to use short paths to predict complex kinetics.

8.4 Validation of Simulation Methodology: Protein Folding Kinetics

To study protein folding kinetics – and especially compare theory to experiment – it is natural to ask which quantities should be compared. The most experimentally accessible quantitative observables of two-state proteins are the folding and unfolding rates from which one can obtain the thermodynamic stability. Thus, it is important to validate any simulation method through quantitative comparison to experiment with proper statistics. As rates and free energies are the natural quantitative experimental measurements, relative or absolute prediction of these quantities is necessary for a direct connection to experiment and a true assessment of theoretical methodology.

8.4.1 Low-viscosity Simulations

We now consider rate predictions made using atomistic potentials based on various approximations of the physics of inter-atomic interactions (including especially solvent-mediated interactions). Caflisch and co-workers have pioneered long atomistic folding simulations using simple, computationally efficient implicit solvent models. By using low (or no) viscosity in their simulations, they accelerate the timescales involved in folding and are able to observe multiple folding transitions in single trajectories. Though not guaranteeing ensemble level convergence, such reversible folding transitions are strong evidence that sampling is sufficient for useful thermodynamic analysis.

For example, two secondary structural motifs were studied by Caflisch et al.: the α-helical Y(MEARA)$_6$ peptide,[54] and Beta3s, a three-stranded antiparallel β-sheet.[55] Surprisingly, the helical peptide, which was shown to contain more helical content (and thus helical stability) than the (AAQAA)$_3$ peptide, folded much more slowly at 300 K, with a mean folding time of $\sim$80 ns. For Beta3s, a mean folding time of 31.8 ns was predicted at 360 K, and a following study predicted a folding time of 39 ns at 330 K,[56] both significantly faster than the $\sim$5 μs timescale reported by De Alba et al. at lower temperatures.[57] Increased sampling of Beta3s in four additional simulations of length 2.7 μs or greater extended the predicted folding time using this model to $\sim$85 ns at 330 K. Additional simulations were also conducted to study the folding of the Beta3s mutant with the two sets of turn GS residues replaced with PG pairs,[38] with the mutant folding three times faster than Beta3s. These inverse folding times thus remain rather high.

Dynamics at low viscosity helps tackle an important challenge of molecular simulations. It is therefore natural to examine the strengths and weaknesses of

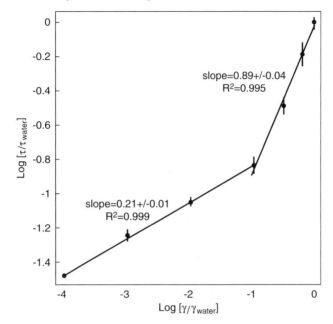

Figure 8.3 Viscosity dependence of the folding time of the Tryptophan Cage molecule in implicit solvent. The folding times and associated errors were calculated using the maximum-likelihood approach. Folding times and viscosities are given relative to the folding time in water and the viscosity of water, respectively. The error bars given are error propagated on the basis of the Cramer–Rao errors for the individual folding times.

this method. A non-linear relationship between folding time and viscosity was reported by Zagrovic et al. for the folding kinetics of a 20-residue tryptophan-cage mini-protein in the GB/SA implicit solvent model of Still et al.[20] under a range of solvent viscosities.[42] Figure 8.3 plots the observed relationship between inverse rate ($\tau = 1/k$) and viscosity ($1/\gamma$) relative to the case for water-like viscosity (i.e. $\gamma_{water} = 91$ ps^{-1}).[58] In the figure it is apparent that linear scaling of the folding time with solvent viscosity holds for viscosities as low as $\sim 1/10$ that of water. However, below this point the folding time scales as $t \sim \gamma^{1/5}$. While applying such scaling rules to the rate predictions of Caflisch and co-workers described above (in low viscosity) would clearly bring their values closer to experimentally established rates for these systems, the precise effect of low viscosity for each of these systems remains unclear.

8.4.2 Estimating Rates with a Two-state Approximation

Including water-like viscosity significantly increases the required sampling time, yet allows absolute folding kinetics to be measured directly. To this end, Pande and co-workers have applied distributed computing to sample trajectory space stochastically and extract rates from an ensemble dynamics (ED) perspective.[21] Two-state behavior is the central concept upon which rates are extracted *via* ED; dwell times in free energy minima of the conformational space are significantly longer than transition times (*i.e.* barrier crossing is much faster than the waiting period). The probability of crossing a barrier separating states A and B by time t is thus given by

$$P(t) = 1 - e^{-kt} \tag{8.6}$$

where k is the folding rate. In the limit of $t \ll 1/k$, this simplifies to $P(t) \approx kt$ and the folding rate (according to the Poisson distribution) is given by

$$k = \frac{N_{folded}}{t \cdot N_{total}} \pm \frac{\sqrt{N_{folded}}}{t \cdot N_{total}} \tag{8.7}$$

For example, if 10 000 simulations are run for 20 ns each and 15 of them cross a given barrier, we obtain a predicted rate of $k = 0.075(\pm 0.019)\,\mu s^{-1}$, corresponding to a folding time of $13.3(\pm 3.4)\,\mu s$. In this way, we can use many short trajectories to investigate the folding behavior of polymers that fold on the microsecond timescale: as we've shown previously, using M processors to simulate folding results in an M-times speedup of barrier crossing events.[59] When $t > 1/k$, as is the case for helix formation and other fast processes, ensemble convergence to absolute equilibrium can be established, and the complete kinetics and thermodynamics can be extracted simultaneously.[60]

In several recent studies, Pande and co-workers have utilized implicit solvent models while maintaining water-like viscosity *via* a Langevin or stochastic dynamics integrator with an inverse relaxation time γ. In the first study,[61] they introduced a method of "coupled ensemble dynamics" as a means to simulate the ensemble folding of the C-terminal β-hairpin of Protein G (1GB1) using the GB/SA continuum solvent model of Still *et al.*[20] and the OPLS united atom force field[16] with water-like viscosity. A total sampling time of $\sim 38\,\mu s$ was obtained, with a calculated inverse folding rate of $4.7(\pm 1.7)\,\mu s$, in good agreement with the experimentally determined value of $6\,\mu s$.[62]

Other hairpin structures have been studied by the Pande group more recently, both in an effort to gain insight into hairpin folding dynamics and for a more thorough comparison to experimental measurements. They reported folding and unfolding rates for three Trp zipper β-hairpins[63] using the methodology described above, including TZ1 (PDBID 1LE0), TZ2 (PDBID 1LE1), and TZ3 (PDBID 1LE0 with G6 replaced by D-proline). The relative inverse folding rates are in good agreement with experimental fluorescence and IR measurements provided by experimental collaborators. Unfolding rates were also predicted with relatively strong agreement.

Beyond these investigations of simple hairpin subunits, several small proteins were studied using an implicit solvent methodology. The first, a 20-residue miniprotein known as the Trp cage, was shown to have an experimental folding time of $\sim 4\,\mu s$. From simulations (totaling $\sim 100\,\mu s$) the folding rate was estimated based on a cutoff parameter in alpha carbon RMSD space: $k_{fold}(3.0\,\text{Å}) = (1.5\,\mu s)^{-1}$, $k_{fold}(2.8\,\text{Å}) = (3.1\,\mu s)^{-1}$, $k_{fold}(2.7\,\text{Å}) = (5.5\,\mu s)^{-1}$, $k_{fold}(2.6\,\text{Å}) = (6.9\,\mu s)^{-1}$, and $k_{fold}(2.5\,\text{Å}) = (8.7\,\mu s)^{-1}$. While the predicted folding time roughly agreed with the experimental value, the calculations illustrated the dependence of rates upon definition of the native state, as was described above (to minimize this dependence cutoffs must be chosen along an optimal reaction coordinate). Post analysis of ensemble folding data is not necessarily trivial unless many folding events are present and a stable native ensemble is easily distinguished from decoys with similar topology. Similar rate predictions were made for two mutants of the 23-residue BBA5 mini-protein and compared to temperature jump measurements made in the Gruebele laboratory.[64] A single mutation replaced F8 with W, which acts as the fluorescent probe, while the double mutant also included a replacement of V3 with Y. The agreement between simulation predictions and experimental measurements was excellent for the double mutant at 6 μs and 7.5($\pm$3.5) μs respectively. The agreement was less striking in the case of the single mutant, where experiment offered an upper limit of 10 μs and simulation predicted 16 μs, with a range of 7 to 43 μs based on the alpha carbon RMSD cutoff used (still a notably accurate prediction).

One of the most notable simulation studies to date was the *tour-de-force* 1-μs trajectory of the villin headpiece conducted by Duan and Kollman.[26] Following the methods described above, Pande and co-workers have simulated the ensemble folding of this 36-residue three-helix bundle (PDBID 1VII) using the GB/SA continuum solvent and the OPLS united atom force field in water-like viscosity.[65] With over 300 μs of simulation time, the folding time was predicted to be 5 μs (1.5–14 μs using alpha carbon RMSD cutoffs of 2.7–3 Å, as described above), which was compared to the 11-μs folding time derived from NMR lineshape analysis. A follow-up study by Eaton and co-workers tested the prediction using temperature-jump fluorescence and found the folding time to be 4.3($\pm$0.6) μs, thereby validating the rate prediction.

To study the formation of more complex protein structure, Pande and co-workers reported unbiased folding simulations of the 23-residue miniprotein BBA5 in explicit solvent.[66] Ten thousand independent MD simulations of the denatured conformation of BBA5 solvated in TIP3P water resulted in an aggregate simulation time of over 100 μs. This sampling yielded 13 complete folding events which, when corrected for the anomalous diffusion constant of the TIP3P model, results in an estimated folding time of 7.5($\pm$4.2) μs. This is in excellent agreement with the experimental folding time of 7.5($\pm$3.5) μs reported by Gruebele and co-workers.[64]

Folding of the villin headpiece was first attempted by Duan and Kollman in 1998.[26] Using TIP3P explicit solvent, their single 1-μs simulation did not show complete folding, which is not surprising given the ~5-μs folding time for that protein. Pande and co-workers have recently reported folding of this protein using the TIP3P water model and the AMBER-GS force field at 300 K,[67,68] thus increasing the maximum sequence size of proteins for which simulated folding has been observed with MD. With a total sampling time of nearly 1 ms, a folding time of 10($\pm$1.7) μs was predicted using a particle mesh Ewald treatment of long range electrostatics. Identical simulations using a reaction field treatment yielded 9.9($\pm$1.5) μs. These values are somewhat slower than the 4.3($\pm$0.6) experimental folding time, which might be due to the slow equilibration previously observed for helix formation under the AMBER-GS potential.[60]

What are the limitations of this two-state method? The direct observation of folding kinetics presents difficulties, especially for larger proteins or those without single exponential behavior. For example, folding ensembles generated from a single unfolded model attempt to populate the unfolded ensemble *and* observe folding. However, the timescale involved for the initial equilibration and the timescale necessary for chain diffusion across the folding barrier scale dramatically with chain length.[69] These factors make it increasingly difficult to observe both equilibration and folding for large proteins. In addition, Paci *et al.* have shown that folding events in extremely short trajectories can proceed from high-energy initial conformations.[41] Deviations from two-state behavior can also make interpretation of ensemble kinetics difficult,[70] and, given the short timescale of current folding simulations (10–1000 ns), any obligate intermediate with an appreciable dwell time (1–100 ns) may represent a sufficient deviation. In a downhill folding scenario, the principal limitation of the ensemble dynamics approach is the potentially lengthy and temperature-dependent timescale for protein conformational diffusion.[71] Fortunately, these challenges may not be intractable: the timescale for downhill equilibration to a relaxed unfolded ensemble may require long simulations,[72] but should be much faster than folding. Also, the detection of intermediates and multiple pathways can be accomplished by the comparison of folding and unfolding ensembles. Finally, these concerns may also be addressed with new Markovian State Model methods,[51–53,73] described in more detail below.

Regardless of the relatively strong agreement between ensemble simulations in implicit solvent and experimental rate measurements, several factors must be considered in interpreting such simulation results. Lacking a discrete representation of water, these studies ignore the potential role that aqueous solvent might play in the folding process. Furthermore, the compact nature of the relaxed unfolded state ensembles observed using the GB/SA solvent model may pose problems for the folding of larger proteins, such as trapping in compact unfolded conformations.

8.4.3 Markovian State Models (MSMs)

The two-state methods described and applied above work well if there are no intermediate states accumulating on timescales comparable to the trajectory length or longer (*e.g.* greater than 20–100 ns) and if the chains are relatively short (*e.g.* less than 50 residues). However, as one examines the folding of larger and more complex proteins, the two-state approximation will surely eventually break down and reaching even just the relaxation time for a given chain will become a challenge. Also, even if the folding is two state, the simple diffusion of

Computer Simulations of Protein Folding

the polymer chain (which scales like the number of residues squared or cubed) will start to require very long trajectories. In anticipation of these problems, we have proposed a new method: Markovian state models.[51–53,73]

Markovian state models transform simulation data gathered from MD trajectories into a kinetic model that includes transition time data. As opposed to traditional transition path sampling analysis,[45,47,74] this method would incorporate all of the simulated data into the results, therefore potentially yielding an increase in efficiency. Our MSM model assumes first-order Markovian transitions between states: simply put, we assume that the next state visited during dynamics will depend solely on the current state and not on previous states visited. Moreover, from an MSM, one can easily calculate any kinetic quantity which can be related to some structural property, such as p_{fold}[75] for all configurations sampled and the mean first passage time (MFPT) from the unfolded state to the folded state. This method also provides a compact representation of the pathways in the system, useful for understanding the mechanisms involved in folding. MSM methods improve on the current graph-based techniques by sampling points using molecular dynamics (MD), thereby greatly increasing the probability that the configurations that are included are kinetically relevant. In addition, the simulation time between points inherently captures transition times, making the direct calculation of folding rates possible.

Early results from MSM methods appear to be promising. Results on a β hairpin[51] and the villin headpiece and protein A[68] find quantitative agreement with experimental folding times, allowing for a quantitative prediction of timescales considerably longer than the individual trajectories used to construct the MSM. Moreover, these methods do not assume two-state behavior and thus can serve as a test of the two-state approximation; the agreement with two-state behavior in these methods supports employing the two-state method in simple proteins, although it is likely that the two-state approximation will break down for larger, more complex proteins or proteins that have unusual kinetics, such as putative downhill folders.

8.4.4 Other Approaches

While the studies described above offer insight into the most elementary events in protein folding, a number of studies have recently been published on the formation and/or denaturation of larger protein structures. Daggett and co-workers have reported *unfolding* rate predictions using explicit solvent models with direct experimental comparisons. The 61-residue engrailed homeodomain (En-HD) forms a three-helix bundle similar to the villin headpiece and is known to undergo thermal denaturation at 373 K with a half-life predicted by long extrapolation of experimental kinetic data at lower temperatures of 4.5 to 25 ns. Mayor *et al.* simulated the thermally induced unfolding of En-HD using the F3C water model[76] in ENCAD[77] at this temperature with an unfolding rate on the tens of nanoseconds timescale.[10,78] The time needed to reach the putative transition state at 75 and 100 °C, 60 ns and 2 ns respectively, was roughly consistent with the extrapolated experimental unfolding rates (precise rates cannot be extracted from a single unfolding event due to the stochastic nature of protein dynamics).

Bolhuis simulated the folding of the C-terminal β-hairpin of protein G using the transition interface sampling method described above to extract transition kinetics.[45] At 300 K, with an equilibrium constant of ~1, the predicted folding time of 5 μs using the TIP3P explicit solvent is in good agreement with the experimental rate of 6 μs[62] as well as the rate predicted by Zagrovic *et al.* using an implicit solvent.[61] The observed agreement suggests that path sampling will be useful in future simulation studies to elucidate the kinetics and mechanisms inherent to protein folding, and it will be interesting to see such methods applied to larger, more complex systems.

Peptides and mini-proteins allow for complete and accurate sampling of folding and unfolding events *via* simulation at biologically relevant temperatures. Pande and co-workers recently studied the helix-coil transition in two 21-residue α-helical sequences and demonstrated complete equilibrium ensemble sampling for multiple variants of the AMBER force field,[11] as shown in Figure 8.4, thus allowing quantitative assessment of the potentials studied. Observing that the previously published AMBER variants resulted in poor equilibrium helix-coil character in comparison to experimental measurements,

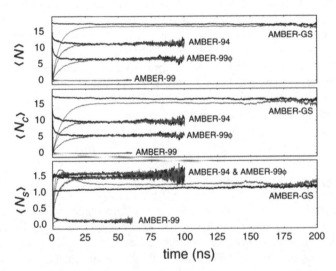

Figure 8.4 Time evolution and convergence of F_s peptide folding ensembles under the AMBER-94, AMBER-GS, AMBER-99, and AMBER-99ϕ potentials. The plots include, from top to bottom, the mean α-helix content, mean contiguous helical length, and mean number of helical segments per conformation according to classical LR counting theory. Native ensembles that converge with corresponding gray folding ensembles are shown in black. Signal noise in the longer time regime is due to fewer simulations reaching that timescale (additional data at long times have been removed for visual clarity).

they tested a new variant denoted AMBER-99ϕ and showed that it more adequately captured the helix-coil dynamics. Based on a multi-state Markovian-based analysis, a primary relaxation time of 151 ns was reported using the more accurate AMBER variant, which agreed well with the 160($\pm$50) ns measured experimentally by Williams *et al.*[79]

Minimalist models have also continued to garner attention recently. It is usually not feasible to obtain direct kinetics information from Go-like models due to difficulty in interpreting the timestep in Go model simulations in terms of a physically measurable quantity. However, it was recently reported that Go model simulations can still be useful in predicting folding timescales of various proteins if the time and temperature are scaled properly to experimental measurements.[80] One caveat in this approach will be the necessity of a rather large training set to obtain calibration data for such scaling. However, considering the tractability for simulation of large systems using minimalist models, it will be interesting to see whether such an approach can be generally applied for other systems.

8.5 Predicting Protein Folding Pathways

8.5.1 Kinetics Simulations

The folding pathway is arguably the most interesting prediction associated with folding simulations. As our ability to observe long-timescale transitions improves, it becomes increasingly important to clearly communicate the observed mechanism. Qualitative descriptions of the folding pathway can only be loosely interpreted in comparison to experiment. First, as mentioned above, results derived from folding simulations can be sensitive to data analysis. For example, Swope and co-workers produced several folding mechanisms for the hairpin from protein G by varying their hydrogen bond definition.[52,73] Second, there are potential semantic issues; a researcher might frame their discussion of β-hairpin folding in terms of zippering, secondary *versus* tertiary contacts, or diffusion-collision versus nucleation-condensation.

The order of "events" is a natural description of a mechanism, but an optimal description of mechanism should account for heterogeneity as well as the interplay between secondary and tertiary contacts. An excellent and recent example comes from protein A. Fersht and co-workers have qualitatively compared several published simulation predictions of the protein A folding pathway to experiment.[81,82] None of the published atomistic simulations were completely consistent with experiment, emphasizing the need for improved simulation predictions of the folding pathway, and improved quantitative means for comparing pathway predictions.

The collaborative effort between the Fersht experimental laboratory and the Daggett simulation laboratory has shed light on an entire family of unfolding mechanisms. The homeodomains, small three-helix proteins, exhibit a spectrum of folding processes, from concurrent secondary and tertiary structure formation (nucleation-condensation mechanism) to sequential secondary and tertiary formation (framework mechanism).[83] They present putative transition state conformations (two each at 373 and 498 K for En-HD; seven at 498 K for c-Myb; and two at 498 K for hTRF1) from high-temperature unfolding for En-HD, c-Myb, and hTRF1, and estimate β_T values (0.83, 0.83, 0.8 respectively) that roughly agree with the experimental β_T values (0.83, 0.79, 0.90). Excluding the mutation of two charged residues, correlation coefficients of 0.79 and 0.74 for En-HD and c-Myb were obtained between the S and Φ values. Gianni et al. report that folding of En-HD resembles the diffusion–collision mechanism more than c-Myb and hTRF1 because the helices are nearly fully formed in the transition state. They do state that movements from diffusion-collision to nucleation-condensation are not detected simply by the helical content of the folding transition states but through analysis of whether the secondary and tertiary structures are formed simultaneously.[83] Given this strategy we feel it is particularly important to generate a statistically meaningful number of transitions to judge the relative timing of events between related molecules.

Through the two-state approximation and distributed computing, the Pande laboratory has examined the folding of several small, two-state proteins. The mechanism found varied with the protein studied. It remains to be seen if a more comprehensive mechanistic survey of many small, two-state proteins will reveal underlying mechanistic similarities or model dependencies. In several cases, distributed computing allowed direct comparison of the performances of different force fields. For example, simulations of the C-terminal β hairpin of protein $G^{35,61}$ found that the initial states of folding were the hydrophobic collapse of the small hydrophobic core, followed by formation of hydrogen bonds.

Simulations of a small zinc finger fold (BBA5) found a different mechanism:[64] the secondary structure formed first and then independently collided to form the folded state, analogous to what one would expect from a diffusion-collision model; this is perhaps not surprising in hindsight, considering that BBA5[84] is a de novo designed protein and its independent elements may be more stable than in typical proteins. Finally, simulations of the villin headpiece found a different mechanism, in which formation of the rough topology was found early, following by the locking in of the side chains.[65]

It is interesting and important to consider the role of force-field variation in the determination of the folding mechanism. Moreover, beyond the force field used to describe protein-protein interactions, one may also expect variations due to the water model chosen, and differences between minimalist models and more detailed, full atomic models. A natural way to quantitatively examine these differences in mechanism is through a correlation of pfold values.[85] As the pfold value gives a quantitative measure of the location of a given conformation along the folding pathway (pfold near 0 means that the conformation is kinetically close to the unfolded state and pfold close to 1 means it is kinetically close to the folded state), a correlation of pfold values between two different models (force fields, solvent models, etc.) yield a quantitative comparison between the kinetic mechanisms that would be predicted.

Upon comparing several different types of explicit water models, implicit water models, and minimalist models (all-atom and Cα Go models), Rhee and Pande[85] found that different explicit models yielded quantitatively similar folding mechanisms. Comparing explicit solvent to implicit solvent models found some greater variation, consistent with other types of comparisons between explicit and implicit solvent.[86,87] When comparing to minimalist models, little correlation was found, indicating that for the protein studied (BBA5), minimalist models could not recapitulate the dynamics described by more detailed models and, moreover, minimalist models did not agree with each other (there was a large discrepancy between all-atom and Cα Go models). While it still remains to be seen if these results will hold for larger, more complex proteins (indeed, BBA5 is a small, human-designed protein and thus may be unusual), these results suggest that there may indeed be differences, as well as laying out a quantitative method for making such comparisons in the future.

8.5.2 Thermodynamics Simulations

The success of thermodynamic methods in the prediction of the relevant folding pathways rests on sampling the entire available phase space. This is because the dominant pathways can be correctly identified only when the relative importance of various intermediates is known. Two major bottlenecks naturally emerge for a correct sampling of the vast phase space: the high dimensionality of protein configuration space and the kinetic trapping during simulations. The following will revisit well-known methods that try to overcome these difficulties.

In the original landscape approach as pioneered by Brooks and co-workers,[6] the free-energy landscape or potential of mean force (PMF) is generated from the equilibrium population distribution. Because it is excessively time consuming to reach equilibrium for high-dimensional protein molecules with conventional molecular dynamics, simulations are performed with umbrella sampling. An additional potential (usually a quadratic or "umbrella" potential) is added to the original Hamiltonian of the system to bias the sampling. By adjusting the bias, the size of the available conformational space can be reduced to expedite the equilibration within the biased Hamiltonian. A series of biased simulations are recombined afterwards to remove the bias in a mathematically strict way using the weighted histogram analysis method.[88] The population distribution $P(q)$ then can be converted to the free energy with $F(q) = -\ln P(q)$. With this approach, Brooks and co-workers have obtained the free-energy landscape and folding dynamics of an α-helical protein (Protein A[89]), an αβ mixed protein (GB1[90,91]), and a mostly β protein (src-SH3[92]) with numerous successful comparisons to experiment. We refer the reader to an excellent review.[6]

Umbrella sampling studies produce informative free-energy landscapes but assume that degrees of freedom orthogonal to the surface equilibrate quickly. The molecular dynamics time needed for significant chain movement could significantly exceed the length of typical umbrella sampling simulations (which are each typically on the nanosecond timescale). However, in spite of this caveat, umbrella sampling approaches have been very successful. One explanation for this success lies in the choice of initial conditions: umbrella sampling simulations employ initial coordinates provided by high-temperature unfolding trajectories. This is a recurring theme: without lengthy simulations, the initial conformations are crucially important, and it appears that unfolding produces reasonable initial models.

Even though umbrella sampling can expedite the sampling by simulating multiple trajectories at the same time, kinetic trapping or slow orthogonal degrees of freedom may still dominate within each umbrella potential. A number of techniques have been developed to overcome this kinetic trapping. Mitsutake et al. have provided an excellent review of these generalized ensemble methods.[93] We will focus on replica exchange molecular dynamics (REMD), which has been widely used in protein-folding simulations. In this approach, a number of simulations ("replicas") are performed in parallel at different temperatures. After a certain time, conformations are exchanged with a Metropolis probability. This criterion ensures that the sampling follows the canonical Boltzmann distribution at each temperature. Kinetic trapping at lower temperatures is avoided by exchanging conformations with higher-temperature replicas. This method is easier to apply than other generalized ensemble methods because it does not require *a priori* knowledge of the population distribution.

After Sugita and Okamoto demonstrated its effectiveness with a gas-phase simulation of the pentapeptide Met-enkephalin,[27] Sanbonmatsu and Garcia obtained the free-energy surface of the same system using explicit water.[28] With 16 parallel replicas they observed enhanced sampling (at least $\sim 5\times$) compared to conventional constant temperature molecular dynamics. Because the method is quite simple and because it is trivially parallelized in low-cost cluster environments, it gained wide application rapidly. Berne and co-workers applied this method to obtain a free-energy landscape for β-hairpin folding in explicit water using 64 replicas with over 4000 atoms.[94] With the equilibrium ensemble and the free-energy landscape in hand, they reported that the β-hairpin population and the hydrogen-bond probability were in agreement with experiments, and proposed that the β strand hydrogen bonds and hydrophobic core form together during the folding pathway.

If care is taken to fully reach equilibrium,[32] REMD becomes powerful for elucidating the folding landscape. For example, Garcia and Onuchic applied the method to a relatively large system, protein A.[29] With 82 replicas for more than 16 000 atoms with temperatures ranging from 277 to 548 K, and with ~ 13 ns molecular dynamics simulations for each replica, they reported convergence to the equilibrium distribution with quantitative determination of the free-energy barrier of folding.

8.6 Conclusions

In the end, an understanding of complex biophysical phenomena will require computer simulation at some level. Most likely, experimental methods will never yield the level of detail that can be reached even today with computer simulations. However, the great challenge for simulations is to prove their validity. Thus, it is naturally the combination of powerful simulations with quantitative experimental validation that will elucidate the nature of how proteins fold.

How well do protein folding kinetics simulations currently compare with experiment? While prediction of relative rates (*e.g.* demonstrating a correlation between experimental and predicted rates) is valuable, prediction of the absolute rate without free parameters is a more stringent test. Though calculation of absolute rates is computationally demanding, we expect such absolute comparisons to become more common (for increasingly complex proteins) with the advent of new methods and increasing computer power. Finally, we stress that a quantitative prediction of rates is not sufficient to guarantee the validity of a model. The ability of fairly different models to quantitatively predict folding rates strongly suggests that more experimental data are needed to further validate simulation. Additionally, several coarse-grained calculations have been employed to study folding and unfolding rates.[80,95,96]

It is also interesting to look to what's on the near horizon. New advances in computational methods have already enabled single trajectories to reach the microsecond timescale routinely, without using a supercomputer, either by using multi-core PCs[97] or streaming processors, such as Graphics Processing Units (GPUs) or the Cell Processor in PS3s.[98] With microsecond length trajectories, fast-folding proteins can now be examined directly, with thousands of trajectories over multiple microsecond timescales directly enabling a full statistical comparison of kinetics between simulation and experiment.[97] Moreover, recent advances in force fields should allow for a significant increase in accuracy, especially with new advances in polarizable force fields.[99,100] The combination of the more advanced computational methods, with modern polarizable force fields, and the sampling power of Markovian state models should yield a potent combination to accurately predict folding properties on the microsecond to millisecond timescale for small, single-domain proteins in the very near future, and likely beyond to the second timescale in the next decade.

The ability to quantitatively predict rates, free energies, and structure from simulations based on physical force fields reflects significant progress made over the last five years. It also draws attention to a new challenge. Even the prediction of experimental observables, such as rates, within experimental uncertainty does not prove that the simulations will yield correct insight into the mechanism of folding. Indeed, recent work suggests that computational models can both agree with experiment, but disagree with each other.[66] Also, observing that a particular residue appears to participate in a non-native contact does not necessarily imply that mutating this position will accelerate folding; for example, Zagrovic and Pande[65] found non-native interactions in their simulations, but did not predict that removing this would necessarily alter the rate (indeed, the simulations performed could not predict a rate change in this case and thus this result is not necessarily in disagreement with the experiment).[101]

However, these sorts of comparisons greatly underscore the need for direct, quantitative comparison between experiment and theory over a broad range of observables as this is the only way to unambiguously test simulation predictions. We must therefore push the link between simulation and experiment further by connecting the two with new observables, multiple techniques, and increasingly strict quantitative comparison and validation of simulation methods. Without more detailed experiments, we may not be able to sufficiently test current simulation methodology and the trustworthiness of refined simulations may remain unclear. Nonetheless, the ability to predict rates, free energies, and structure of small proteins is a significant advance for simulation, likely heralding even more significant advances over the next five years.

References

1. C. M. Dobson, *Trends Biochem. Sci.*, 1999, **24**(9), 329–232.
2. C. M. Dobson and A. Sali *et al.*, *Angew. Chem., Int. Ed. Engl.*, 1998, **37**, 868–893.
3. V. S. Pande and A. Grosberg *et al.*, *Curr. Opin. Struct. Biol.*, 1998, **8**(1), 68–79.
4. V. Grantcharova and E. J. Alm *et al.*, *Curr. Opin. Struct. Biol.*, 2001, **11**(1), 70–82.
5. M. Levitt, *Nature Str. Biol.*, 2001, **8**, 392–393.
6. J. E. Shea and C. L. Brooks III, *Annu. Rev. Phys. Chem.*, 2001, **52**, 499–535.
7. D. Baker and W. A. Eaton, *Curr. Opin. Struct. Biol.*, 2004, **14**(1), 67–69.
8. H. Abe and N. Go, *Biopolymers*, 1981, **20**, 1013.
9. V. S. Pande and D. S. Rokhsar, *Proc. Natl. Acad. Sci. USA*, 1998, **95**(4), 1490–1494.
10. U. Mayor and N. R. Guydosh *et al.*, *Nature*, 2003, **421**(6925), 863–867.
11. E. J. Sorin and V. S. Pande, *J. Comput. Chem.*, 2005, **26**(7), 682–690.
12. P. J. Steinbach and B. R. Brooks, *J. Comput. Chem.*, 1994, **15**(7), 667–683.
13. I. G. Tironi and R. Sperb *et al.*, *J. Chem. Phys.*, 1995, **102**(13), 5451–5459.
14. W. D. Cornell and P. Cieplak *et al.*, *J. Am. Chem. Soc.*, 1995, **117**, 5179–5197.
15. B. R. Brooks and R. E. Bruccoleri *et al.*, *J. Comp. Chem.*, 1983, **4**, 187–217.
16. W. L. Jorgensen and J. Tirado-Rives, *J. Am. Chem. Soc.*, 1988, **110**, 1657–1666.
17. R. Zhou and X. Huang *et al.*, *Science*, 2004, **305**(10), 1605–1609.
18. E. J. Sorin and Y. M. Rhee *et al.*, *J. Mol. Biol.*, 2006, **356**, 248–256.
19. J. A. Grant and B. T. Pickup *et al.*, *J. Comput. Chem.*, 2000, **22**(6), 608–640.
20. D. Qiu and P. S. Shenkin *et al.*, *J. Phys. Chem. A*, 1997, **101**, 3005–3014.
21. V. S. Pande and I. Baker *et al.*, *Biopolymers*, 2003, **68**, 91–109.
22. M. S. Cheung and A. E. Garcia *et al.*, *Proc. Natl. Acad. Sci. USA*, 2002, **99**(2), 685–690.

23. M. R. Shirts and J. Pitera *et al.*, *J. Chem. Phys.*, 2003, **119**, 5740.
24. M. R. Shirts and V. S. Pande, *J. Chem. Phys.*, 2005, **122**(13), 134508.
25. V. Tsui and D. A. Case, *Biopolymers*, 2001, **56**, 275–291.
26. Y. Duan and P. A. Kollman, *Science*, 1998, **282**, 740–744.
27. Y. Sugita and Y. Okamoto, *Chem. Phys. Lett.*, 1999, **314**, 141–151.
28. K. Y. Sanbonmatsu and A. E. Garcia, *Proteins*, 2002, **46**(2), 225–234.
29. A. E. Garcia and J. N. Onuchic, *Proc. Natl. Acad. Sci. USA*, 2003, **100**(24), 13898–13903.
30. S. Gnanakaran and H. Nymeyer *et al.*, *Curr. Opin. Struct. Biol.*, 2003, **13**(2), 168–174.
31. A. K. Felts and Y. Harano *et al.*, *Proteins*, 2004, **56**(2), 310–321.
32. Y. M. Rhee and V. S. Pande, *Biophys. J.*, 2003, **84**, 775–786.
33. V. Daggett and M. Levitt, *J. Mol. Biol.*, 1993, **232**, 600–619.
34. D. O. Alonso and V. Daggett, *J. Mol. Biol.*, 1995, **247**, 501–520.
35. V. S. Pande and D. S. Rokhsar, *Proc. Natl. Acad. Sci. USA*, 1999, **96**(16), 9062–9067.
36. V. Daggett and A. R. Fersht, *Trends Biochem. Sci.*, 2003, **28**, 18–25.
37. J. Ervin and M. Gruebele, *J. Biol. Phys.*, 2002, **28**(2), 0092–0606.
38. P. Ferrara and A. Caflisch, *J. Mol. Biol.*, 2001, **306**(4), 837–850.
39. J. Gsponer and A. Caflisch, *J. Mol. Biol.*, 2001, **309**(1), 285–298.
40. A. Caflisch, *Trends Biotechnol.*, 2003, **21**(10), 423–425.
41. E. Paci and A. Cavalli *et al.*, *Proc. Natl. Acad. Sci. USA*, 2003, **100**(14), 8217–8222.
42. B. Zagrovic and V. Pande, *J. Comput. Chem.*, 2003, **24**(12), 1432–1436.
43. J. N. Onuchic and Z. Luthey-Schulten *et al.*, *Annu. Rev. Phys. Chem.*, 1997, **48**, 545–600.
44. E. J. Sorin and B. J. Nakatani *et al.*, *J. Mol. Biol.*, 2004, **337**(4), 789–797.
45. P. G. Bolhuis, *Proc. Natl. Acad. Sci. USA*, 2003, **100**(21), 12129–12134.
46. D. Chandler, *Introduction to Modern Statistical Mechanics*, Oxford University Press, 1987.
47. P. G. Bolhuis, *Biophys. J.*, 2005, **88**(1), 50–61.
48. M. S. Apaydin and D. L. Brutlag *et al.*, *J. Comput. Biol.*, 2003, **10**(3–4), 257–281.
49. G. Song and S. Thomas *et al.*, *Proceedings of the Pacific Symposium on Biocomputing*, 2003.
50. G. Hummer, *J. Chem. Phys.*, 2004, **120**(2), 516–523.
51. N. Singhal and C. D. Snow *et al.*, *J. Chem. Phys.*, 2004, **121**(1), 415–425.
52. W. C. Swope and J. W. Pitera *et al.*, *J. Phys. Chem. B*, 2004, **108**(21), 6571–6581.
53. N. Singhal and V. S. Pande, *J. Chem. Phys.*, 2005, **123**(20), 204909.
54. A. Hiltpold and P. Ferrara *et al.*, *J. Phys. Chem. B*, 2000, **104**, 10080–10086.
55. P. Ferrara and A. Caflisch, *Proc. Natl. Acad. Sci. USA*, 2000, **97**(20), 10780–10785.
56. A. Cavalli and P. Ferrara *et al.*, *Proteins Struct. Funct. Genet.*, 2002, **47**(3), 305–314.

57. E. De Alba and J. Santoro et al., Protein Sci., 1999, **8**, 854–865.
58. S. Yun-yu and L. Wang et al., Mol. Simul., 1988, **1**, 369–383.
59. M. R. Shirts and V. S. Pande, Phys. Rev. Lett., 2001, **86**(22), 4983–4987.
60. E. J. Sorin and V. S. Pande, Biophys. J., 2005, **88**(4), 2472–2493.
61. B. Zagrovic and E. J. Sorin et al., J. Mol. Biol., 2001, **313**, 151–169.
62. V. Munoz and P. A. Thompson et al., Nature, 1997, **390**(6656), 196–198.
63. C. D. Snow and L. Qiu et al., Proc. Natl. Acad. Sci. USA, 2004, **101**(12), 4077–4082.
64. C. Snow and H. Nguyen et al., Nature, 2002, **420**, 102–106.
65. B. Zagrovic and C. Snow et al., J. Mol. Biol., 2002, **323**, 927–937.
66. Y. M. Rhee and E. J. Sorin et al., Proc. Natl. Acad. Sci. USA, 2004, **101**(17), 6456–6461.
67. G. Jayachandran and V. Vishal et al., J. Chem. Phys., 2006, **124**(16), 164902.
68. G. Jayachandran and V. Vishal et al., J. Struct. Biol., 2007, **157**(3), 491–499.
69. F. Krieger and B. Fierz et al., J. Mol. Biol., 2003, **332**(1), 265–274.
70. A. R. Fersht, Proc. Natl. Acad. Sci. USA, 2002, **99**(22), 14122–14125.
71. A. N. Naganathan, J. M. Sanchez-Ruiz and V. Munoz, J. Am. Chem. Soc., 2005, **127**, 17970–17971.
72. N. J. Marianayagam and N. L. Fawzi et al., Proc. Natl. Acad. Sci. USA, 2005, **102**(46), 16684–16689.
73. W. C. Swope and J. W. Pitera et al., J. Phys. Chem. B, 2004, **108**, 6582–6594.
74. C. Dellago and P. G. Bolhuis et al., J. Chem. Phys., 1998, **108**, 1964–1977.
75. R. Du and V. S. Pande et al., J. Chem. Phys., 1998, **108**, 334–350.
76. M. Levitt and M. Hirshberg et al., J. Phys. Chem. B, 1997, **101**, 5051–5061.
77. M. Levitt, ENCAD, Energy Calculations and Dynamics, Palo Alto, CA, Molecular Applications Group, 1990.
78. U. Mayor and C. M. Johnson et al., Proc. Natl. Acad. Sci. USA, 2000, **97**, 13518–13522.
79. S. Williams and T. P. Causgrove et al., Biochemistry, 1996, **35**, 691–697.
80. L. L. Chavez and J. N. Onuchic et al., J. Am. Chem. Soc., 2004, **126**, 8426–8432.
81. S. Sato and T. L. Religa et al., Proc. Natl. Acad. Sci. USA, 2004, **101**(18), 6952–6956.
82. P. G. Wolynes, Proc. Natl. Acad. Sci. USA, 2004, **101**(18), 6837–6838.
83. S. Gianni and N. R. Guydosh et al., Proc. Natl. Acad. Sci. USA, 2003, **100**(23), 13286–13291.
84. J. J. Ottesen and B. Imperiali, Nat. Struct. Biol., 2001, **8**(6), 535–539.
85. Y. M. Rhee and V. S. Pande, Chem. Phys. Lett., 2006, **323**, 66–77.
86. J. Wagoner and N. A. Baker, J. Comput. Chem., 2004, **25**(13), 1623–1629.
87. J. A. Wagoner and N. A. Baker, Proc. Natl. Acad. Sci. USA, 2006, **103**(22), 8331–8336.

88. A. M. Ferrenberg and R. H. Swendsen, *Phys. Rev. Lett.*, 1989, **63**(12), 1195–1198.
89. E. M. Boczko and C. L. Brooks III, *Science*, 1995, **269**(5222), 393–396.
90. F. B. Sheinerman and C. L. Brooks, *J. Mol. Biol.*, 1998, **278**, 439–456.
91. F. B. Sheinerman and C. L. Brooks, *Proc. Natl. Acad. Sci. USA*, 1998, **95**(4), 1562–1567.
92. J. E. Shea and J. N. Onuchic *et al.*, *Proc. Natl. Acad. Sci. USA*, 2002, **99**(25), 16064–16068.
93. A. Mitsutake and Y. Sugita *et al.*, *Biopolymers*, 2001, **60**(2), 96–123.
94. R. H. Zhou and B. J. Berne *et al.*, *Proc. Natl. Acad. Sci. USA*, 2001, **98**(26), 14931–14936.
95. V. Munoz and W. A. Eaton, *Proc. Natl. Acad. Sci. USA*, 1999, **96**, 11311–11316.
96. D. N. Ivankov and A. V. Finkelstein, *Proc. Natl. Acad. Sci. USA*, 2004, **101**, 8942–8944.
97. D. L. Ensign and P. M. Kasson *et al.*, *J. Mol. Biol.*, 2007, **374**(3), 806–816.
98. E. Elsen and M. Houston *et al.*, *Proceedings of the 2006 ACM/IEEE Conference on Supercomputing*, 2006, 188.
99. P. E. Lopes and G. Lamoureux *et al.*, *J. Phys. Chem. B*, 2007, **111**(11), 2873–2885.
100. M. J. Schnieders and N. A. Baker *et al.*, *J. Chem. Phys.*, 2007, **126**(12), 124114.
101. J. Kubelka and W. A. Eaton *et al.*, *J. Mol. Biol.*, 2003, **329**(4), 625–630.

CHAPTER 9
Protein Design: Tailoring Sequence, Structure, and Folding Properties

ANDREAS LEHMANN, CHRISTOPHER J. LANCI,
THOMAS J. PETTY II, SEUNG-GU KANG
AND JEFFERY G. SAVEN

Makineni Theoretical Laboratories, Department of Chemistry, University of Pennsylvania, Philadelphia, PA 19104, USA

9.1 Introduction

Protein design algorithms identify protein sequences consistent with a particular fold, and often simultaneously quantify the many subtle, non-covalent interactions that govern protein folding, stability and function. Efforts in protein design stand to advance our knowledge of protein folding and function and also can identify new proteins with applications to biotechnology, catalysis, and materials research. Here, recent developments in protein design are discussed with a focus on features common to many of the computational design methods. A sampling of studies is presented in which computationally designed proteins have been experimentally realized, exemplifying what may be learned and accomplished with protein design.

Advances in protein design inform our understanding of the molecular basis of life processes and provide tools for new applications in biotechnology. Proteins are molecular workhorses, and they play central roles in cellular functions such as cytoskeleton assembly, transport, signaling, bioenergetics, metabolism and gene regulation. Structural proteins (*e.g.* actin, microtubules, collagen) are vital for maintaining the morphological properties of organelles, cells, and tissues. Enzymes catalyze reactions selectively and efficiently, and protein-based hormones and receptors are critical for inter- and intracellular communication. Thus proteins exhibit a multitude of functions, and this versatility can potentially be leveraged. Polypeptide synthesis and protein overexpression are often straightforward, facilitating the realization of natural and non-natural sequences. Designed proteins obtained using such methods provide systems for critically testing our present understanding of the molecular features most relevant to protein folding, stability, and function. In addition, novel proteins can provide new biotechnological applications, such as selective catalysts and sensors.

Protein folding enables function to be encoded in sequence. Most polymers can take on a large number of conformations in dilute solution. Folding, however, implies that a protein has a well-defined three-dimensional structure under physiological conditions, a structure that is usually requisite for the protein's function. Anfinsen[1] showed that ribonuclease could be denatured and refolded without loss of enzymatic activity. This led to the general acceptance of the "thermodynamic hypothesis," which states that "the three-dimensional structure of a native protein in its normal physiological milieu ... is the one in which the Gibbs free energy of the whole system is lowest; that is, that the native conformation is determined by the totality of inter-atomic interactions and hence by the amino acid sequence, in a given environment."[1] While there are certainly exceptions to this rule, the three-dimensional structures of most proteins are encoded in their amino acid sequences. Due to the complexity of proteins and the many possible compact structures to which they can fold, protein structure prediction from amino acid sequence remains one of the fundamental open problems of molecular science, but much progress has been made in recent years.[2] In addition, protein folding dynamics remains an active area of research. Such studies are motivated in part by Levinthal's paradox:[3] how is it that proteins having exponentially large numbers of conformations are able to fold on timescales of minutes or less? There has been much recent development of mechanisms and models that quantitatively describe folding kinetics,[4-9] including the energy landscape theory of protein folding[10-20] (see Chapter 3 by Wolynes for a detailed account of the energy landscape approach to protein folding).

Proteins are involved in an increasing number of industrial applications such as chemical production, pharmaceuticals and fine chemicals, pulp and paper, food, textiles, and energy.[21-23] Exploiting the structural and functional features of proteins stands also to give rise to new biomaterials. Peptides may respond structurally and functionally to environmental changes in pH, temperature, pressure, salt concentration, UV or visible light exposure. Such "smart-material" peptides may be used as building blocks for filaments and fibrils, scaffolds, hydrogels, and surfactants.[24] These protein and peptide-based systems have applications in tissue and surface engineering, as drug or cell carriers, for patterning of targeted cell growth, as miniaturized solar cells or in optical and electronic devices.[25-31] Commercial biocatalysts have been obtained from natural enzymes or *via* the directed mutagenesis of such enzymes.[22] Advances in genomics, directed evolution, and bioinformatics[21] have led to the development of recombinant enzymes and biomolecular pathways.[32–36] For example, a recently developed industrial process uses a re-engineered and extended *E. coli* pathway to convert D-glucose into the polymer precursor 1,3-propanediol within a single organism;[36] such efforts illustrate the "green chemistry"[37] possible with biomolecular systems. Protein design efforts have also led to the discovery of new sensors and enzymes.[38–42] Therapeutic proteins may affect intercellular communication and the physiology of the human immune system, and there remains enormous potential for the development of therapeutic agents for autoimmune diseases, cancer, infections, and inflammatory diseases.[43–45] Thus, the ability to reliably design structure and function into proteins stands to inform our basic understanding of biomolecular function and can lead to a wide variety of biotechnological and biomedical applications.

Although realizing a particular protein sequence may be straightforward (problems with protein synthesis or expression notwithstanding), protein design is non-trivial. Proteins are large macromolecules that range in length from tens to tens of thousands of amino acid residues. The number of possible sequences is exponentially large. For a protein with 100 variable residues, there are 20^{100} possible sequences. In addition, many amino acids have multiple possible side-chain conformations, further increasing the complexity of the search for sequences consistent with a desired structure. Non-covalent interactions such as van der Waals forces, hydrogen bonding, salt bridges, and solvation effects stabilize the protein structure. Many of these interactions are coupled and interdependent in protein structures, and their parameterization in the form of a molecular energy function is necessarily approximate. As a result, accurate determination of the stabilities of proteins using molecular simulations is often impractical and remains computationally intensive.[46,47] The difficulties of protein structure prediction suggest that the mapping from sequence to structure is subtle.[2,48] Ultimately, the best assessment of the quality of a particular protein design effort is to create and study the resulting sequences. This validation of design is typically more experimentally intensive than structure prediction, where thousands of structures already in the protein structural database may serve as test cases.

9.2 Empirical Approaches to Protein Design

9.2.1 Hierarchical Protein Design

Early work in protein design, also referred to as the inverse folding problem,[49] applied knowledge gleaned from biochemical experiments and structural databases to the construction of small proteins. Often a hierarchical approach was used, where sequences likely to form particular substructures or secondary structures were assembled with a particular tertiary structure in mind.[50,51] Principles guiding the design process included the trend to have largely hydrophobic amino acids within the interior of the folded protein,[52,53] and the propensities of individual amino acids to appear in particular secondary structures such as α-helices or β-sheets.[54–56] Early design efforts often focused on α-helical proteins and did not necessarily use all 20 amino acids.[57–61] Electrostatic interactions were found to play a smaller role in determining topology than the relative positioning of interior hydrophobic residues.[62] *Via* such qualitative protein design, it was found that two proteins having 50% or greater sequence identity could have different folds.[63,64]

Helical proteins have been targets of many design efforts. Designed, α-helical peptides have been constructed to solubilize membrane proteins[65] and to retard the HIV-1 infection of human T cells.[66] Rational, structure-informed *de novo* design based on α-helical bundles has been used to design proteins that bind metal ions such as zinc,[67–70] iron,[71,72] copper,[73] cadmium,[74] mercury,[75] and calcium,[68] as well as proteins that bind metal-containing cofactors such as heme.[76–78] The B1 domain of Streptococcal IgG-binding protein G was also used as a template for designing a zinc binding protein.[79] A four-helix peptide (maquette) was shown to efficiently incorporate an iron-sulfur cluster as a tetramer and exhibit properties typical of natural ferredoxins.[80] While successfully realizing some of the targeted functionality and structure, often these designed proteins did not have well-defined tertiary structures. These proteins exhibited more mobility in the interior than native proteins, and in many cases had the features of a molten-globule-like state.[50,51,81] The approaches used often consider the secondary structure propensities of the amino acids and the appropriate patterning of hydrophobic residues for a particular tertiary structure. These methods do not typically consider, however, the complementarity of steric and other inter-atomic interactions (*e.g.* hydrogen bonding) observed in the structures of natural proteins. As a result, such designed proteins may be compact and have a large degree of the appropriate secondary structure but may not form well-defined tertiary structures.

9.2.2 Combinatorial Methods

Generating and screening combinatorial protein libraries for variants with new or improved function or stability has become an established method for protein engineering.[82] Diverse, partially random libraries are experimentally obtained by using degenerate oligonucleotides during gene assembly,[83] by performing the polymerase chain reaction under mutagenic conditions,[84] or by using DNA shuffling.[85,86] In directed evolution protocols, these methods are used as part of an iterative mutation-selection-enrichment cycle. The use of such protocols has been spurred by the development of high-throughput assays and the availability of various library platforms for expressing and displaying proteins. Phage display libraries are popular tools in directed evolution,[87] but bacterial,[88] yeast,[89] and ribosomal[90] display systems are also widely used.

Combinatorial libraries have been used to examine protein folding and stability.[91–94] Proper patterning of hydrophobicity was found to be a key determinant of whether variants of λ repressor are compatible with the wild-type fold,[91] and multiple substitutions in the helix-turn-helix region of λ repressor are largely additive with regard to their impact on folded state stability of the protein.[92] Combinatorial libraries have been constructed to investigate the utility of binary patterning of hydrophobic and hydrophilic amino acids in a manner consistent with predetermined secondary and tertiary structures. Such efforts have identified native-like protein structures that fold into well-ordered four-helix bundles and beta proteins.[95–102]

9.2.3 Directed Evolution

Directed evolution and related methods have identified proteins with a variety of functionally important properties. Partial randomization of sequence followed by selection is a powerful, evolutionarily inspired, tool for introducing new function into proteins. Through directed evolution, protein function can be engineered if a suitable selection assay is available. Applications include the ability to maintain binding affinity when removing segments non-essential for a protein,[103] improve binding affinity,[104] evolve RNA polymerase from DNA polymerase,[105] endow an antibody with catalytic activity,[106–109] and accelerate the maturation of a red fluorescent protein.[110] Retroviruses have been reengineered to greatly enhance their spreading efficiency through human fibrosarcoma cells for possible use in gene therapy.[111] New biosynthetic pathways have been engineered in *E. coli* for the production of non-native carotenoids,[33] and a new genetic circuit has been evolved.[112] These studies show that directed evolution methods have a broad reach into exploring and tailoring the functions of proteins. Complex functional properties may be engineered without requiring a detailed molecular understanding as to how these are achieved. The development of a selection method yielding proteins with the desired properties is one of the key features of such methods. The particular selection strategy usually is dependent on the desired function. Affinity columns with immobilized ligands are useful for selecting tight binding proteins. For more subtle and complex functions, such as catalysis, more sophisticated selections must be used and often are tied to cell (or phage) viability. In addition, the targeted functions must be accessible *via* evolution methods where usually only a few mutations are accumulated per generation.

9.2.4 Intrinsic Limitations

Despite some of the striking successes of hierarchical protein design, combinatorial, and directed evolution methods, these techniques are not without shortcomings. Optimizing the many interactions within a particular folded state structure can be difficult by inspection alone and, as mentioned, many designed proteins do not have well-defined tertiary or oligomeric structures[50] While combinatorial methods can address large numbers of sequences – 10^3 to 10^6 for high-throughput screening and 10^{12} for display methods – these numbers are far smaller than the number of possible sequences. Thus, such approaches may miss proteins with the desired properties and structures. Directed evolution methods often require a well-folded protein as a starting point for evolving new structures and functions,[82,113] and as a result these methods are usually limited to the redesign of natural proteins. DNA shuffling protocols often require high degrees of sequence homology (60%) for appreciable rates of recombination.[114] Several approaches have been developed to overcome this limitation,[115,116] but functional hybrids often become more sparse.

It is of interest to explore sequence and structure more extensively and to arrive at proteins having novel structures that differ from natural proteins. The development of methods that allow simultaneous consideration of the myriad of interactions and levels of structure present in proteins quantitatively permit detailed predictions about structure, stability, and sequence variability that may be rigorously tested. Such studies can further our physical chemical understanding of the stability, folding, functions, and dynamics of proteins.

9.3 Computational Approaches to Structured-based Design

Computational protein design involves the search for sequences compatible with a given fold with the aid of computer modeling methods to address and quantify protein structure and amino acid variability. The template fold is often represented in atomistic detail and, as opposed to more qualitative design methods, inter-atomic interactions involving variable residues are explicitly quantified and evaluated. Computational design involves simultaneous consideration of multiple interacting residues. A complete enumeration of all possible sequences, however, is only possible when only a few residues are varied. As a result, powerful methods for sampling or characterizing sequences consistent with a particular fold must be employed when large numbers of residues are varied. Such methods may be used to design particular sequences or to guide the construction of combinatorial libraries and directed evolution experiments.

Although there are different approaches to computational protein design, most make use of similar methods for specifying the properties of the target protein and for quantifying the physical and chemical interactions that stabilize structure and confer functionality. A target polypeptide backbone structure serves as a template to guide the selection of sequences. At each variable position, residue degrees of freedom include the allowed amino acids and their side-chain conformations, which are usually treated as a discrete set of rotamer states. The interactions between all residues are quantified through energy or scoring functions, which often contain terms representing such effects as hydrogen bonding, van der Waals interactions, electrostatic interactions, and solvation. Energy functions are often used to arrive at foldability criteria, which take unfolded states into account and quantify the degree to which a particular sequence is likely to fold into the target structure. There have been several studies that examine protein design and variability through the use of sequence-alignment methods,[117,118] but herein we focus on the design of structure and sequence based upon the physico-chemical properties of the amino acids.

9.3.1 Backbone Structure and Sequence Constraints

A backbone structure specifies the coordinates of the main chain atoms, the bonded series of carbon, nitrogen, and oxygen atoms associated with each amino acid residue in a protein. This structure serves as the target for computational protein design. The backbone coordinates may be obtained from a known structure[119] or from modeling novel structures.[120–123] Naturally occurring protein structures are often used, as many such structures support a wide variety of biological functions, *e.g.* the TIM barrel superfamily.[124] This specification of the template structure partially defines the design problem, reduces the number of degrees of freedom, and avoids some of the difficulties of structure prediction. The fixed-backbone approximation has been successfully used in computational protein design, but flexibility of the main chain may be included to accommodate backbone readjustments that result from changes in sequence.[125,126] While flexible-backbone protein design is more computationally intensive than fixed-backbone design, recent studies have shown that such methods can yield well-structured proteins.[121,127]

9.3.2 Residue Degrees of Freedom

Since the backbone structure is largely predetermined, the degrees of freedom in protein design mainly involve the distinguishable states of the amino acid residues. These residue degrees of freedom include both the allowed amino acids at each variable position as well as the side-chain conformations of these amino acids. The naturally occurring 20 amino acids are most often used, but this number may be reduced, as in the patterning of residue properties,[95,128] or expanded with the inclusion of non-natural amino acids. Most amino acids may assume multiple, distinguishable side-chain conformations (rotamers).[129–131] Libraries of allowed rotamer states reduce the complexity of the side-chain states to a discrete set of side-chain conformations. Such rotamer states have been statistically deduced from protein structure databases, and usually are consistent with bond and torsional angles corresponding to local energy minima.[131] Depending on the size and topology of an amino acid, the number of possible rotamers can range from one (glycine and alanine) to as many as 80–100 (or more) for larger side chains. Rotamer libraries have been developed that are backbone-independent or are sensitive to the local backbone and secondary structure.[129,132–143] Atomistic representations of side chains and their conformations enable the design of well-packed protein interiors. Rotamer libraries may also be developed for non-biological amino acids by identifying local torsional minima using a molecular mechanics force field.[144,145]

In order to make the design of larger proteins more tractable, approaches have been introduced that reduce the residue degrees of freedom. Such methods include reducing the number of allowed amino acids, often in a patterned or site-specific manner,[95,128] or simplifying the representation of the amino acid side chains. Studies with limited numbers of amino acids can suggest the minimal set of amino acids necessary for certain structures. For example, a 108-residue, four-helix bundle structure has been constructed using only 7 of the 20 possible amino acids.[146] Similarly, only five amino acids have been used to reconstruct large portions of an SH3 domain structure.[147] Simplified models of side chains have been successfully used in the design of proteins, using an energy landscape approach.[148]

9.3.3 Energy Function

Quantification of intra- and intermolecular interactions is critical to computational protein design. The energy functions used in protein design are often similar to those used in molecular modeling and simulation,[149] such as the atom-based molecular mechanics force fields Amber,[150] CHARMm,[151] and Gromos.[152] Energy functions like these may comprise contributions arising from deformation of bond lengths, bond angles, and dihedral angles, as well as non-bonding interactions arising from van der Waals, electrostatic, and hydrogen bonding interactions (see Chapter 8 by Pande for more on force fields in protein folding simulations). In protein design, the non-bonding interaction terms often dominate, since the approximation of discrete rotamer states and the rigid backbone largely fixes bond lengths and bond angles. So as not to overestimate the repulsive energies of van der Waals interactions, which may often be readily alleviated by slight backbone adjustments, the van der Waals radii are often uniformly diminished using *ad hoc* scaling factors.[153,154]

In addition to atom-based physico-chemical energy functions, effective energies quantifying the structural propensities of the amino acids can be included using scoring functions. Statistical analysis of protein structure databases reveals that the relative frequencies of amino acids may depend on the local structural environment, and experimental studies have quantified the degree to which different amino acids destabilize secondary structures.[55,56,149,155-157] Such experimental or database studies can yield effective scoring functions. The individual terms in the energy function may be weighted in order to combine molecular mechanics potentials and empirical scoring functions. This weighting may be subtle, and is often accomplished by comparison with known sequences and structures,[158] or with training sets of randomized sequences.[159]

9.3.4 Solvation

Solvation and hydrophobic effects play critical roles in protein folding.[160] Hydrophobic residues tend to be sequestered in the interior of the protein, while hydrophilic residues are found more frequently on the exterior. In protein design, evaluating free energies of solvation through explicit modeling of solvent is computationally prohibitive. Solvation effects, and indirectly the hydrophobic effect, are often approximated using energies expressed in terms of the solvent accessible area of each atom. The corresponding free energy cost per unit area exposed is often parameterized using a structural protein database, or known free energies of transfer between water and either a vacuum or organic phase.[161–164] Although much simpler than the explicit modeling of solvent, calculating such surface areas may still be computationally expensive,[165] particularly since these areas are sequence dependent. As an alternative method, a statistical potential may be introduced that quantifies the propensities of the amino acids to reside in buried and exposed local environments.[166]

9.3.5 Foldability Criteria and Negative Design

A designed protein should fold into a unique three-dimensional structure defined as the "native" state. Non-target conformations of the protein should not be appreciably populated. In order to achieve this, the conformational energy landscape should have "funnel" shape, with the folded state at the free-energy minimum[11–13,167] (see Chapter 3 by Wolynes). Including information about stabilization with respect to misfolded structures is often referred to as "negative design".[60] In order to achieve structural specificity, the target structure should correspond to an energetic ground state that has an energy gap separating the target from other competing structures.[13,168,169]

Many approaches to computational protein design focus on energy minimization (through variation of sequence) as the foldability criterion. However, the notion that decreasing energy is correlated with improved foldability can be problematic, particularly for models involving reduced representations of the amino acids.[170] The absence of explicit negative design may result in proteins that populate multiple topologies.[171] For simple models of proteins, other foldability criteria that more accurately approximate the free energy of folding and/or address unfolded structures explicitly may be used. Such criteria can act as objective functions in sequence design[169,172–174] Such quantities include Δ/Γ, where Δ is the energy gap between the target structure and the average energy of other competing, unfolded structures and Γ^2 is the variance of the energy averaged over this same ensemble of unfolded structures.[169,172–175]

For atomistic representations of proteins, however, energy minimization appears to be a viable strategy. This is not unreasonable, given that most design algorithms yield structures that are sterically and energetically self-consistent, in keeping with what is observed in natural structures. For proteins comprising a single chain, such tightly packed sequences are specific to the target backbone structure, and it is unlikely that the same interior packing could be observed in alternative backbone conformations. Viewed in another way, the use of explicit side-chain conformations in protein design increases the effective number of monomer types by associating a set of rotamers with each amino acid, and tailoring a sequence for a particular tertiary structure becomes more straightforward with the expanded monomer set.[13,176,177] Elements of negative design may also already be involved in the design process in an indirect manner, *via* the use of effective energies of the residues in unfolded structures (*i.e.* reference energies),[120,178] by imposition of composition constraints on the numbers of each amino acid,[179,180] or by application of hydrophobic and hydrophilic patterning.[128] Explicit negative design can become crucial in cases where degenerate or low-energy competing structures are likely to compete with the target structure, such as may be the case in low-resolution protein models (coarse-grained or simplified representations of amino acids)[148] or in cases involving protein–protein interfaces, which may have smooth energy landscapes supporting multiple possible orientations of the associating proteins.[181]

Implementations of negative design have yielded well-folded proteins. Qualitative use of negative design based on the patterning of hydrophobic and hydrophilic amino acids enabled the conversion of a designed amyloid-forming protein into a monomeric β-sheet protein.[102] In the redesign of a three-helix bundle topology, an ensemble of denatured structures from folding simulations was used as a set of competing unfolded structures.[148] Optimal sequences were selected using a modified Δ/Γ score, and one such sequence appeared native-like upon experimental characterization. Explicit negative design has also been implemented in an algorithm for designing coiled-coil interfaces,[181] where sequence specificity was selected by comparing energies of the target structure with non-target, misfolded, homodimeric and heterodimeric states.

9.3.6 Search and Characterization of Sequence Ensembles

Protein design involves identifying viable sequences subject to imposed constraints on structure, sequence, and function. Various algorithms can be utilized for the identification of sequences consistent with the target structure and target protein properties. These approaches may be grouped into two categories: directed or search-based methods that seek to identify sequences optimizing a particular scoring or energy function; and probabilistic approaches that seek to characterize the properties of the ensemble sequences likely to fold to the desired structure.

In an optimization approach, the goal is identification of a high-scoring (low-energy) sequence for the target using energy or other scoring functions. Since exhaustive enumeration of sequences and rotamer positioning are only tractable for cases involving just a few variable residues, approaches such as genetic algorithms,[125] simulated annealing,[158,182] and Monte Carlo methods[183] are used. Alternatively, elimination and pruning methods identify global optima by successively removing residue states that cannot be a part of the optimal solution.[153,184–186]

A statistical approach estimates the site-specific probabilities of the amino acids among sequences consistent with the targeted structure and other desired properties.[166,174] Such a probabilistic approach is motivated by several considerations. Nature often provides multiple sequences that fold to the same structure, so there are usually multiple possible solutions in protein design. Probabilistic information regarding the likelihoods of the amino acids is a natural input into combinatorial studies of proteins. Many aspects involved in design, such as parameterized energy functions, discrete side-chain conformations, fixing backbone atoms, and effective solvation energies, involve approximations. The sequences identified by optimization-based approaches are likely to be sensitive to the details of the energy function used and to the nature of these approximations, whereas statistical features may be more robust. Site-specific amino acid probabilities can highlight the allowed mutations at each location, and provide a broad characterization of the ensemble of sequences. Such methods are implemented in two complementary approaches: maximization of an effective entropy to determine the most likely set of site-specific amino acid probabilities,[120,166] and sampling of sequences using Monte Carlo methods.[183,187] These probabilities may then be used to determine specific protein sequences or to specify the composition in a combinatorial library. This approach has been termed a statistical, computationally assisted design strategy (SCADS).[120,188]

9.4 Recent Successes in Protein Design

De novo designed proteins have appeared within recent years that make use of advances in computational design methods. We discuss only a few here, but other noteworthy design achievements include biocatalysts,[42,189,190] sensors,[191] and protein–protein interactions.[192] Several recent reviews have also appeared that detail further exploration of this field.[126,127,169,176,177,193-195]

Verification of a designed protein sequence is best accomplished by experimental realization and characterization. Empirical structure determination efforts, *via* X-ray crystallography or NMR structure determination, provide demanding but time-intensive assessments of design. Often stringent biophysical and functional assays are used. The examples discussed here involve the computational design of proteins that have been characterized experimentally, affirming the theoretical methodology employed.

9.4.1 Tailored Mutations for Ultrafast Folding

Computational protein design has been used to probe folding kinetics and to engineer ultrafast (microsecond) folding mutants of small proteins (see Chapter 6 by Gruebele for thorough discussion on fast folding). Protein engineering *via* mutation is a common tool for investigating folding dynamics.[7,8,196] Small, ultrafast-folding proteins are of interest because they lend themselves to comparison with atomistic molecular dynamics simulations[197] (see Chapter 8 by Pande) and with theories of protein folding[13,15,198] (see Chapter 3 by Wolynes). Additionally, such fast-folding proteins are ideal for examining how the folded state structure may influence folding rate.[196,199–201] Designed proteins with well-packed hydrophobic interiors are among the fastest-folding proteins, with folding times of 1–30 μs.[202–204]

Zhu et al.[205] explored the use of computationally designed mutations for kinetics studies using a 47-residue three-helix bundle albumen-binding protein, which has a folding time of $t_f = 6$ μs, where $t_f = 1/k_f$ and k_f results from a two-state analysis of temperature jump folding studies. SCADS was used both to identify frustrated sites, sites where other amino acids are more favorable than the wild type, and to suggest mutations at these positions that are structurally consistent with the native protein. The stability and folding kinetics of four suggested single mutants and one double mutant were then analysed using laser-induced T-jump infrared techniques (see Chapter 6 by Gruebele). One designed mutant had a folding time of $t_f = 1$ μs, placing it among the fastest folding proteins known to date. A linear correlation was observed in which the maximal folding rate decreased with the overall hydrophobicity of the protein, suggesting that tailored hydrophobic interactions can leverage the rapid rate of hydrophobic collapse[206] and lead to ultrafast-folding proteins.

Monte Carlo sampling methods were used in a probabilistic context to design an ultrafast-folding mutant of the 20-residue Trp-cage protein.[207] These calculations identified a P12W mutation, termed the Trp^2-cage due to the second tryptophan. The Trp^2-cage mutant was more stable and folded more rapidly compared to the wild-type structure: the folding time obtained from a kinetic two-state analysis is $t_f = 0.94$ μs, compared to 4.1 μs for the wild type.[207,208] These findings make the Trp^2-cage mutant one of the fastest-folding proteins characterized to date, and an ideal model system for further computational and experimental kinetic studies.

9.4.2 Designing Structure and Sequence

Protein design can suggest not only variants of naturally occurring proteins but also lead to de novo designed proteins as well, where structure, sequence, and even function are elements of the design process. A small protein based on a zinc finger topology has been successfully computationally designed.[153] Novel sequences that fold into desired target structures have been identified by cycling between sequence design and backbone optimization.[158,179,209–211] A computationally designed 97-residue α/β protein, Top7, based upon a topology not found in isolation, has recently been realized.[121,212] This design was achieved using backbone templates assembled from structural fragments with subsequent design of sequence.[158] These findings suggest that additional protein folds not yet found in nature may be physically possible (Figure 9.1).

Probabilistic methods have also been successful in large-scale protein design. SCADS has been applied to the de novo design of DFsc, a 114-residue four-helix bundle containing a di-iron center.[120] The backbone template was based on the crystal structure of a previously designed dimeric helix protein DF1.[213] The topology of DF1 was re-engineered by altering the interhelical turns in an

Protein	Structure	Residues Designed | Total	Design Approach	Reference
Phage 434 Cro		7 | 63	De novo design of hydrophobic core	Desjarlais, J.R. et al. Protein Science 1995
FSD-1		28 | 28	Fully automated design	Dahiyat, B.I. et al. Science 1997
Coiled coil		4 | 31	Design of specificity and recognition	Havranek, J.J. et al. Nature Structural Biology 2003
Albumen binding domain	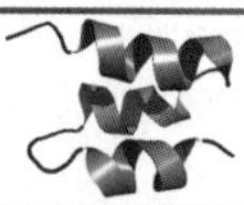	47 | 47	Energy landscape approach with negative design	Jin, W.Z. et al. Structure 2003
Top7		93 | 93	Fold and sequence design	Kuhlman, B. et al. Science 2003
DFsc		114 | 114	Designer monomeric dinuclear metalloprotein	Calhoun, J.R. et al. J. Molecular Biology 2003
WSK-3		33 | 104*	Water-soluble analog of transmembrane potassium channel	Slovic, A.M. et al. PNAS 2004
Four helix bundle	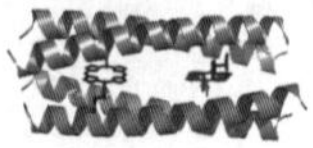	34 | 34*	Specific binding of nonbiological cofactor	Cochran, F.V. et al. J. Am. Chem. Soc. 2005
Trp2-cage		1 | 20	Ultrafast folding protein (microsecond timescale)	Bunagan, M.R. et al. J. Phys. Chem. B 2006
Dps		10 | 167*	Hydrophobic interior of ferritin-like protein	Swift, J. et al. J. Am. Chem. Soc. 2006
PAsc	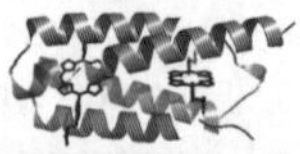	108 | 108	De novo design of a single-chain diphenyl-porphyrin metalloprotein	Bender, G.M. et al. J. Am. Chem. Soc. 2007

Figure 9.1 Examples of computationally designed proteins. The column labeled "Residues" indicates the number of designed, variable residues over the total length of the protein. The asterisks indicate homo-oligomeric protein complexes, and the number of variable residues and total length of each subunit are indicated. WSK-3 and the four-helix bundle with a non-biological co-factor are tetramers; Dps is a dodecamer.

effort to arrive at a stable and monomeric (single-chain) protein. Residue identities and conformations were fixed at 26 positions to confer metal binding, provide access to the active site, and initiate helix formation. SCADS was used to identify the remaining 88 residues. The structure of the final sequence has been characterized using CD and NMR spectroscopy. DFsc is well structured and may be catalytically active.[120]

Computational methods have also been extended to the design of β-sheet metalloproteins.[123] The rubredoxin protein family consists of simple iron sulfur beta proteins having a redox active Fe(II)/Fe(III) ion in a tetra-cysteine binding site.[213] A novel backbone was obtained by simplifying the metal binding site of rubredoxin to a pair of pseudo-equivalent β-hairpins. The two-stranded sheet was extended into a three-stranded structure, resulting in a dimer that was connected with a tryptophan zipper hairpin.[214] The resulting structure contained only 40 residues, compared to the 54 in rubredoxin, and consisted of an entirely different overall topology. Some residues were fixed prior to computational design including the Trpzip linker, the four metal binding cysteines, two glycine residues that adopt an α_L conformation, and an isoleucine to shield the active site from solvent. The remaining residues were chosen based on the highest probabilities from SCADS calculations, resulting in the protein RM1. Experimental studies confirmed that the metal ions bound in the proper geometry with the expected stoichiometry, and that RM1 is monomeric and properly folded with the expected β-sheet structure, both with and without metal ions. RM1 could be reversibly reduced and oxidized over 16 cycles under aerobic conditions, suggesting a protein that is significantly more robust electrochemically than previously designed metalloproteins.[71,215]

9.4.3 Facilitating the Study of Membrane Proteins

Integral membrane proteins comprise roughly 30% of the human proteome, including many important drug targets. There is much interest in functional and structural studies of such proteins. Transmembrane proteins have large numbers of hydrophobic amino acids on their exteriors, which often serve to anchor the protein in a lipid bilayer. As a result, these proteins are problematic since they are aggregation prone and do not usually express at high levels. It is thus difficult to characterize their biophysical properties and to generate high-quality crystals for structure determination. One approach to overcome these problems is the creation of soluble variants of membrane proteins that maintain structure and functionally related properties. As an example of such an approach, the hydrophobic transmembrane domain of the potassium channel KcsA was redesigned to yield a water-soluble variant of the protein.[188] The backbone template used in this study was the high-resolution structure of KcsA.[216,217] Thirty-five surface residues were targeted for mutation based on solvent accessibility. The calculations were constrained by fixing the solvation energy to match that of a water-soluble protein with similar size.[166] One resulting protein, WSK-3, shared many properties with the parent KcsA. WSK-3 showed the correct helical structure, formed predominantly tetramers, and bound both agitoxin2 and small molecule channel blockers.[188,218] This work exemplifies the solubilization of a membrane protein by computational sequence redesign, a method that may be extended to other membrane proteins, possibly even those with unknown structure through the use of homology modeling.

9.4.4 Proteins with Non-biological Components

Nature often overcomes the physico-chemical limitations of the natural amino acids through the incorporation of co-factors, which can provide structural and electronic properties required for light harvesting or catalytic activity that may not be accessible using only the amino acids. Co-factors take the form of metal ions, organometallic compounds, and organic molecules. The ability to design proteins that can selectively bind non-biological co-factors can lead to novel proteins with functionalities not seen in nature. Instead of redesigning a natural protein, Cochran *et al.* created a novel protein scaffold.[122] Such design of structure "from scratch" is necessary when the co-factor molecules differ substantially from those seen in nature. The targeted complex comprised a homotetrameric helical bundle that contained a pair of synthetic iron porphyrin co-factors. Subject to constraints on residues involved in metal coordination, the identities of the remaining unconstrained positions were determined using SCADS. CD spectroscopy, size-exclusion chromatography, and analytical ultracentrifugation revealed that the 34-residue peptide goes from a partially disordered monomer to an α-helical tetramer upon co-factor binding.[122] Importantly, the spectra of the Soret band indicate that the complex selectively binds the target co-factor, and not other iron-containing porphyrins. The strategy has been extended to design a 108-residue single chain protein that binds two non-biological porphyrin-based co-factors.[233] The success of this study indicates that computational design methods can lead to the discovery of protein systems that selectively form complexes with large non-biological components.

9.4.5 Symmetric Structures

Since many large protein systems are polymeric with well-defined quaternary structures, it is of interest to develop computational methods for the design of such protein assemblies. For homo-oligomers or crystalline systems, the system's symmetry facilitates design calculations. Fu *et al.*[219] developed a symmetry assumption that greatly reduced the complexity of the calculations for symmetric complexes of identical chains. This assumption allows for calculation of the site-specific probabilities using only a single variable chain, rather than an entire multi-subunit complex.

Swift *et al.* recently applied this methodology to redesign a ferritin-like protein, Dps (DNA binding protein from starved cell). Dps is a dodecamer of four-helix bundle proteins that is important for iron storage and preventing DNA damage.[220] The goal of the study was to explore the degree to which the hydrophobicity of the interior protein cavity could be modified through mutation, yielding a system having potential application to the encapsulation of hydrophobic molecules. The mutation sites were identified among the exposed hydrophilic residues on the surface of the interior cavity. Using the calculated site-specific amino acid probabilities, three different variants were designed having three, seven, and ten hydrophobic mutations residues per subunit. The stability with respect to chemical denaturation was found to decrease with the increasing number of hydrophobic mutations. All of the mutants, however, exhibited high melting temperatures ($T_m = 74$–$90\ °C$). Despite the large number of hydrophobic mutations, the rate of iron oxidation and mineralization of the variant with seven mutations was comparable to that of wild-type Dps.[220] The results confirm that the ferritin family of proteins is a robust scaffold for engineering nanoscale molecular containers.

9.4.6 Computational Methods for Directed Evolution

Computational design methods may be used to guide the combinatorial design of proteins.[176,177,194,221,222] In addition, such methods can aid the design of directed evolution methods, which have been successfully applied to engineer and optimize enzymes and other proteins.[223] Due to the random nature of mutagenesis and recombination techniques, only a small fraction of the possible mutations are typically explored in most experiments, and many of the mutants will be unstable, poorly folded, or not functional. Often the structures of the proteins under study are known, however, and this information can be leveraged to improve the efficiency of such experiments. Computational algorithms have been developed in an effort to optimize mutant libraries by targeting mutagenesis to specific residues and designing recombination strategies.

Computational methods have been applied to assist *in vitro* recombination experiments at the DNA level. Maranas and co-workers[224] developed a computational framework, eCodonOpt, to remove the inherent bias in combinatorial libraries by optimizing codon usage for DNA shuffling. DNA shuffling leads to crossover positions that tend to be biased toward regions having high sequence similarity with the parent strand. This bias can lead to libraries with little sequence diversity. Optimizing codon usage of the parental DNA, through the use of eCodonOpt, increases the average number of crossovers per recombination.

Optimization of crossover points can also lead to more successful recombination approaches. In protein evolution natural recombination is often thought to occur at regions of well-defined substructures or domains. This concept has been harnessed by the computer algorithm SCHEMA.[225] SCHEMA calculates the interactions between residues and determines how many stabilizing interactions will be broken during recombination. A resulting profile for each residue is generated, and crossovers corresponding to minima involving regions where the largest number of stabilizing interactions is preserved. The SCHEMA algorithm has been tested using earlier recombination experiments,[225] and nearly all of the crossovers that improved stability or function occurred near minima in the schema profiles.

The computational method SIRCH has been developed to evaluate the functionality of protein hybrids generated by recombination.[226] SIRCH identifies residue–residue clashes in a given library. Based on the number and severity of the clashes, the hybrids can be classified according to their functional potential. The results of SIRCH have been compared to functional crossover positions identified in the recombination of human and *E. Coli* glycinamide ribonucleotide (GAR) transformylases.[115,227–230] The experimentally determined positions were consistent with the computational results, and SIRCH was able to distinguish crossover directionality (*i.e.* an A-B *versus* a B-A crossover).

9.5 Outlook

The design of a novel protein may appear to be a daunting process, given the many degrees of freedom and myriads of subtle interactions that guide folding. Computational methods have been successful, however, in overcoming these hurdles by leveraging some of the fundamental rules governing protein folding and by addressing many coupled degrees of (sequence) freedom simultaneously. Through the use of computational design methods it is now feasible to sample, search, and characterize sequences for a variety of target proteins. Several challenges still remain. It will be of interest to quantify sought-after properties in a manner consistent with protein design algorithms. Such properties include solubility, cellular toxicity, or lack thereof, high-affinity ligand binding, selective protein–protein association, and specificity in enzyme catalysis. With continued efforts in protein design, a greater understanding of these properties will become available as well as an improved understanding of protein folding and assembly.

Although the field of computational design is still under development, it has already proven successful in generating new proteins with a diverse range of structures and functions. Computational design methods may also be extended to proteins containing unnatural amino acids and non-biological folding systems or "foldamers."[231,232] Aided by computational methods, novel molecular "machinery" comprising proteins (or other polymers) may potentially be designed that can carry out desired functions with the same specificity and selectivity of natural proteins.

Acknowledgements

The authors gratefully acknowledge support from the US Department of Energy (DE-FG02-04ER46156), the University of Pennsylvania's Nano/Bio

Interface Center through the National Science Foundation (NSF) NSEC DMR-0425780. Support is also acknowledged from the National Institutes of Health (GM61267, GM71628), and Laboratory for Research on the Structure of Matter through NSF MRSEC DMR05-20020. The figures were rendered using PyMol (DeLano Scientific LLC).

References

1. C. B. Anfinsen, *Science*, 1973, **181**, 223–230.
2. J. J. Vincent, C. H. Tai, B. K. Sathyanarayana and B. Lee, *Proteins*, 2005, **61**(7), 67–83.
3. C. Levinthal, in *Mössbauer Spectroscopy in Biological Systems*, ed. P. Debrunner, J. C. M. Tsibiris, E.M. Münck, University of Illinois Press, Urbana, 1969, pp. 22–24.
4. O. B. Ptitsyn and A. A. Rashin, *Biophys. Chem.*, 1975, **3**, 1–20.
5. P. S. Kim and R. L. Baldwin, *Annu. Rev. Biochem.*, 1982, **51**, 459–489.
6. P. S. Kim and R. L. Baldwin, *Annu. Rev. Biochem.*, 1990, **59**, 631–660.
7. A. R. Fersht, *Proc. Natl. Acad. Sci. USA*, 1995, **92**, 10869–10873.
8. A. R. Fersht, *Curr. Opin. Struct. Biol.*, 1997, **7**, 3–9.
9. E. Shakhnovich, V. Abkevich and O. Ptitsyn, *Nature*, 1996, **379**, 96–98.
10. J. D. Bryngelson and P. G. Wolynes, *Proc. Natl. Acad. Sci. USA*, 1987, **84**, 7524–7528.
11. J. D. Bryngelson, J. N. Onuchic, N. D. Socci and P. G. Wolynes, *Proteins Struct. Funct. Genet.*, 1995, **21**, 167–195.
12. J. N. Onuchic, P. G. Wolynes, Z. Lutheyschulten and N. D. Socci, *Proc. Natl. Acad. Sci. USA*, 1995, **92**, 3626–3630.
13. J. N. Onuchic, Z. LutheySchulten and P. G. Wolynes, *Annu. Rev. Phys. Chem.*, 1997, **48**, 545–600.
14. B. A. Shoemaker and P. G. Wolynes, *J. Mol. Biol.*, 1999, **287**, 657–674.
15. J. N. Onuchic and P. G. Wolynes, *Curr. Opin. Struct. Biol.*, 2004, **14**, 70–75.
16. A. Akmal and V. Muñoz, *Proteins Struct. Funct. Bioinformatics*, 2004, **57**, 142–152.
17. W. A. Eaton, V. Muñoz, S. J. Hagen, G. S. Jas, L. J. Lapidus, E. R. Henry and J. Hofrichter, *Annu. Rev. Biophys. Biomol. Struct.*, 2000, **29**, 327–359.
18. V. Muñoz, *Int. J. Quantum Chem.*, 2002, **90**, 1522–1528.
19. A. N. Naganathan, J. M. Sanchez-Ruiz and V. Muñoz, *J. Am. Chem. Soc.*, 2005, **127**, 17970–17971.
20. V. Muñoz and J. M. Sanchez-Ruiz, *Proc. Natl. Acad. Sci. USA*, 2004, **101**, 17646–17651.
21. A. T. Bull, A. C. Ward and M. Goodfellow, *Microbiol. Mol. Biol. Rev.*, 2000, **64**, 573–606.
22. H. E. Schoemaker, D. Mink and M. G. Wubbolts, *Science*, 2003, **299**, 1694–1697.
23. A. J. Straathof, S. Panke and A. Schmid, *Curr. Opin. Biotechnol.*, 2002, **13**, 548–556.

24. D. E. Wagner, C. L. Phillips, W. M. Ali, G. E. Nybakken, E. D. Crawford, A. D. Schwab, W. F. Smith and R. Fairman, *Proc. Natl. Acad. Sci. USA*, 2005, **102**, 12656–12661.
25. R. Fairman and K. S. Akerfeldt, *Curr. Opin. Struct. Biol.*, 2005, **15**, 453–463.
26. S. A. Maskarinec and D. A. Tirrell, *Curr. Opin. Biotechnol.*, 2005, **16**, 422–426.
27. R. Langer and D. A. Tirrell, *Nature*, 2004, **428**, 487–492.
28. V. P. Torchilin and A. N. Lukyanov, *Drug Discov. Today*, 2003, **8**, 259–266.
29. B. D. Ratner and S. J. Bryant, *Annu. Rev. Biomed. Eng.*, 2004, **6**, 41–75.
30. M. E. Davis, P. C. Hsieh, A. J. Grodzinsky and R. T. Lee, *Circ. Res.*, 2005, **97**, 8–15.
31. F. Hudecz, Z. Banoczi and G. Csik, *Med. Res. Rev.*, 2005, **25**, 679–736.
32. C. W. Wang, M. K. Oh and J. C. Liao, *Biotechnol. Bioeng.*, 1999, **62**, 235–241.
33. C. Schmidt-Dannert, D. Umeno and F. H. Arnold, *Nature Biotechnol.*, 2000, **18**, 750–753.
34. A. Crameri, G. Dawes, E. Rodriguez Jr., S. Silver and W. P. Stemmer, *Nat. Biotechnol.*, 1997, **15**, 436–438.
35. E. J. A. X. Van de Sandt and E. De Vroom, *Chimica Oggi-Chemistry Today*, 2000, **18**, 72–75.
36. C. E. Nakamura and G. M. Whited, *Curr. Opin. Biotechnol.*, 2003, **14**, 454–459.
37. J. Fahrenkamp-Uppenbrink, *Science*, 2002, **297**, 798.
38. K. R. Rogers, *Mol. Biotechnol.*, 2000, **14**, 109–129.
39. D. V. Lim, J. M. Simpson, E. A. Kearns and M. F. Kramer, *Clin. Microbiol. Rev.*, 2005, **18**, 583–607.
40. M. Allert, S. S. Rizk, L. L. Looger and H. W. Hellinga, *Proc. Natl. Acad. Sci. USA*, 2004, **101**, 7907–7912.
41 A. Korkegian, M. E. Black, D. Baker and B. L. Stoddard, *Science*, 2005, **308**, 857–860.
42. M. A. Dwyer, L. L. Looger and H. W. Hellinga, *Science*, 2004, **304**, 1967–1971.
43. O. H. Brekke and I. Sandlie, *Nat. Rev. Drug Discov.*, 2003, **2**, 52–62.
44. G. A. Lazar, S. A. Marshall, J. J. Plecs, S. L. Mayo and J. R. Desjarlais, *Curr. Opin. Struct. Biol.*, 2003, **13**, 513–518.
45. A. P. Vasserot, C. D. Dickinson, Y. Tang, W. D. Huse, K. S. Manchester and J. D. Watkins, *Drug Discov. Today*, 2003, **8**, 118–126.
46. M.R. Shirts and V.S. Pande, *J. Chem. Phys.*, 2005, **122**, 134508.
47. M. R. Shirts, J. W. Pitera, W. C. Swope and V. S. Pande, *J. Chem. Phys.*, 2003, **119**, 5740–5761.
48. A. Tramontano, *FEBS J.*, 2007, **274**, 1651–1654.
49. C. Pabo, *Nature*, 1983, **301**, 200.
50. J. W. Bryson, S. F. Betz, H. S. Lu, D.J. Suich, H. X. X. Zhou, K. T. O'Neil and W. F. Degrado, *Science*, 1995, **270**, 935–941.

51. W. F. Degrado, C. M. Summa, V. Pavone, F. Nastri and A. Lombardi, *Annu. Rev. Biochem.*, 1999, **68**, 779–819.
52. G. D. Rose, A. R. Geselowitz, G. J. Lesser, R. H. Lee and M. H. Zehfus, *Science*, 1985, **229**, 834–838.
53. S. Miller, J. Janin, A. M. Lesk and C. Chothia, *J. Mol. Biol.*, 1987, **196**, 641–656.
54. K. T. O'Neil, H. R. Wolfe, S. Ericksonviitanen and W. F. Degrado, *Science*, 1987, **236**, 1454–1456.
55. D. L. Minor and P. S. Kim, *Nature*, 1994, **367**, 660–663.
56. D. L. Minor and P. S. Kim, *Nature*, 1994, **371**, 264–267.
57. D. Eisenberg, W. Wilcox, S. M. Eshita, P. M. Pryciak, S. P. Ho and W. F. DeGrado, *Proteins*, 1986, **1**, 16–22.
58. S. P. Ho and W. F. Degrado, *J. Am. Chem. Soc.*, 1987, **109**, 6751–6758.
59. L. Regan and W. F. Degrado, *Science*, 1988, **241**, 976–978.
60. M. H. Hecht, J. S. Richardson, D. C. Richardson and R. C. Ogden, *Science*, 1990, **249**, 884–891.
61. C. P. Hill, D. H. Anderson, L. Wesson, W. F. Degrado and D. Eisenberg, *Science*, 1990, **249**, 543–546.
62. B. Lovejoy, S. Choe, D. Cascio, D. K. Mcrorie, W. F. Degrado and D. Eisenberg, *Science*, 1993, **259**, 1288–1293.
63. G. D. Rose and T. P. Creamer, *Proteins Struct. Funct. Genet.*, 1994, **19**, 1–3.
64. S. Dalal, S. Balasubramanian and L. Regan, *Nat. Struct. Biol.*, 1997, **4**, 548–552.
65. C. E. Schafmeister, L. J. W. Miercke and R. M. Stroud, *Science*, 1993, **262**, 734–738.
66. M. J. Root, M. S. Kay and P. S. Kim, *Science*, 2001, **291**, 884–888.
67. L. Regan and N. D. Clarke, *Biochemistry*, 1990, **29**, 10878–10883.
68. W. D. Kohn, C. M. Kay, B. D. Sykes and R. S. Hodges, *J. Am. Chem. Soc.*, 1998, **120**, 1124–1132.
69. T. M. Handel, S. A. Williams and W. F. Degrado, *Science*, 1993, **261**, 879–885.
70. C. Sissi, P. Rossi, F. Felluga, F. Formaggio, M. Palumbo, P. Tecilla, C. Toniolo and P. Scrimin, *J. Am. Chem. Soc.*, 2001, **123**, 3169–3170.
71. E. Farinas and L. Regan, *Protein Science*, 1998, **7**, 1939–1946.
72. A. Lombardi, C. M. Summa, S. Geremia, L. Randaccio, V. Pavone and W. F. DeGrado, *Proc. Natl. Acad. Sci. USA*, 2000, **97**, 6298–6305.
73. R. Schnepf, P. Horth, E. Bill, K. Wieghardt, P. Hildebrandt and W. Haehnel, *J. Am. Chem. Soc.*, 2001, **123**, 2186–2195.
74. X. Q. Li, K. Suzuki, K. Kanaori, K. Tajima, A. Kashiwada, H. Hiroaki, D. Kohda and T. Tanaka, *Protein Science*, 2000, **9**, 1327–1333.
75. G. R. Dieckmann, D. K. McRorie, D. L. Tierney, L. M. Utschig, C. P. Singer, T. V. O'Halloran, J. E. PennerHahn, W. F. DeGrado and V. L. Pecoraro, *J. Am. Chem. Soc.*, 1997, **119**, 6195–6196.
76. D. E. Robertson, R. S. Farid, C. C. Moser, J. L. Urbauer, S. E. Mulholland, R. Pidikiti, J. D. Lear, A. J. Wand, W. F. Degrado and P. L. Dutton, *Nature*, 1994, **368**, 425–431.

77. H. K. Rau, N. DeJonge and W. Haehnel, *Proc. Natl. Acad. Sci. USA*, 1998, **95**, 11526–11531.
78. H. K. Rau, N. DeJonge and W. Haehnel, *Angew. Chem. Int. Ed.*, 2000, **39**, 250–253.
79. M. Klemba, K. H. Gardner, S. Marino, N. D. Clarke and L. Regan, *Nat. Struct. Biol.*, 1995, **2**, 368–373.
80. S. E. Mulholland, B. R. Gibney, F. Rabanal and P. L. Dutton, *Biochemistry*, 1999, **38**, 10442–10448.
81. D. P. Raleigh and W. F. Degrado, *J. Am. Chem. Soc.*, 1992, **114**, 10079–10081.
82. L. Yuan, I. Kurek, J. English and R. Keenan, *Microbiol. Mol. Biol. Rev.*, 2005, **69**, 373–392.
83. A. Knappik, L. M. Ge, A. Honegger, P. Pack, M. Fischer, G. Wellnhofer, A. Hoess, J. Wolle, A. Pluckthun and B. Virnekas, *J. Mol. Biol.*, 2000, **296**, 57–86.
84. M. Zaccolo, D. M. Williams, D. M. Brown and E. Gherardi, *J. Mol. Biol.*, 1996, **255**, 589–603.
85. W. P. C. Stemmer, *Nature*, 1994, **370**, 389–391.
86. W. P. C. Stemmer, *Proc. Natl. Acad. Sci. USA*, 1994, **91**, 10747–10751.
87. R. H. Hoess, *Chem. Rev.*, 2001, **101**, 3205–3218.
88. T. N. Nguyen, M. Hansson, S. Stahl, T. Bachi, A. Robert, W. Domzig, H. Binz and M. Uhlen, *Gene*, 1993, **128**, 89–94.
89. E. T. Boder and K. D. Wittrup, *Nat. Biotechnol.*, 1997, **15**, 553–557.
90. J. Hanes, C. Schaffitzel, A. Knappik and A. Pluckthun, *Nat. Biotechnol.*, 2000, **18**, 1287–1292.
91. W. A. Lim and R. T. Sauer, *Nature*, 1989, **339**, 31–36.
92. L. M. Gregoret and R. T. Sauer, *Proc. Natl. Acad. Sci. USA*, 1993, **90**, 4246–4250.
93. A. R. Davidson and R. T. Sauer, *Proc. Natl. Acad. Sci. USA*, 1994, **91**, 2146–2150.
94. A. R. Davidson, K. J. Lumb and R. T. Sauer, *Nat. Struct. Biol.*, 1995, **2**, 856–864.
95. S. Kamtekar, J. M. Schiffer, H. Y. Xiong, J. M. Babik and M. H. Hecht, *Science*, 1993, **262**, 1680–1685.
96. D. A. Moffet, L. K. Certain, A. J. Smith, A. J. Kessel, K. A. Beckwith and M. H. Hecht, *J. Am. Chem. Soc.*, 2000, **122**, 7612–7613.
97. M. W. West, W. X. Wang, J. Patterson, J. D. Mancias, J. R. Beasley and M. H. Hecht, *Proc. Natl. Acad. Sci. USA*, 1999, **96**, 11211–11216.
98. B. M. Broome and M. H. Hecht, *J. Mol. Biol.*, 2000, **296**, 961–968.
99. D. A. Moffet and M. H. Hecht, *Chem. Rev.*, 2001, **101**, 3191–3203.
100. Y. N. Wei, S. Kim, D. Fela, J. Baum and M. H. Hecht, *Proc. Natl. Acad. Sci. USA*, 2003, **100**, 13270–13273.
101. Y. N. Wei, T. Liu, S. L. Sazinsky, D. A. Moffet, I. Pelczer and M. H. Hecht, *Protein Science*, 2003, **12**, 92–102.
102. W. X. Wang and M. H. Hecht, *Proc. Natl. Acad. Sci. USA*, 2002, **99**, 2760–2765.

103. A. C. Braisted and J. A. Wells, *Proc. Natl. Acad. Sci. USA*, 1996, **93**, 5688–5692.
104. L. Giver, A. Gershenson, P. O. Freskgard and F. H. Arnold, *Proc. Natl. Acad. Sci. USA*, 1998, **95**, 12809–12813.
105. G. Xia, L. J. Chen, T. Sera, M. Fa, P. G. Schultz and F. E. Romesberg, *Proc. Natl. Acad. Sci. USA*, 2002, **99**, 6597–6602.
106. S. J. Pollack, J. W. Jacobs and P. G. Schultz, *Science*, 1986, **234**, 1570–1573.
107. A. Tramontano, K. D. Janda and R. A. Lerner, *Science*, 1986, **234**, 1566–1570.
108. P. G. Schultz and R. A. Lerner, *Science*, 1995, **269**, 1835–1842.
109. K. M. Shokat, M. K. Ko, T. S. Scanlan, L. Kochersperger, S. Yonkovich, S. Thaisrivongs and P. G. Schultz, *Angew. Chem. Int. Ed.*, 1990, **29**, 1296–1303.
110. B. J. Bevis and B. S. Glick, *Nat. Biotechnol.*, 2002, **20**, 83–87.
111. R. M. Schneider, Y. Medvedovska, I. Hartl, B. Voelker, M. P. Chadwick, S. J. Russell, K. Cichutek and C. J. Buchholz, *Gene Therapy*, 2003, **10**, 1370–1380.
112. Y. Yokobayashi, R. Weiss and F. H. Arnold, *Proc. Natl. Acad. Sci. USA*, 2002, **99**, 16587–16591.
113. F. H. Arnold, *Nature*, 2001, **409**, 253–257.
114. G. L. Moore and C. D. Maranas, *AIChE J.*, 2004, **50**, 262–272.
115. M. Ostermeier, J. H. Shim and S. J. Benkovic, *Nat. Biotechnol.*, 1999, **17**, 1205–1209.
116. V. Sieber, C. A. Martinez and F. H. Arnold, *Nat. Biotechnol.*, 2001, **19**, 456–460.
117. W. P. Russ, D. M. Lowery, P. Mishra, M. B. Yaffe and R. Ranganathan, *Nature*, 2005, **437**, 579–583.
118. M. Socolich, S. W. Lockless, W. P. Russ, H. Lee, K. H. Gardner and R. Ranganathan, *Nature*, 2005, **437**, 512–518.
119. H. M. Berman, J. Westbrook, Z. Feng, G. Gilliland, T. N. Bhat, H. Weissig, I. N. Shindyalov and P. E. Bourne, *Nucleic Acids Res.*, 2000, **28**, 235–242.
120. J. R. Calhoun, H. Kono, S. Lahr, W. Wang, W. F. DeGrado and J. G. Saven, *J. Mol. Biol.*, 2003, **334**, 1101–1115.
121. B. Kuhlman, G. Dantas, G. C. Ireton, G. Varani, B. L. Stoddard and D. Baker, *Science*, 2003, **302**, 1364–1368.
122. F. V. Cochran, S. P. Wu, W. Wang, V. Nanda, J. G. Saven, M. J. Therien and W. F. DeGrado, *J. Am. Chem. Soc.*, 2005, **127**, 1346–1347.
123. V. Nanda, M. M. Rosenblatt, A. Osyczka, H. Kono, Z. Getahun, P. L. Dutton, J. G. Saven and W. F. DeGrado, *J. Am. Chem. Soc.*, 2005, **127**, 5804–5805.
124. N. Nagano, C. A. Orengo and J. M. Thornton, *J. Mol. Biol.*, 2002, **321**, 741–765.
125. J. R. Desjarlais and T. M. Handel, *Protein Science*, 1995, **4**, 2006–2018.

126. C. A. Floudas, H. K. Fung, S. R. McAllister, M. Monnigmann and R. Rajgaria, *Chem. Eng. Sci.*, 2006, **61**, 966–988.
127. C. M. Kraemer-Pecore, J. T. J. Lecomte and J. R. Desjarlais, *Protein Science*, 2003, **12**, 2194–2205.
128. S. A. Marshall and S. L. Mayo, *J. Mol. Biol.*, 2001, **305**, 619–631.
129. R. L. Dunbrack and F. E. Cohen, *Protein Science*, 1997, **6**, 1661–1681.
130. S. C. Lovell, J. M. Word, J. S. Richardson and D. C. Richardson, *Proteins Struct. Funct. Genet.*, 2000, **40**, 389–408.
131. R. L. Dunbrack, *Curr. Opin. Struct. Biol.*, 2002, **12**, 431–440.
132. R. Chandrasekaran and G. N. Ramachandran, *Int. J. Protein Res.*, 1970, **2**, 223–233.
133. J. Janin, S. Wodak, M. Levitt and B. Maigret, *J. Mol. Biol.*, 1978, **125**, 357–386.
134. T. N. Bhat, V. Sasisekharan and M. Vijayan, *Int. J. Pept. Protein Res.*, 1979, **13**, 170–184.
135. M. N. G. James and A. R. Sielecki, *J. Mol. Biol.*, 1983, **163**, 299–361.
136. E. Benedetti, G. Morelli, G. Nemethy and H. A. Scheraga, *Int. J. Pept. Protein Res.*, 1983, **22**, 1–15.
137. J. W. Ponder and F. M. Richards, *J. Mol. Biol.*, 1987, **193**, 775–791.
138. M. J. McGregor, S. A. Islam and M. J. E. Sternberg, *J. Mol. Biol.*, 1987, **198**, 295–310.
139. P. Tuffery, C. Etchebest, S. Hazout and R. Lavery, *J. Biomol. Struct. Dynam.*, 1991, **8**, 1267–1289.
140. R. L. Dunbrack and M. Karplus, *J. Mol. Biol.*, 1993, **230**, 543–574.
141. H. Schrauber, F. Eisenhaber and P. Argos, *J. Mol. Biol.*, 1993, **230**, 592–612.
142. H. Kono and J. Doi, *J. Comput. Chem.*, 1996, **17**, 1667–1683.
143. M. DeMaeyer, J. Desmet and I. Lasters, *Folding & Design*, 1997, **2**, 53–66.
144. J. W. Chin, T. A. Cropp, J. C. Anderson, M. Mukherji, Z. W. Zhang and P. G. Schultz, *Science*, 2003, **301**, 964–967.
145. D. Datta, P. Wang, I. S. Carrico, S. L. Mayo and D. A. Tirrell, *J. Am. Chem. Soc.*, 2002, **124**, 5652–5653.
146. C. E. Schafmeister, S. L. LaPorte, L. J. W. Miercke and R. M. Stroud, *Nat. Struct. Biol.*, 1997, **4**, 1039–1046.
147. D. S. Riddle, J. V. Santiago, S. T. BrayHall, N. Doshi, V. P. Grantcharova, Q. Yi and D. Baker, *Nat. Struct. Biol.*, 1997, **4**, 805–809.
148. W. Z. Jin, O. Kambara, H. Sasakawa, A. Tamura and S. Takada, *Structure*, 2003, **11**, 581–590.
149. D. B. Gordon, S. A. Marshall and S. L. Mayo, *Curr. Opin. Struct. Biol.*, 1999, **9**, 509–513.
150. S. J. Weiner, P. A. Kollman, D. A. Case, U. C. Singh, C. Ghio, G. Alagona, S. Profeta and P. Weiner, *J. Am. Chem. Soc.*, 1984, **106**, 765–784.
151. B. R. Brooks, R. E. Bruccoleri, B. D. Olafson, D. J. States, S. Swaminathan and M. Karplus, *J. Comput. Chem.*, 1983, **4**, 187–217.
152. J. Hermans, H. J. C. Berendsen, W. F. Vangunsteren and J. P. M. Postma, *Biopolymers*, 1984, **23**, 1513–1518.

153. B. I. Dahiyat and S. L. Mayo, *Science*, 1997, **278**, 82–87.
154. A. E. Keating, V. N. Malashkevich, B. Tidor and P. S. Kim, *Proc. Natl. Acad. Sci. USA*, 2001, **98**, 14825–14830.
155. J. S. Richardson and D. C. Richardson, *Science*, 1988, **240**, 1648–1652.
156. K. T. O'Neil and W. F. Degrado, *Science*, 1990, **250**, 646–651.
157. A. G. Street and S. L. Mayo, *Proc. Natl. Acad. Sci. USA*, 1999, **96**, 9074–9076.
158. B. Kuhlman and D. Baker, *Proc. Natl. Acad. Sci. USA*, 2000, **97**, 10383–10388.
159. T. L. Chiu and R. A. Goldstein, *Protein Eng.*, 1998, **11**, 749–752.
160. K. A. Dill, *Biochemistry*, 1990, **29**, 7133–7155.
161. J. L. Fauchere and V. Pliska, *Eur. J. Med. Chem.*, 1983, **18**, 369–375.
162. D. Eisenberg and A. D. McLachlan, *Nature*, 1986, **319**, 199–203.
163. T. Ooi, M. Oobatake, G. Nemethy and H. A. Scheraga, *Proc. Natl. Acad. Sci. USA*, 1987, **84**, 3086–3090.
164. K. A. Sharp, A. Nicholls, R. Friedman and B. Honig, *Biochemistry*, 1991, **30**, 9686–9697.
165. A. G. Street and S. L. Mayo, *Folding & Design*, 1998, **3**, 253–258.
166. H. Kono and J. G. Saven, *J. Mol. Biol.*, 2001, **306**, 607–628.
167. P. G. Wolynes, J. N. Onuchic and D. Thirumalai, *Science*, 1995, **267**, 1619–1620.
168. E. I. Shakhnovich, *Fold Des.*, 1998, **3**, R45–R58.
169. J. G. Saven, *Chem. Rev.*, 2001, **101**, 3113–3130.
170. K. Yue, K. M. Fiebig, P. D. Thomas, H. S. Chan, E. I. Shakhnovich and K. A. Dill, *Proc. Natl. Acad. Sci. USA*, 1995, **92**, 325–329.
171. J. W. Bryson, J. R. Desjarlais, T. M. Handel and W. F. DeGrado, *Protein Sci.*, 1998, **7**, 1404–1414.
172. A. R. Dinner, V. Abkevich, E. Shakhnovich and M. Karplus, *Proteins Struct. Funct. Genet.*, 1999, **35**, 34–40.
173. D. K. Klimov and D. Thirumalai, *J. Chem. Phys.*, 1998, **109**, 4119–4125.
174. J. M. Zou and J. G. Saven, *J. Mol. Biol.*, 2000, **296**, 281–294.
175. R. A. Goldstein, Z. A. Lutheyschulten and P. G. Wolynes, *Proc. Natl. Acad. Sci. USA*, 1992, **89**, 9029–9033.
176. S. Park, Y. Xi and J. G. Saven, *Curr. Opin. Struct. Biol.*, 2004, **14**, 487–494.
177. S. Park, X. Fu Stowell, W. Wang, X. Yang and J. G. Saven, *Annu. Rep. Prog. Chem., Sect. C*, 2004, **100**, 195–236.
178. L. Wernisch, S. Hery and S. J. Wodak, *J. Mol. Biol.*, 2000, **301**, 713–736.
179. P. Koehl and M. Levitt, *J. Mol. Biol.*, 1999, **293**, 1161–1181.
180. P. Koehl and M. Levitt, *J. Mol. Biol.*, 1999, **293**, 1183–1193.
181. J. J. Havranek and P. B. Harbury, *Nat. Struct. Biol.*, 2003, **10**, 45–52.
182. H. W. Hellinga and F. M. Richards, *Proc. Natl. Acad. Sci. USA*, 1994, **91**, 5803–5807.
183. X. Yang and J. G. Saven, *Chem. Phys. Lett.*, 2005, **401**, 205–210.
184. R. F. Goldstein, *Biophys. J.*, 1994, **66**, 1335–1340.
185. D. B. Gordon and S. L. Mayo, *J. Comput. Chem.*, 1998, **19**, 1505–1514.

186. C. A. Voigt, D. B. Gordon and S. L. Mayo, *J. Mol. Biol.*, 2000, **299**, 789–803.
187. J. Zou and J. G. Saven, *J. Chem. Phys.*, 2003, **118**, 3843–3854.
188. A. M. Slovic, H. Kono, J. D. Lear, J. G. Saven and W. F. DeGrado, *Proc. Natl. Acad. Sci. USA*, 2004, **101**, 1828–1833.
189. D. N. Bolon, C. A. Voigt and S. L. Mayo, *Curr. Opin. Chem. Biol.*, 2002, **6**, 125–129.
190. J. Kaplan and W. F. Degrado, *Proc. Natl. Acad. Sci. USA*, 2004, **101**, 11566–11570.
191. L. L. Looger, M. A. Dwyer, J. J. Smith and H. W. Hellinga, *Nature*, 2003, **423**, 185–190.
192. T. Kortemme, L. A. Joachimiak, A. N. Bullock, A. D. Schuler, B. L. Stoddard and D. Baker, *Nat. Struct. Mol. Biol.*, 2004, **11**, 371–379.
193. L. Baltzer and W. F. DeGrado, *Curr. Opin. Struct. Biol.*, 2004, **14**, 455–457.
194. S. Park, H. Kono, W. Wang, E. T. Boder and J. G. Saven, *Comp. Chem. Eng.*, 2005, **29**, 407–421.
195. O. Schueler-Furman, C. Wang, P. Bradley, K. Misura and D. Baker, *Science*, 2005, **310**, 638–642.
196. H. Nguyen, M. Jager, J. W. Kelly and M. Gruebele, *J. Phys. Chem. B*, 2005, **109**, 15182–15186.
197. C. D. Snow, E. J. Sorin, Y. M. Rhee and V. S. Pande, *Annu. Rev. Biophys. Biomol. Struct.*, 2005, **34**, 43–69.
198. V. Muñoz and W. A. Eaton, *Proc. Natl. Acad. Sci. USA*, 1999, **96**, 11311–11316.
199. K. W. Plaxco, K. T. Simons and D. Baker, *J. Mol. Biol.*, 1998, **277**, 985–994.
200. W. Y. Yang and M. Gruebele, *Nature*, **423**, 193–197.
201. M. Gruebele, *Curr. Opin. Struct. Biol.*, 2002, **12**, 161–168.
202. B. Gillespie, D. M. Vu, P. S. Shah, S. A. Marshall, R. B. Dyer, S. L. Mayo and K. W. Plaxco, *J. Mol. Biol.*, 2003, **330**, 813–819.
203. Y. Zhu, D. O. Alonso, K. Maki, C. Y. Huang, S. J. Lahr, V. Daggett, H. Roder, W. F. DeGrado and F. Gai, *Proc. Natl. Acad. Sci. USA*, 2003, **100**, 15486–15491.
204. J. Kubelka, T. K. Chiu, D. R. Davies, W. A. Eaton and J. Hofrichter, *J. Mol. Biol.*, 2006, **359**, 546–553.
205. Y. Zhu, X. Fu, T. Wang, A. Tamura, S. Takada, J. G. Saven and F. Gai, *Chem. Phys.*, 2004, **307**, 99.
206. M. Sadqi, L. J. Lapidus and V. Muñoz, *Proc. Natl. Acad. Sci. USA*, 2003, **100**, 12117–12122.
207. M. R. Bunagan, X. Yang, J. G. Saven and F. Gai, *J. Phys. Chem. B*, 2006, **110**, 3759–3763.
208. L. L. Qiu, S. A. Pabit, A. E. Roitberg and S. J. Hagen, *J. Am. Chem. Soc.*, 2002, **124**, 12952–12953.
209. A. Jaramillo, L. Wernisch, S. Hery and S. J. Wodak, *Proc. Natl. Acad. Sci. USA*, 2002, **99**, 13554–13559.

210. K. Raha, A. M. Wollacott, M. J. Italia and J. R. Desjarlais, *Protein Sci.*, 2000, **9**, 1106–1119.
211. A. Su and S. L. Mayo, *Protein Sci.*, 1997, **6**, 1701–1707.
212. B. Kuhlman and D. Baker, *Curr. Opin. Struct. Biol.*, 2004, **14**, 89–95.
213. L. Di Costanzo, H. Wade, S. Geremia, L. Randaccio, V. Pavone, W. F. DeGrado and A. Lombardi, *J. Am. Chem. Soc.*, 2001, **123**, 12749–12757.
214. A. G. Cochran, N. J. Skelton and M. A. Starovasnik, *Proc. Natl. Acad. Sci. USA*, 2001, **98**, 5578–5583.
215. D. E. Benson, M. S. Wisz, W. Liu and H. W. Hellinga, *Biochemistry*, 1998, **37**, 7070–7076.
216. D. A. Doyle, J. Morais Cabral, R. A. Pfuetzner, A. Kuo, J. M. Gulbis, S. L. Cohen, B. T. Chait and R. MacKinnon, *Science*, 1998, **280**, 69–77.
217. Y. F. Zhou, J. H. Morais-Cabral, A. Kaufman and R. MacKinnon, *Nature*, 2001, **414**, 43–48.
218. J. Bronson, O. S. Lee and J. G. Saven, *Biophys. J.*, 2006, **90**, 1156–1163.
219. X. R. Fu, H. Kono and J. G. Saven, *Protein Eng.*, 2003, **16**, 971–977.
220. J. Swift, W. A. Wehbi, B. D. Kelly, X. Fu Stowell, J. G. Saven, and I. J. Dmochowski, *J. Am. Chem. Soc.*, 2006, **128**, 6611–6619.
221. J. G. Saven, *Curr. Opin. Struct. Biol.*, 2002, **12**, 453–458.
222. R. J. Hayes, J. Bentzien, M. L. Ary, M. Y. Hwang, J. M. Jacinto, J. Vielmetter, A. Kundu and B. I. Dahiyat, *Proc. Natl. Acad. Sci. USA*, 2002, **99**, 15926–15931.
223. J. D. Bloom, M. M. Meyer, P. Meinhold, C. R. Otey, D. MacMillan and F. H. Arnold, *Curr. Opin. Struct. Biol.*, 2005, **15**, 447–452.
224. G. L. Moore and C. D. Maranas, *Nucleic Acids Res.*, 2002, **30**, 2407–2416.
225. C. A. Voigt, C. Martinez, Z. G. Wang, S. L. Mayo and F. H. Arnold, *Nat. Struct. Biol.*, 2002, **9**, 553–558.
226. G. L. Moore and C. D. Maranas, *Proc. Natl. Acad. Sci. USA*, 2003, **100**, 5091–5096.
227. S. Lutz, M. Ostermeier, G. L. Moore, C. D. Maranas and S. J. Benkovic, *Proc. Natl. Acad. Sci. USA*, 2001, **98**, 11248–11253.
228. S. Lutz, M. Ostermeier and S. J. Benkovic, *Nucleic Acids Res.*, 2001, **29**, E16.
229. M. Ostermeier, A. E. Nixon, J. H. Shim and S. J. Benkovic, *Proc. Natl. Acad. Sci. USA*, 1999, **96**, 3562–3567.
230. M. Ostermeier and S. J. Benkovic, *Nat. Biotechnol.*, 1999, **17**, 639–640.
231. S. H. Gellman, *Acc. Chem. Res.*, 1998, **31**, 173–180.
232. D. J. Hill, M. J. Mio, R. B. Prince, T. S. Hughes and J. S. Moore, *Chem. Rev.*, 2001, **101**, 3893–4012.
233. G. M. Bender, A. Lehmann, H. Zou, H. Cheng, H. C. Fry, D. Engel, M. J. Therien, J. K. Blasie, H. Roder, J. G. Saven and W. F. DeGrado, *J. Am. Chem. Soc.*, 2007, **129**, 10732–10740.

CHAPTER 10
Protein Misfolding and β-Amyloid Formation

ALEXANDRA ESTERAS-CHOPO, MARIA TERESA PASTOR AND LUIS SERRANO

European Molecular Biology Laboratory, Meyerhofstrasse 1, D-69117 Heidelberg, Germany

10.1 Introduction

One of the most basic biological processes is the folding of a linear sequence of amino acids into the three-dimensional structure of a functional protein. As there are 20 different types of amino acids and due to the wide range of protein sizes, the number of possible random amino acid sequences exceeds the estimates of the total number of atoms of the universe. Therefore, natural proteins are a very select group of molecules, with very special characteristics that differentiate them from the random amino acid sequences. Two of their most important features are their ability to fold in unique structures and their ability to generate a wide range of functions. Protein functions include the control and regulation of essentially every chemical process which our lives depend on, as well as the provision of key components to virtually all the structural frameworks within our bodies. Although the code that governs protein folding remains a mystery, we do know that primary sequence is subject to evolutionary pressure to maintain functionality, and by extension to fold into a stable structure. In the last few years, this structure–function dogma has been expanded by the discovery of functional unstructured proteins and of unstructured regions in many functional proteins, both linked to very important cellular processes.[1] Many of these intrinsically unfolded proteins or regions undergo transitions to more structured states upon binding to their target.

RSC Biomolecular Sciences
Protein Folding, Misfolding and Aggregation: Classical Themes and Novel Approaches
Edited by Victor Muñoz
© Royal Society of Chemistry 2008

However, under certain circumstances, some of the members of this limited group can fail in their correct folding process, leading to a functional deficit that can have serious consequences at the organism level. More recently, an emerging class of late-onset, slow-progressing diseases appears to result from a gain (rather than a deficit) of function associated with an abnormally folded form of the protein. These diseases, which are characterized by ordered, fibrillar aggregates of protein known as amyloid fibrils, include common neurodegenerative pathologies such as Alzheimer's (AD) and Parkinson's disease (PD) as well as many rare systemic diseases such as familial amyloid polyneuropathy (Table 10.1). Interestingly, some intrinsically disordered proteins, like the Aβ

Table 10.1 A summary of the main human amyloidoses, the organs affected, the proteins or peptides involved, and the localization of the deposits associated with every disease.

Clinical syndrome	Organ affected	Plaque components	Cellular localization
Alzheimer's disease	Brain: cerebral cortex, hippocampus	Amyloid β peptide	Extracellular
		Tau protein	Tangles in neuronal cytoplasm
Parkinson's disease	Brain: substantia nigra, hypothalamus	α-Synuclein	Neuronal cytoplasm
Polyglutamine expansion disease (*e.g.* Huntington's disease)	Brain: striatum, cerebral cortex	Long glutamine stretches within certain proteins *e.g.* Huntington	Neuronal nuclei and cytoplasm
Spongiform encephalopathy	Brain: cortex, thalamus, brain stem, cerebellum	Prion protein	Extra- and intracellular
Type II diabetes	Pancreas, insulinomas	Islet amyloid polypeptide	Extracellular
Familial amyloidotic polyneuropathy 1	Systemic; peripheral nerves, heart, vitreous	Mutant transthyretin and fragments thereof	Extracellular
Senile systemic amyloidosis	Systemic; cardiac	Wild-type transthyretin and fragments thereof	Extracellular
Haemodialysis-related amyloidosis	Systemic; joints, bones, liver, tongue, lungs	β$_2$-microglobulin	Extracellular
Finnish hereditary amyloidosis	Systemic; ocular	Fragments of mutant gelsolin	Extracellular
Hereditary systemic amyloidosis	Systemic: renal and visceral disease; organomegaly	Mutant lysozyme	Extracellular

peptide, islet amyloid polypeptide (IAPP), tau and α-synuclein, can also enter this misfolding pathway and form ordered amyloid fibrils.

Despite active research in the last few years, there are still many open questions regarding amyloid fibril formation such as the factors that determine amyloid formation by normally soluble proteins, the mechanism of toxicity associated with these diseases, and the structure of the mature amyloid fibrils and intermediate species appearing in this process. Efforts from different areas of life and medical sciences are being directed at creating an integrated picture of the different processes associated with these diseases with the final aim of identifying potential therapeutical strategies. Here we discuss the molecular aspects of protein misfolding and amyloid formation with an emphasis on the experimental approaches (see Chapter 11 by Dima and co-workers for an account of results obtained with computational methods). In the first section we describe the general principles underlying protein aggregation into β-amyloids. The second section focuses on some of our own findings obtained with a minimalist approach consisting in studying the *in vitro* aggregation of very small peptides as model systems for amyloid formation.

10.2 General Principles of Amyloid Formation

10.2.1 Historical Perspective

The understanding of amyloid fibril formation has advanced as better technology has become available. Amyloid deposits were detected for the first time and named by Rudolph Virchow in 1854 by iodine staining of brain sections with an abnormal macroscopic appearance.[2] Initially, light microscopy and histopathological dyes such as Thioflavin T (ThT) and Congo red (CR) were used in their characterization. The phenomenon of positive birefringence of the amyloid deposits upon Congo red binding introduced the possibility of an ordered submicroscopic structure. In 1959, electron microscopy studies of different amyloid tissues demonstrated that all of them exhibit a comparable fibrillar structure in fixed tissue sections.[3] The association of a fibrillar structure with the tinctorial properties exhibited by amyloid deposits provided the basis for amyloid fibril isolation. The first isolated amyloid fibrils showed morphology and properties similar to the ones observed in tissues.[4,5] This finding, together with the improvement of the extraction methods and introduction of techniques to solve the proteins forming the fibrils, opened the era of biochemical studies. Sequencing of the proteins isolated from different amyloid deposits demonstrated that each of the clinical syndromes is associated with a different protein (Table 10.1). The discovery that amyloid fibrils formed *in vitro* from synthetic or recombinant polypeptides and proteins are similar to the ones formed *in vivo* by natural amyloids unlocked the possibilities of carrying out biophysical characterization of the process and tackling the challenge of unraveling the structure and organization of amyloid fibrils.[2]

10.2.2 Molecular Basis of Amyloidosis: Protein Misfolding

10.2.2.1 In vitro Studies

In the early 1970s, the discovery that lysosomal extracts are sufficient to convert amyloidogenic proteins into amyloid fibrils introduced the assumption that proteolytic processing was the amyloidogenic determinant.[6,7] However, based on many evidences from independent investigations, it is now widely accepted that protein misfolding is the molecular basis of all the amyloid related disorders.[8–10] Conformational changes of the natively folded or unfolded protein have been demonstrated to precede the formation of amyloid structures. Nevertheless, proteolysis still plays an important role in the generation of aberrant protein fragments highly prone to amyloid formation. For example, the amyloid precursor protein (APP) does not undergo fibril formation while its proteolysis product, the Aβ peptide, is the major component of the extracellular amyloid plaques associated with AD.[11]

Under certain conditions, all the amyloidogenic proteins can adopt an amyloid prone state that favors the intermolecular interactions leading to the formation of oligomeric species. A natively folded polypeptide chain loses or is unable to attain its closely packed three-dimensional structure, populating non-correctly folded states in equilibrium with each other (Figure 10.1).[9] It has been proposed that natively unfolded proteins associated with amyloid diseases (Aβ, tau, α-synuclein, IAPP) can adopt a metastable, partially structured conformation that is stabilized by oligomerization.[10] In these amyloid prone states, the protein can nucleate initial oligomeric assemblies where the content of secondary β structure is generally increased. These "seeds" or "nuclei" provide a sort of template where other misfolded or partially folded molecules are recruited, thereby increasing the size of the assemblies that finally give rise to the fibrils.[13]

10.2.2.2 Protein Misfolding in the Cell

To put the misfolding and conformational hypothesis in a cellular context some other cellular events associated with the life of a protein have to be taken into account (Figure 10.1). In eukaryotic cells, the vast majority of proteins fold either in the cytosol or in the endoplasmic reticulum. These environments are much more complex compared to the one associated with most experiments conducted *in vitro*. For example, the cytosol is so densely packed with all the molecular components that are needed for survival and replication that the macromolecular concentration can exceed $350\,\text{mg}\,\text{ml}^{-1}$.[14] This high degree of molecular crowding means that incompletely or improperly folded molecules will probably aggregate with each other or associate improperly with other cellular components. To avoid this problem, a series of auxiliary systems have evolved to assist proteins to fold efficiently.[15] These species involve "folding catalysts" that accelerate slow steps in protein folding, such as the formation of disulfide bonds or the isomerization of X_{aa}-proline peptide bonds, and "molecular chaperones" that act to avoid the consequences of protein misfolding

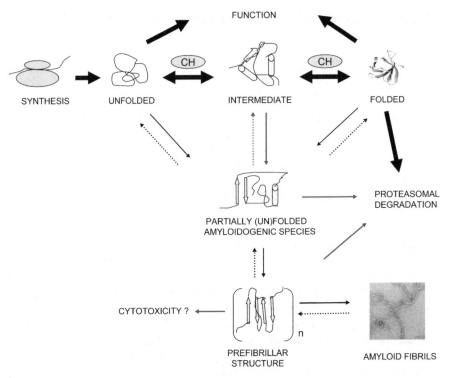

Figure 10.1 Schematic representation of protein (mis)folding in the cell. Thick arrows represent physiological processes leading to normal protein function. CH: chaperones and folding catalysts. Thin arrows correspond to the events involved in amyloid formation. Grey thin arrows indicate steps of the amyloid process that are not yet well established as the identity of the amyloidogenic precursor species, the cytotoxic species, or the species that can be targeted for degradation. Dashed arrows indicate non-favored processes. For the sake of simplicity, formation of disordered aggregates has not been included. Adapted from Ref. 12.

and aggregation. In addition, there is also a series of quality control mechanisms to check whether proteins are correctly folded. Most of these take place in the endoplasmic reticulum, the major folding compartment of eukaryotic cells.[16] Protein molecules that do not meet the quality requirements are targeted for destruction. The best-characterized degradation mechanisms are part of the "unfolded protein response," which involves ubiquitination of proteins destined for disposal, followed by their destruction in the cytosol by the proteasome.[17]

10.2.2.3 Conditions Promoting Amyloid Formation in vivo *and* in vitro

In the cellular environment, the onset of aggregation may be triggered by any factor that results in a rise of the concentration of the amyloidogenic precursor, thereby shifting the equilibrium between correctly and partially folded molecules (Figure 10.1). Protein mutations, environmental changes, and chemical modifications have been reported to favor amyloid formation in several pathological proteins.

Mutations could modulate the extent of amyloid formation by reducing the conformational stability of the protein, or by specifically increasing the amyloid propensity of the polypeptide sequence. For example, two mutations in lysozyme associated with hereditary systemic amyloidosis[18] seem to facilitate amyloid formation by reducing the stability of the globular protein, as revealed by a series of *in vitro* studies.[19] Mutations in the Aβ peptide associated with AD phenotypes have been demonstrated to accelerate amyloid formation *in vitro*.[20] Other mutations in the APP increase the Aβ peptide production roughly twofold, as has been shown in transfected cells, transgenic mice, and in the affected individuals themselves.[21] The total amount of peptide would increase, along with the population comprising partially folded molecules. Also, inherently amyloidogenic proteins at normal concentration can lead to amyloid deposits after very prolonged periods of time. This is the case of transthyretin (TTR) in senile systemic amyloidosis. This disease occurs with increasing frequency after the age of 70 years, becoming almost universal above the age of 90 years.[22]

Interaction with cellular components may also favor amyloid formation. Many proteins have been reported to be associated with amyloid deposits found in AD and some other central nervous system (CNS) and systemic disorders. *In vivo* and *in vitro* studies have shown that many of these elements can regulate Aβ amyloid formation (reviewed in reference 23). There is some evidence that implicates proteoglycans in the pathophysiology of amyloid, probably through the promotion and stabilization of fibrils.[24] Some lines of evidence propose that Aβ production takes place in the late endosomal/lysosomal system.[25] The pH of these compartments (late endosome 5-5.5, lysosome 4.5) fits very well with the observation that Aβ amyloidogenesis is favored at acidic pH.[26] Metal ions also seem to be related with an acceleration of fibril formation *in vivo*. The Aβ peptide has been shown to bind to Cu^{2+}, Fe^{2+}, and Zn^{2+}. Metal binding induces a β-sheet-like conformational change in Aβ, resulting in enhanced fibril formation.[11]

Another factor that can favor amyloid formation is covalent post-translational modifications of proteins. Amyloid formation by two intrinsically disordered proteins, tau and α-synuclein, seems to be promoted by phosphorylation. Analysis of the deposits formed by both proteins has shown that they contain mainly the hyperphosphorylated form.[27]

Amyloid formation could be also favored by conditions that increase the accumulation of misfolded intermediates, such as impairment of the quality control machinery. It has been demonstrated that inactivating mutations of any of the components of the quality control or harsh environmental conditions such as heat shock, oxidative stress, or chemical modification may impair the activity of the clearing machinery and/or increase the number of misfolded or unfolded proteins, overwhelming the capacity of molecular chaperones and the proteasome.[17] Changes in the activity levels of this quality control associated with ageing could be also associated with the TTR deposits mentioned above.

Experimental conditions can be designed to force the self-assembly of different non-pathogenic proteins *in vitro*.[28] This can happen under conditions which are partially destabilizing (acidic pH values, high temperature, lack of ligands, or moderate concentrations of salts or co-solvents such as trifluoroethanol) where the tertiary interactions are destabilized, whereas the secondary contacts are still favored. Nevertheless, many of these conditions are quite unlikely to occur *in vivo*.

10.2.2.4 The Mechanisms of Amyloid Formation

Amyloids are considered the final state of a nucleated polymerization process that correlates macroscopically with the formation of insoluble fibrillar structures. The process of amyloid fibril formation is characterized by three different phases (Figure 10.2A):

a) *Slow nucleation phase* or *lag phase*. During this time, the conformational transition of the soluble precursor is supposed to take place to generate the amyloid prone species (see Chapter 11 by Dima *et al.* for more details on nucleation and related conformational changes).

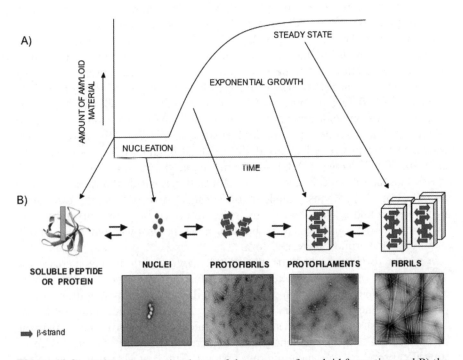

Figure 10.2 A) The three main phases of the process of amyloid formation and B) the different species associated: schematic representation and electron micrographs.

Protein Misfolding and β-Amyloid Formation

b) *Growth phase*, in which the nucleus grows to form larger polymers. This step does not occur until the amyloid prone species is at concentrations above a certain level (*i.e.* the critical concentration). Addition of exogenous nuclei or seeds can accelerate the normally slow nucleation step. Furthermore, increasing protein concentration can reduce as well the time span before amyloid formation commences.

c) A *steady-state phase*, in which the ordered fibril and the monomer appear to be in dynamic equilibrium.[13]

The study of amyloid fibril formation from different amyloid proteins and peptides has revealed the presence of intermediate species along the pathway. Electron microscopy (EM), sedimentation, and atomic force microscopy (AFM) have been used for the characterization of these intermediate species.[29] Figure 10.2B shows schematic representations and EM micrographs of the different species associated with every phase of the nucleation dependent process. The first stage can be depicted as "seeds" or nuclei and these are usually described as being spherical or globular in appearance.[30] These globular particles associate over time to form protofibrils. Protofibrils were identified to appear transiently during Aβ peptide fibrillogenesis[31,32] and seem to be precursors of full-length fibrils. These precursors generally have a curved appearance and are shorter than the mature fibrils. Interest in the structure of these early species has grown in recent years, since they appear to be directly involved in the mechanism of toxicity (see below). These protofibrils evolve to form protofilaments that pack together forming the final mature amyloid fibrils. All these species can co-exist in solution, and only at the plateau phase is the equilibrium of the mature fibrillar form and monomeric forms reached. It is important to remark that protein self-association into β-sheets can lead to products of different quaternary structure. Amyloid formation refers to the formation of ordered fibrillar material with β-sheet structure. Protein aggregation refers to non-ordered β-sheet assemblies with amorphous morphology. Both processes can co-occur in amyloid-related diseases since nuclei could also be considered as amorphous aggregates but they are not necessarily linked. Recent studies have aimed at revealing the sequence determinants directing a protein sequence towards the amyloid or the amorphous aggregation pathway.[33–35]

10.2.3 The Structural Architecture of Amyloid Fibrils

Amyloid fibrils from *in vivo* deposits or even fibrils that have been assembled *in vitro* are difficult to study using conventional structural techniques. Methods such as single crystal X-ray crystallography and solution nuclear magnetic resonance (NMR) cannot be used on fibrils since they are insoluble and non-crystalline. Hence, high-resolution structures of amyloid fibrils still remain an open challenge. The strategy used in the field has been the combination of low- to moderate-resolution data from different techniques to build models of the structure of mature fibrils. All mature amyloid fibrils exhibit a similar hierarchical organization that can be dissected into different levels.

10.2.3.1 Supramolecular Structure

Simple EM micrographs of mature amyloid fibrils reveal the presence of fibrillar subunits denominated protofilaments, which pack up to constitute the fibril. More sophisticated techniques such as image reconstruction of cross-sectional EM images[36,37] or a combination of cryo-electron microscopy with single particle averaging methods[38–40] have been used to study in greater detail the protofibrillar structure of a wide range of *ex vivo* and synthetic fibrils. In all the models proposed a variable number of protofilaments twist around each other. However, the number and the arrangement (*e.g.* left or right winding) of the protofilaments might differ even within the same sample.

10.2.3.2 The Protofilament Core

X-Ray fibre diffraction has been used to examine the internal structure of amyloid fibrils. It has been shown that all amyloid fibrils possess a characteristic cross-β structure (Figure 10.3). The β-sheets are hydrogen bonded along the length of the fibrils, and the β-strands run perpendicularly to the long axis of the fibril. Whether these are in parallel or anti-parallel arrangement is still under debate. However, the detailed analysis of X-ray diffraction patterns for different *ex vivo* and synthetic fibril preparations has allowed the proposal of a model for the organization of the protofilament core.[41,42] In this model the protofilaments are composed of four β-sheets (the number of β-sheets might change depending on the amyloid protein) running parallel to the axis of the protofilament, whereas their component β-strands are almost orthogonal to the axis.

Another intriguing question regarding amyloid structure is how peptides and proteins of very different lengths ranging from six to hundreds of amino acids can self-assemble into similar amyloid structures. Several models propose that the whole amino acid sequence refolds into a parallel β-helix to form the core of the amyloid structure (Figure 10.4A).[43–46] Adjacent strands of the helix are connected by H-bonds as in a normal β-sheet, while any part of the sequence incompatible with the architecture can be accommodated as loops between the core forming β-strands.[47,48]

Other models propose that only a segment of the protein, the so-called *amyloid domain*, forms an *amyloid core* that is surrounded by the rest of the protein (Figure 10.4B). The folding state of the globular appendages is unknown and might differ depending on the amyloid protein. These models are based on the hypothesis that amyloid formation is the result of a "gain of interaction" (GOI) due to the formation of a new intermolecular bond, which is contributed by a region of the protein.[49] Observations of 3D domain swapping led Eisenberg and co-workers to propose a particular case of this GOI model

Protein Misfolding and β-Amyloid Formation

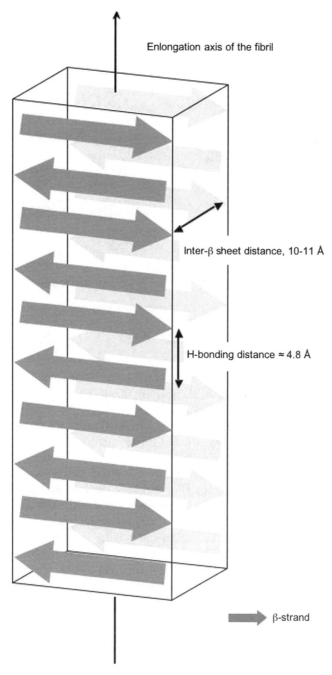

Figure 10.3 Schematic representation of the cross-β motif. The arrangement of the β-strands and β-sheets can differ depending on the amyloid sequence.

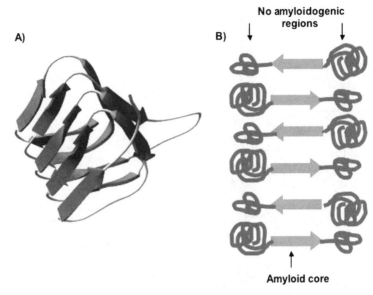

Figure 10.4 Models of the primordial structure of amyloid fibrils. A) β-helix structure. B) Schematic representation of the "zipper spine" model proposed by Eisenberg and co-workers. Adapted from reference 52.

called the "zipper model".[50] In this model a specific region of a protein binds to the same region in another molecule, forming the β-sheet spine of the amyloid, while the rest of the molecule decorates the periphery.[51,52]

Recent work by the Eisenberg group has resulted in the first crystal structure of an amyloid fibril, in this case formed by a heptapeptide (GNNQQNY) and a related hexapeptide (NNQQNY) from the yeast prion protein Sup35.[53] The structure shows a β-sandwich formed by parallel β-sheets with anti-parallel arrangement. The face-to-face interface between the β-sheets is described as a dry interface, where the side chains are interdigitated with their counterparts in the mating β-sheet. According to the "zipper model" this structure should represent the organization of the β-sheet spine of the amyloid fibrils. This study has been expanded with the structure of more amyloid forming fragments that also show steric zippers but with variations depending on the sequence.[54] These works represent a major step towards the determination of a high-resolution structure of amyloid fibrils.

10.2.4 Amyloid Induced Toxicity

Nowadays, there is strong evidence indicating that amyloid formation is the cause or at least a central event in the pathogenesis of amyloidoses, not a consequence of the disease.[27] During the past two decades, several animal models such as the nematode *Caenorhabditis elegans*,[55] the fruit fly *Drosophila*

Protein Misfolding and β-Amyloid Formation

melanogaster,[56,57] and the mouse *Mus musculus*[57–59] have been used to study the expression of the wild type and mutant versions of the principal amyloidogenic proteins. These animal models recapitulate some of the symptoms associated with those diseases such as neuron death, movement disorders, and learning impairment. Genetic evidence also linked mutations in some genes with familial forms of these diseases.[27] In fact, point mutations of amyloidogenic proteins associated with early onset of AD have been shown to accelerate amyloid formation *in vitro*.[20]

10.2.4.1 Pathogenic Species

The common presence of amyloid fibril deposits associated with all amyloid-related pathologies initially suggested that mature amyloid fibrils were the species responsible for cellular impairment and cell death. However, studies with *post-mortem* material have shown that accumulation of amyloid fibrils in neurons does not always correlate with cell degeneration and clinical symptoms.[60,61] Based on this and some other related results, mature fibrils are currently regarded as inert end products of the amyloid fibril formation reaction.[62] However, given that the process of amyloid formation is associated with amyloid disorders, the toxic agents must occur at a certain stage of this process. Hence, one or several of the fibrillation intermediates could be the species responsible for amyloid toxicity. This emerging idea is supported by several observations. For example, incubation of cell cultures with different intermediates of the fibrillation pathway resulted in different rates of cell death.[63–65] Microinjection of oligomers of the Aβ peptide disrupted cognitive function in mice.[66] Furthermore, additional support has arrived from non-pathogenic proteins that are able to form amyloid like fibrils *in vitro*. Small prefibrillar aggregates formed during their self-assembly were shown to be toxic in cell cultures. These cytotoxic species showed common structural features suggesting that toxicity could be inherent to these prefibrillar aggregates, thus implying a common pathogenic mechanism for all amyloidoses.[63,67]

10.2.4.2 Proposed Mechanisms of Amyloid Toxicity

Many authors believe that the shared structural features of amyloid aggregates, both at the level of protofibrils and mature fibrils, could be reflected into common toxicity mechanisms. Although considerable efforts are currently directed towards deciphering the cytotoxic amyloid induced pathway, it remains still unknown.

As previously mentioned, there is significant evidence suggesting that ordered intermediates on the pathway to fibril formation are responsible for cell dysfunction and death.[63,65,66,68–70] One hypothesis about the mechanism of amyloid-induced toxicity proposes that the pathogenecity of these species would arise from a gain of function of the misfolded protein. In this mechanism, the toxic intermediates can interact with cellular components such as the proteasome,[71] chaperones,[72] or the plasma membrane. In fact, the current main view in the field points to the plasma membrane as the primary target of amyloid toxicity.[64,73,74] Several mechanisms have been proposed to explain how the toxic species might be causing membrane perturbation. The "channel hypothesis" postulates that protofibrils would form non-specific membrane pores resulting in unbalance of the cellular ion content.[75] Recently, several authors have proposed that membrane perturbation could also take place through an independent pore formation mechanism, such as change in membrane fluidity[73,76,77] or in membrane conductance.[74] Another related pathway is based on the finding that interactions of Aβ oligomers with Cu^+ or Fe^{2+} generates H_2O_2.[78] Lipid and protein peroxidation induced by this process could impair the normal function of membrane proteins such as ATPases, or glucose and glutamate transporters.[79] All these different toxic pathways could be acting synergistically. By disturbing both ion homeostasis and energy metabolism, relatively low levels of membrane-associated oxidative stress can render neurons vulnerable to cytotoxicity and apoptosis.

In addition to mechanisms related to the gain of function, amyloid pathogenesis could be caused as well by the loss of function of the soluble protein or even by fibril accumulation. It has been described that fibril accumulation can produce chronic inflammatory responses,[80] as well as damage to organs by interfering with the proper flow of nutrients to the cells or by sheer weight.[22] All these hypotheses, and many others not described here, might simply be different consequences of the formation of toxic amyloid species. Furthermore, the dominant mechanism or mechanisms may depend on the cellular type or tissue affected.

10.2.5 Experimental Techniques to Study Amyloid Formation

To be considered amyloid, any proteinaceous fibrillar material must fulfill the following characteristics:

a) Fibrils show straight unbranched fibrillar morphology as detected by EM, widths of ~7–12 nm and indeterminate length
b) Fibrils show a cross-β X-ray diffraction pattern
c) Fibrils bind dyes, such as CR and ThT
d) Protein solutions display a polymeric β-sheet CD signature.

None of these features alone is sufficient to unambiguously determine the amyloid nature of the aggregates. A combination of techniques is advisable to increase the accuracy of the classification. The identification and characterization of amyloid material is usually performed by microscopy (EM and AFM) and spectroscopic techniques. All these techniques provide low-resolution data about amyloid formation, but are generally easy and inexpensive. The choice of the appropriate technique depends on several factors, such as protein availability, solubility, concentration requirements, access to instrumentation, and

Protein Misfolding and β-Amyloid Formation 227

expertise. These are the most commonly used techniques in biochemistry laboratories:

Electron Microscopy: EM, which has a resolution of approximately 2 nm, is typically used for the identification and morphological characterization of the nuclei, filament, and fibrils that appear during amyloid fibril growth. Negative staining techniques do not require long sample preparation.[81] However, the sample is in a dry state that might affect fibril morphology.

Atomic Force Microscopy: AFM is a method of imaging surfaces with nanometre resolution. It requires even less sample preparation than EM and can be used in a continuous mode. *In situ* AFM of Aβ (1–42) fibrillation has allowed us to follow the time course of fibril formation in buffered solutions.[82] This study also shows how different surfaces (hydrophilic mica or hydrophobic graphite) can affect the behavior of Aβ in different environments, a factor that must be taken into account during evaluation of the results.

Circular Dichroism: CD spectroscopy measures the difference in absorbance of right- and left-circularly polarized light as a function of the wavelength. Far-UV CD spectroscopy (180–250 nm) provides a very convenient method to monitor secondary structure in solution. Its application to the study of amyloid formation relies on the common conformational transition of amyloid proteins from their native structures to a polymeric β-sheet (Figure 10.5A). Increasing β-sheet content is associated with fibrillar morphology, relative insolubility, and protease resistance. The polymeric β-sheet signature is characterized by a minimum between 215 and 220 nm and a maximum between 195 and 202 nm (Figure 10.5A). Near-UV CD (250–320 nm) detects primarily the presence of tertiary interactions involving aromatic residues. It can be used to obtain

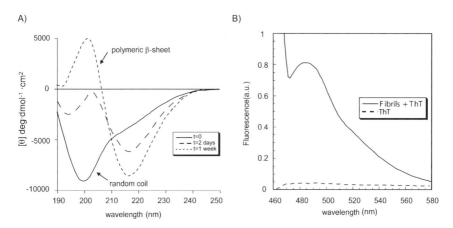

Figure 10.5 Spectroscopic techniques to study amyloid fibril formation. A) Conformational transition followed by far-UV CD of the amyloidogenic sequence STVIIE from random coil ($t = 0$) to the polymeric β-sheet spectra ($t = 1$ week). B) Enhancement of the ThT fluorescent emission spectra upon binding to mature STVIIE fibrils.

information about changes in the folding state of the globular domain of proteins upon fibril formation.

CD has many advantages to monitor β-sheet self-association as it is fast and allows recovery of the sample, which can afterwards be analysed by other techniques. However, some of the experimental conditions normally used to induce amyloid formation *in vitro* such as high salt concentrations or organic solvents, *e.g.* DMSO, can prevent CD measurements. High scattering from already formed fibrils and fibril deposition during the acquisition time are other factors that can make CD measurements of amyloid formation difficult.

Dye binding techniques: CR and ThT are two dyes widely used to diagnose the amyloid nature in protein deposits *ex vivo*. Their use has also been extended to monitor the formation of amyloid structures in solution. CR binding to the cross-β structure of the amyloid fibril induces a change in the UV absorption spectrum of the dye. Upon binding to amyloid fibrils, there is an increase of CR absorbance around 540 nm.[83] The binding of ThT to amyloid fibrils leads to a strong enhancement of the ThT fluorescence emission at 485 nm following excitation at 440 nm (Figure 10.5B).[84] As dye binding depends on the amount of amyloid fibrils, both assays can also be used to quantify amyloid formation *in vitro*.

Nevertheless, it has to be taken into consideration that the magnitude of the effect varies as a function of fibril morphology, and even non-fibrillar aggregates can show dye binding.[85] Therefore, they cannot be considered as a definitive proof of amyloid nature, but rather as a quantitative measure of this process once the amyloid nature had been proven by some other technique.

10.2.6 Cytotoxicity Studies

To study amyloid-induced toxicity, a simple and easy system that is able to reproduce the pathological effects observed in patients would be desirable. Although animal models of amyloid disease are very useful tools to study amyloid diseases, obtaining these model systems is complicated and time consuming. An alternative is the use of cellular systems such as primary neuron cultures to study neurodegenerative disorders.[75] However, obtaining and maintaining these cultures is also cumbersome. To avoid these problems, PC12 cells have become a widely used alternative system to study amyloid toxicity since they are easily handled. The PC12 pheochromocytoma line is a clonal line derived from a rat adrenal medullary tumor. In the presence of nerve growth factor (NGF), PC12 cells cease to multiply and differentiate into sympathetic-like neurons that allow their use as a useful model system to study amyloid-induced toxicity. Conclusions obtained from animal models and neuron cultures correlate well with the results obtained from PC12 assays, like, for example, the non-pathogenic role of mature fibrils or the toxicity of oligomeric species.[66,75,86] These results support the use of PC12 cell cultures as a suitable model system to study amyloid-induced toxicity.

The degree of amyloid-induced toxicity is usually quantified using the MTT assay (Figure 10.6A). This is a commercial kit based on the cleavage of yellow

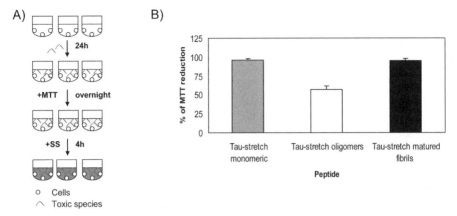

Figure 10.6 A) Schematic representation of MTT test used to quantify the toxicity induced by amyloid related species. B) Example of cytotoxicity of monomers, oligomers, and mature fibrils by the hexapeptide sequence KVQIIN, derived from the tau protein (tau stretch). Monomers and mature fibrils do not affect cell viability, while oligomers reduce the percentage of MTT reduction and therefore the rate of cell survival.

MTT (3-[4,5,dimethylthiazol-2-yl]-2,5-diphenyltetrazolium bromide) by metabolically active cells to yield a dark blue formazan product that can be spectrophotometrically measured (wavelength between 550 and 600 nm). Since only living cells can cleave the MTT product, the absorbance measurement is correlated to the percentage of cell survival (Figure 6B).

10.3 Experimental Studies on Amyloid Model Systems

10.3.1 Diversity and Commonalities in the Amyloid Protein Family

Up to now, there are more than 20 different proteins or peptides that have been associated with human amyloidosis. In the last few years the amyloid field has widened with the discovery of non-pathogenic proteins[28,87,88] and designed peptides and proteins[81,89,90] that can form amyloid fibrils *in vitro*. Such *in vitro* fibrils are morphologically and structurally undistinguishable from those formed *in vivo*. Therefore, the group of peptides and proteins able to form amyloid fibrils *in vitro* and/or *in vivo* is very heterogeneous, not sharing any apparent fold or sequence patterns. In their non-fibrillar form, amyloidogenic proteins display a wide range of native folds. For example, transthyretin is a homotetramer with high β-sheet structure, and only one helix per monomer.[91] In contrast, the native form of lysozyme is predominantly α-helix with a small amount of β-structure.[92] Many amyloidogenic polypeptides consist primarily of random-coil structures in their native, soluble states. These include the Aβ peptide,[32] IAPP,[93] α-synuclein,[94] and tau.[95]

X-Ray fiber diffraction data indicate, however, that all amyloid fibrils share a cross-β structure, regardless of the sequence or native fold of the soluble precursor.[42] Also, for many amyloidogenic proteins it has been demonstrated that amyloid formation involves a loss of native structure and an increase of β-sheet population. These results provide strong support for the increasingly adopted view in the field that the ability to form amyloid fibrils is a general property of the polypeptide backbone, and that there might be certain general principles governing protein fibrillization.[28,96] According with this hypothesis, the propensity to form amyloid structures is related to some physico-chemical properties of the polypeptide chain such as charge, hydrophobicity, and secondary structure propensity, together with a consideration of the distribution of hydrophobic and polar residues[81,97,98] (see Chapter 11 by Dima et al. for more on this issue).

10.3.2 Protein Amyloidogenic Regions

It is well established in the field that a protein has to be partially or fully unfolded to form amyloid fibrils.[9,10,99] However, most natively unfolded proteins do not undergo fibril formation *in vivo*,[100] indicating that unfolding is necessary, but not sufficient to promote protein polymerization. Hence, there must be some sequence motifs that are more prone to self-assembly into amyloid material than others. Very strong evidence in favor of this idea has been provided by recent work in our lab.[34,81,101,102]

Recent investigations also indicate that the amyloidogenic capability of a protein seems to be concentrated in particular protein regions and, more specifically, in small sequence fragments therein.[51,103–107] It has been shown, for example, that a hexamer of human IAPP (residues 22–27, NFGAIL) and even a pentamer (residues 23–27, FGAIL) are already sufficient for amyloid formation and cytotoxicity.[104] Recent studies on protein self-assembly indicate that only six residues of the molecule need to be ordered to give rise to an ordered filament and point mutations at the hexapeptide region prevent protein aggregation.[107] In the case of the Alzheimer peptide (Aβ 1–40), the sequence KLVFFA (residues 16–21) has been identified as the shortest sequence able to form fibrils *in vitro*, but is not toxic in cell culture.[105] The comparison of two homologous proteins that differ in amyloidogenic properties has served to identify short divergent sequence fragments that once swapped into the non-amyloidogenic protein can trigger amyloid formation.[51,106]

10.3.2.1 Experimental Mapping of Protein Amyloidogenic Regions

As a protein must be at least partially unfolded to undergo amyloid fibril formation,[10,99] the regions of a protein more prone to unfolding or with higher flexibility are the most suitable to interact intermolecularly. Protein engineering experiments also indicate that there is a good correlation between the solvent exposed regions of a protein and those shown to be critical in the rate-determining steps of protein self-association.[108] Such exposed and/or highly flexible regions of a protein can be identified by limited proteolysis.[109,110] Once they have been isolated and purified, the ability of these fragments to form fibrils can be tested using the standard set of experimental techniques (see above).

NMR studies on amyloid precursor states can also be conducted to identify the key amyloidogenic regions of a protein.[111] Analysis of peptides consisting of these regions can further confirm their amyloidogenic properties.[103] Minimal amyloid sequences can be searched within these fragments by spotting overlapping peptides on a membrane and by assaying their binding capabilities to the full-length protein.[112]

10.3.2.2 Development of Peptide Model Systems of Amyloid Formation

The research carried out in our group during the past years has mainly focused on establishing the sequence and structural bases of amyloid formation. Since amyloid deposits formed by natural proteins are difficult to study using standard biophysical techniques, our strategy has consisted in devising peptide and protein model systems that possess suitable physical properties such as reversibility, good solubility, *etc.* to allow biophysical characterization, and that are small enough to permit a detailed analysis of the aggregation process.[81,113] If amyloid formation is actually driven by short fragments of a misfolded protein, small model peptides should be more suitable than proteins to investigate those elements in sequences that favor amyloid formation. Furthermore, whereas a mutation would alter just the self-assembly properties of a small peptide, it might lead to protein destabilization, complicating the extraction of pure sequence propensities to form amyloid fibrils.

Initially, we reported the computer-aided design of a hexapeptide-based model system for amyloidogenesis.[81] Its simplicity served to highlight that fibril formation is due to a very delicate balance between specific side-chain and electrostatic interactions within a sequence and to propose a structural model of the fibril that is consistent with the organization of the protofilament core of naturally amyloidogenic proteins (see Section 6.2). This result validates its use to study amyloid fibril formation and structure.

The small size of this model peptide system has been exploited to determine how exact sequence details modulate, or completely disrupt, the apparent generality of amyloid fibril formation in proteins.[34] This question has been addressed by systematically replacing the residues of a *de novo* designed amyloid peptide with all natural amino acids. Previously, only alanine[114] and proline[107] scannings had been carried out in amyloid fragments found experimentally to determine the role of each residue.

From this saturation mutagenesis experiment, a sequence pattern to identify amyloidogenic stretches in proteins was extracted.

Acidic pH   {P}-{PKRHW}-[VLSCWFNQE]-[ILTYWFNE]-[FIY]-{PKRH}
Neutral pH   {P}-{PKRHW}-[VLSCWFNQ]-[ILTYWFN]-[FIY]-{PKRH}

PROSITE syntax (http://www.expasy.org/prosite/). "[ ]"residues allowed at the position; "{ }" residues forbidden at the position; "-" separates each pattern element.

The positional scanning mutagenesis revealed that there is a position dependence in the sensitivity of amyloid fibril formation to mutation, and also that mutationally very tolerant (*edges*) and restrictive (*core*) positions can be found within an amyloid sequence.[34] This amyloidogenic pattern has been successfully tested experimentally and also by *in silico* sequence scanning of amyloid proteins and protein databases. Analysis of protein databases has shown that highly amyloidogenic sequences matching the pattern are less frequent in proteins than innocuous amino acid combinations. Furthermore, when present, such amyloidogenic sequences are surrounded by amino acids that disrupt their aggregating capability (*amyloid breakers*).

As a final test, a set of amyloid peptides and proteins that form amyloid fibrils *in vitro* and/or *in vivo* and whose amyloid regions have been investigated experimentally were scanned with our amyloid pattern. The results suggest that the pattern is able to detect hexapeptide stretches that overlap with the amyloid regions found experimentally. For example, for the Aβ (1–42) peptide, the region that agrees with the pattern consists of residues 16–21 ($_{16}$KLVFFA$_{21}$). Interestingly, this region overlaps with the minimal sequence shown to be essential for Aβ polymerization, residues 16–20.[105]

10.3.2.3 Testing the Amyloid Stretch Hypothesis

The results mentioned above suggest that even a short amino acid stretch bearing a highly amyloidogenic motif could provide the driving force needed to trigger the self-assembly process of a protein. This is referred to as the *amyloid stretch hypothesis*. Previously, it has been shown that insertion of long amyloid domains (30–80 amino acids) from naturally occurring amyloid proteins can trigger amyloid formation of some non-amyloidogenic proteins both *in vitro*[115,116] and *in vivo*.[117] Other studies have used shorter amyloidogenic fragments (10-mer and 7-mer) from the N-terminus region of the yeast prion Sup35 with the same aim.[118,119] But since the target protein was already amyloidogenic[118] and/or they did not reach clear conclusions,[119] none of these studies provided compelling evidences in favor of the *amyloid stretch hypothesis*.

Our strategy to demonstrate the *amyloid stretch hypothesis* was the conversion of a non-amyloidogenic protein into an amyloid-prone molecule by inserting just a 6-residue amyloidogenic stretch.[101] We chose the α-spectrin SH3 domain (α-SH3) as a target non-amyloidogenic protein. This protein has been shown not to be amyloidogenic under any conditions tested in our laboratory.[113] As amyloidogenic sequences, we selected a *de novo* designed peptide, STVIIE, highly amyloidogenic *in vitro*.[81] To validate the *amyloid pattern* within the context of a protein, we designed point mutants of this sequence at different position categories (*core* and *edges*). We also introduced an amyloidogenic 6-residue fragment of the Aβ amyloid peptide identified with the *amyloid pattern*, $_{16}$KLVFFA$_{21}$.[34] Sequences were inserted at different positions of the protein. The N-terminus of the SH3 domain is natively disordered while the C-terminus is structured into a β-strand.[120] This difference provides a convenient framework to assess the influence of the structural environment of the sequence on its amyloidogenic capabilities.

The initial conformation of the variants carrying the amyloidogenic insertions was investigated using far- and near-UV spectroscopy. In nearly all the cases they showed CD signatures similar to the ones of the WT SH3 domain. Hence, we could conclude that the insertions do not affect the folding of the domain. In order to prove that the amyloid insertions do not affect the stability of the globular domain, we estimated the stability of the α-SH3 variants using thermally induced unfolding monitored by CD.

Next, we carried out fibrillation assays to test the amyloidogenic properties of the amyloid SH3 variants. Samples were set up at two different protein concentrations and checked for amyloid formation by CD and EM at different time points. Modified versions of the α-SH3 carrying these short amyloidogenic sequences in the N-terminus are as stable as the WT protein, but they fibrillate under conditions where the original domain still remains soluble. Thus, the amyloidogenic behavior shown by these proteins is not due to an extra destabilization of the protein, but rather to the amyloidogenic properties of the inserted sequence. The amyloidogenicity shown by the variants bearing the insertion at the N-terminus compares to the amyloidogenic behavior exhibited by the peptides.[34] The N-terminus variant carrying the highly amyloidogenic sequence STVIIE (STVIIE-SH3, **1**-SH) forms abundant amyloid material after three months (Figure 10.7). Mutation Glu6Thr at position #6 of the more tolerant *edges* (STVII<u>T</u>-SH3, **2**-SH) keeps the amyloidogenic feature of the **1**-SH variant while mutation Ile5Lys at the most restrictive position of the *core* completely abolishes amyloid fibril formation (STVIKT-SH, **3**-SH). These results suggest that the amyloidogenic properties of proteins containing 6-residue stretches that match the *amyloid pattern* can be modified by designing mutations according to the amino acid tolerance at each position provided by the pattern. Our results also show that the hexapeptide sequence derived from the Aβ peptide, $_{16}$KLVFFA$_{21}$, is also able to trigger amyloid formation by a completely unrelated domain (KLVFFA-SH3, **Aβ**-SH).

Despite being slightly destabilized the C-terminal variants do not form any kind of amyloid material under the same conditions in which the N-terminal ones do. As we have mentioned before the N-terminus of α-SH3 is disordered, while the C-terminus is structured into a β-strand. Therefore, the amyloid tail should be more exposed in mutants with the insertion at the N-terminus, and thus more accessible to intermolecular interactions such as those involved in

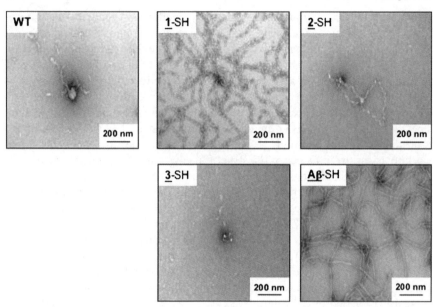

Figure 10.7 Amyloid formation by the different N-terminus α-SH3 variants. All the proteins were incubated under similar conditions (pH 2.6, $c \sim 300\,\mu M$, $t = 3$ months and room temperature). $\underline{2}$-SH was assayed only at the lowest concentration ($c \sim 100\,\mu M$) because of solubility problems. Taken from reference 101. © The National Academy of Sciences of the USA.

amyloid fibril formation.[120] The result highlights that the structural environment of the amyloidogenic stretch plays a fundamental role in whether or not an amyloid-prone sequence can productively trigger the amyloid self-assembly process. In order to be amyloidogenic a protein must carry an appropriate amyloid stretch (*sequence determinant*) that must be or become locally unfolded to initiate the process of amyloid formation (*structural determinant*).

10.3.2.4 Studying Amyloid Cytotoxicity with Short Peptides

Due to the complexity of the events involved in the pathogenicity of amyloid formation, simplified models to study the molecular bases of this process are desirable. Peptide model systems have been very helpful to provide outstanding knowledge about the underlying factors in amyloid formation.[121] Therefore, short peptides capable of polymerizing into fibrils with properties similar to those of natural amyloid proteins could be a successful alternative.[81,89] Based on that, we scanned all the human amyloidogenic proteins described so far with the amyloidogenic pattern described by our group.[34] Peptide fibrillation assays showed that the amyloid stretches identified are able to form amyloid-like fibrils. We found that the fibrils formed by these hexapeptides are not pathogenic in PC12 cell culture while prefibrillar ordered aggregates of amyloid stretches were toxic.

Interestingly, all of the toxic oligomers formed by sequence-unrelated hexapeptides displayed identical morphology by EM. These toxic oligomers correspond to the same intermediate species of the amyloid formation pathway, namely protofibrils. These results suggest that sequence does not play a general role in the toxicity mechanism, which seems to depend exclusively on aggregate structure. This finding was further confirmed by the observation that D- and L-versions of the same sequence exhibit similar toxicities. Furthermore, the same toxic species were identified as responsible for the toxic effects of both the full-length $A\beta_{1-42}$ and the $A\beta$ hexapeptide stretch identified by the amyloid pattern. We also took advantage of these model peptides to explore the mechanism by which these prefibrillar aggregates impair cell function and trigger cell death. Analysis of fluorescently labeled peptides showed attachment of the prefibrillar structures to the cell membrane, indicating that the plasma membrane is the primary target of amyloid induced toxicity. Also, we showed that cell death induced by toxic prefibrillar aggregates is mediated by apoptosis. Based on these findings, we concluded that self-assembly of putative amyloid fragments into prefibrillar aggregates impairs cell functions and triggers cell death in the same way full-length proteins do.[122]

The demonstration that short amyloid sequences can trigger amyloid formation in a soluble domain and that they can be used as a successful model system to study amyloid cytotoxicity may have significant impact in facilitating the development of anti-amyloid therapeutics. Amyloid-prone protein regions identified with the pattern can be synthesized and used to screen for molecules that not only block amyloid formation but also amyloid-induced cytotoxicity.[123]

Acknowledgements

We would like to thank Mjriam Mayer for critical reading of the manuscript. We acknowledge support from an EU grant to AEC (RTN project 2001-00364 "Protein Folding, Misfolding and Disease") and by an EU grant (APOPIS) to MTP.

References

1. H. J. Dyson and P. E. Wright, *Nat. Rev. Mol. Cell Biol.*, 2005, **6**, 197–208.
2. J. D. Sipe and A. S. Cohen, *J. Struct. Biol.*, 2000, **130**, 88–98.
3. A. S. Cohen and E. Calkins, *Nature*, 1959, **183**, 1202–1203.
4. A. S. Cohen and E. Calkins, *J. Cell. Biol.*, 1964, **21**, 481–486.
5. T. Shirahama and A. S. Cohen, *J. Cell. Biol.*, 1967, **33**, 679–708.
6. G. G. Glenner, D. Ein, E. D. Eanes, H. A. Bladen, W. Terry and D. L. Page, *Science*, 1971, **174**, 712–714.
7. T. Shirahama and A. S. Cohen, *Am. J. Pathol.*, 1975, **81**, 101–116.

8. J. W. Kelly, *Curr. Opin. Struct. Biol.*, 1996, **6**, 11–17.
9. J. W. Kelly, *Curr. Opin. Struct. Biol.*, 1998, **8**, 101–106.
10. J. C. Rochet and P. T. Lansbury Jr., *Curr. Opin. Struct. Biol.*, 2000, **10**, 60–68.
11. E. Bossy-Wetzel, R. Schwarzenbacher and S. A. Lipton, *Nat. Med.*, 2004, **10**(Suppl.), S2–S9.
12. M. Stefani, *Biochim. Biophys. Acta*, 2004, **1739**, 5–25.
13. J. D. Harper and P. T. Lansbury Jr., *Annu. Rev. Biochem.*, 1997, **66**, 385–407.
14. R. J. Ellis, *Curr. Opin. Struct. Biol.*, 2001, **11**, 114–119.
15. F. U. Hartl and M. Hayer-Hartl, *Science*, 2002, **295**, 1852–1858.
16. R. Sitia and I. Braakman, *Nature*, 2003, **426**, 891–894.
17. A. L. Goldberg, *Nature*, 2003, **426**, 895–899.
18. M. B. Pepys, P. N. Hawkins, D. R. Booth, D. M. Vigushin, G. A. Tennent, A. K. Soutar, N. Totty, O. Nguyen, C. C. Blake and C. J. Terry et al., *Nature*, 1993, **362**, 553–557.
19. D. R. Booth, M. Sunde, V. Bellotti, C. V. Robinson, W. L. Hutchinson, P. E. Fraser, P. N. Hawkins, C. M. Dobson, S. E. Radford, C. C. Blake and M. B. Pepys, *Nature*, 1997, **385**, 787–793.
20. C. Nilsberth, A. Westlind-Danielsson, C. B. Eckman, M. M. Condron, K. Axelman, C. Forsell, C. Stenh, J. Luthman, D. B. Teplow, S. G. Younkin, J. Naslund and L. Lannfelt, *Nat. Neurosci.*, 2001, **4**, 887–893.
21. D. J. Selkoe, *Nat. Cell Biol.*, 2004, **6**, 1054–1061.
22. M. B. Pepys, *Philos. Trans. R. Soc. London, Ser. B*, 2001, **356**, 203–210, discussion 210–201.
23. J. McLaurin, D. Yang, C. M. Yip and P. E. Fraser, *J. Struct. Biol.*, 2000, **130**, 259–270.
24. J. D. Sipe, *Annu. Rev. Biochem.*, 1992, **61**, 947–975.
25. S. H. Pasternak, J. W. Callahan and D. J. Mahuran, *J. Alzheimers Dis.*, 2004, **6**, 53–65.
26. Y. Su and P. T. Chang, *Brain Res.*, 2001, **893**, 287–291.
27. C. A. Ross and M. A. Poirier, *Nat. Med.*, 2004, **10**(Suppl.), S10–S17.
28. F. Chiti, P. Webster, N. Taddei, A. Clark, M. Stefani, G. Ramponi and C. M. Dobson, *Proc. Natl. Acad. Sci. U.S.A.*, 1999, **96**, 3590–3594.
29. L. C. Serpell, *Biochim. Biophys. Acta*, 2000, **1502**, 16–30.
30. B. Seilheimer, B. Bohrmann, L. Bondolfi, F. Muller, D. Stuber and H. Dobeli, *J. Struct. Biol.*, 1997, **119**, 59–71.
31. J. D. Harper, S. S. Wong, C. M. Lieber and P. T. Lansbury, *Chem. Biol.*, 1997, **4**, 119–125.
32. D. M. Walsh, D. M. Hartley, Y. Kusumoto, Y. Fezoui, M. M. Condron, A. Lomakin, G. B. Benedek, D. J. Selkoe and D. B. Teplow, *J. Biol. Chem.*, 1999, **274**, 25945–25952.
33. A. M. Fernandez-Escamilla, F. Rousseau, J. Schymkowitz and L. Serrano, *Nat. Biotechnol.*, 2004, **22**, 1302–1306.
34. M. Lopez de la Paz and L. Serrano, *Proc. Natl. Acad. Sci. U.S.A.*, 2004, **101**, 87–92.

35. A. P. Pawar, K. F. Dubay, J. Zurdo, F. Chiti, M. Vendruscolo and C. M. Dobson, *J. Mol. Biol.*, 2005, **350**, 379–392.
36. L. C. Serpell, M. Sunde, P. E. Fraser, P. K. Luther, E. P. Morris, O. Sangren, E. Lundgren and C. C. Blake, *J. Mol. Biol.*, 1995, **254**, 113–118.
37. L. C. Serpell, M. Sunde, M. D. Benson, G. A. Tennent, M. B. Pepys and P. E. Fraser, *J. Mol. Biol.*, 2000, **300**, 1033–1039.
38. J. L. Jimenez, J. I. Guijarro, E. Orlova, J. Zurdo, C. M. Dobson, M. Sunde and H. R. Saibil, *Embo. J.*, 1999, **18**, 815–821.
39. J. L. Jimenez, G. Tennent, M. Pepys and H. R. Saibil, *J. Mol. Biol.*, 2001, **311**, 241–247.
40. J. L. Jimenez, E. J. Nettleton, M. Bouchard, C. V. Robinson, C. M. Dobson and H. R. Saibil, *Proc. Natl. Acad. Sci. U.S.A.*, 2002, **99**, 9196–9201.
41. C. Blake and L. Serpell, *Structure*, 1996, **4**, 989–998.
42. M. Sunde, L. C. Serpell, M. Bartlam, P. E. Fraser, M. B. Pepys and C. C. Blake, *J. Mol. Biol.*, 1997, **273**, 729–739.
43. M. Hoshino, H. Katou, Y. Hagihara, K. Hasegawa, H. Naiki and Y. Goto, *Nat. Struct. Biol.*, 2002, **9**, 332–336.
44. A. Kishimoto, K. Hasegawa, H. Suzuki, H. Taguchi, K. Namba and M. Yoshida, *Biochem. Biophys. Res. Commun.*, 2004, **315**, 739–745.
45. M. F. Perutz, J. T. Finch, J. Berriman and A. Lesk, *Proc. Natl. Acad. Sci. U.S.A.*, 2002, **99**, 5591–5595.
46. H. Wille, M. D. Michelitsch, V. Guenebaut, S. Supattapone, A. Serban, F. E. Cohen, D. A. Agard and S. B. Prusiner, *Proc. Natl. Acad. Sci. U.S.A.*, 2002, **99**, 3563–3568.
47. R. W. Pickersgill, *Structure (Cambridge, MA, U.S.)*, 2003, **11**, 137–138.
48. R. Wetzel, *Structure (Cambridge, MA, U.S.)*, 2002, **10**, 1031–1036.
49. J. S. Elam, A. B. Taylor, R. Strange, S. Antonyuk, P. A. Doucette, J. A. Rodriguez, S. S. Hasnain, L. J. Hayward, J. S. Valentine, T. O. Yeates and P. J. Hart, *Nat. Struct. Biol.*, 2003, **10**, 461–467.
50. Y. Liu, G. Gotte, M. Libonati and D. Eisenberg, *Nat. Struct. Biol.*, 2001, **8**, 211–214.
51. M. I. Ivanova, M. R. Sawaya, M. Gingery, A. Attinger and D. Eisenberg, *Proc. Natl. Acad. Sci. U.S.A.*, 2004, **101**, 10584–10589.
52. S. Sambashivan, Y. Liu, M. R. Sawaya, M. Gingery and D. Eisenberg, *Nature*, 2005, **437**, 266–269.
53. R. Nelson, M. R. Sawaya, M. Balbirnie, A. O. Madsen, C. Riekel, R. Grothe and D. Eisenberg, *Nature*, 2005, **435**, 773–778.
54. M. R. Sawaya, S. Sambashivan, R. Nelson, M. I. Ivanova, S. A. Sievers, M. I. Apostol, M. J. Thompson, M. Balbirnie, J. J. Wiltzius, H. T. McFarlane, A. Madsen, C. Riekel and D. Eisenberg, *Nature*, 2007, **447**, 453–457.
55. Y. Wu and Y. Luo, *Curr. Alzheimer Res.*, 2005, **2**, 37–45.
56. N. M. Bonini and M. E. Fortini, *Annu. Rev. Neurosci.*, 2003, **26**, 627–656.

57. F. Chen, D. David, A. Ferrari and J. Gotz, *Curr. Drug Targets*, 2004, **5**, 503–515.
58. S. L. Karsten and D. H. Geschwind, *Cell*, 2005, **120**, 572–574.
59. H. Y. Zoghbi and J. Botas, *Trends Genet.*, 2002, **18**, 463–471.
60. S. Kuemmerle, C. A. Gutekunst, A. M. Klein, X. J. Li, S. Li, M. F. Beal, S. M. Hersch and R. J. Ferrante, *Ann. Neurol.*, 1999, **46**, 842–849.
61. R. D. Terry, E. Masliah, D. P. Salmon, N. Butters, R. DeTeresa, R. Hill, L. A. Hansen and R. Katzman, *Ann. Neurol.*, 1991, **30**, 572–580.
62. B. Caughey and P. T. Lansbury, *Annu. Rev. Neurosci.*, 2003, **26**, 267–298.
63. M. Bucciantini, E. Giannoni, F. Chiti, F. Baroni, L. Formigli, J. Zurdo, N. Taddei, G. Ramponi, C. M. Dobson and M. Stefani, *Nature*, 2002, **416**, 507–511.
64. A. Demuro, E. Mina, R. Kayed, S. C. Milton, I. Parker and C. G. Glabe, *J. Biol. Chem.*, 2005, **280**, 17294–17300.
65. N. Reixach, S. Deechongkit, X. Jiang, J. W. Kelly and J. N. Buxbaum, *Proc. Natl. Acad. Sci. U.S.A.*, 2004, **101**, 2817–2822.
66. J. P. Cleary, D. M. Walsh, J. J. Hofmeister, G. M. Shankar, M. A. Kuskowski, D. J. Selkoe and K. H. Ashe, *Nat. Neurosci.*, 2005, **8**, 79–84.
67. M. Bucciantini, G. Calloni, F. Chiti, L. Formigli, D. Nosi, C. M. Dobson and M. Stefani, *J. Biol. Chem.*, 2004, **279**, 31374–31382.
68. M. P. Lambert, A. K. Barlow, B. A. Chromy, C. Edwards, R. Freed, M. Liosatos, T. E. Morgan, I. Rozovsky, B. Trommer, K. L. Viola, P. Wals, C. Zhang, C. E. Finch, G. A. Krafft and W. L. Klein, *Proc. Natl. Acad. Sci. U.S.A.*, 1998, **95**, 6448–6453.
69. H. Mukai, T. Isagawa, E. Goyama, S. Tanaka, N. F. Bence, A. Tamura, Y. Ono and R. R. Kopito, *Proc. Natl. Acad. Sci. U.S.A.*, 2005, **102**, 10887–10892.
70. D. M. Walsh, I. Klyubin, J. V. Fadeeva, W. K. Cullen, R. Anwyl, M. S. Wolfe, M. J. Rowan and D. J. Selkoe, *Nature*, 2002, **416**, 535–539.
71. N. F. Bence, R. M. Sampat and R. R. Kopito, *Science*, 2001, **292**, 1552–1555.
72. P. J. Muchowski and J. L. Wacker, *Nat. Rev. Neurosci.*, 2005, **6**, 11–22.
73. X. Hou, S. J. Richardson, M.-I. Aguilar and D. H. Small, *Biochemistry*, 2005, **44**, 11618–11627.
74. R. Kayed, Y. Sokolov, B. Edmonds, T. M. McIntire, S. C. Milton, J. E. Hall and C. G. Glabe, *J. Biol. Chem.*, 2004, **279**, 46363–46366.
75. J. C. Rochet, T. F. Outeiro, K. A. Conway, T. T. Ding, M. J. Volles, H. A. Lashuel, R. M. Bieganski, S. L. Lindquist and P. T. Lansbury, *J. Mol. Neurosci.*, 2004, **23**, 23–34.
76. K. Wong, Y. Qiu, W. Hyun, R. Nixon, J. VanCleff, J. Sanchez-Salazar, S. B. Prusiner and S. J. DeArmond, *Neurology*, 1996, **47**, 741–750.
77. M. Xiaocui, Y. Sha, K. Lin and S. Nie, *Protein Pept. Lett.*, 2002, **9**, 173–178.
78. D. A. Butterfield, J. Drake, C. Pocernich and A. Castegna, *Trends Mol. Med.*, 2001, **7**, 548–554.
79. M. P. Mattson, *Physiol. Rev.*, 1997, **77**, 1081–1132.

80. C. Soto, *Nat. Rev. Neurosci.*, 2003, **4**, 49–60.
81. M. Lopez de la Paz, K. Goldie, J. Zurdo, E. Lacroix, C. M. Dobson, A. Hoenger and L. Serrano, *Proc. Natl. Acad. Sci. U.S.A.*, 2002, **99**, 16052–16057.
82. T. Kowalewski and D. M. Holtzman, *Proc. Natl. Acad. Sci. U.S.A.*, 1999, **96**, 3688–3693.
83. W. E. Klunk, J. W. Pettegrew and D. J. Abraham, *J. Histochem. Cytochem.*, 1989, **37**, 1273–1281.
84. H. LeVine III, *Protein Sci.*, 1993, **2**, 404–410.
85. S. J. Wood, B. Maleeff, T. Hart and R. Wetzel, *J. Mol. Biol.*, 1996, **256**, 870–877.
86. B. A. Chromy, R. J. Nowak, M. P. Lambert, K. L. Viola, L. Chang, P. T. Velasco, B. W. Jones, S. J. Fernandez, P. N. Lacor, P. Horowitz, C. E. Finch, G. A. Krafft and W. L. Klein, *Biochemistry*, 2003, **42**, 12749–12760.
87. M. Fandrich, M. A. Fletcher and C. M. Dobson, *Nature*, 2001, **410**, 165–166.
88. J. I. Guijarro, M. Sunde, J. A. Jones, I. D. Campbell and C. M. Dobson, *Proc. Natl. Acad. Sci. U.S.A.*, 1998, **95**, 4224–4228.
89. Y. Fezoui, D. M. Hartley, D. M. Walsh, D. J. Selkoe, J. J. Osterhout and D. B. Teplow, *Nat. Struct. Biol.*, 2000, **7**, 1095–1099.
90. M. W. West, W. Wang, J. Patterson, J. D. Mancias, J. R. Beasley and M. H. Hecht, *Proc. Natl. Acad. Sci. U.S.A.*, 1999, **96**, 11211–11216.
91. C. C. Blake, M. J. Geisow, S. J. Oatley, B. Rerat and C. Rerat, *J. Mol. Biol.*, 1978, **121**, 339–356.
92. P. J. Artymiuk and C. C. Blake, *J. Mol. Biol.*, 1981, **152**, 737–762.
93. R. Kayed, J. Bernhagen, N. Greenfield, K. Sweimeh, H. Brunner, W. Voelter and A. Kapurniotu, *J. Mol. Biol.*, 1999, **287**, 781–796.
94. P. H. Weinreb, W. Zhen, A. W. Poon, K. A. Conway and P. T. Lansbury Jr., *Biochemistry*, 1996, **35**, 13709–13715.
95. O. Schweers, E. Schonbrunn-Hanebeck, A. Marx and E. Mandelkow, *J. Biol. Chem.*, 1994, **269**, 24290–24297.
96. C. M. Dobson, *Biochem. Soc. Symp.*, 2001, 1–26.
97. F. Chiti, M. Calamai, N. Taddei, M. Stefani, G. Ramponi and C. M. Dobson, *Proc. Natl. Acad. Sci. U.S.A.*, 2002, **99**(4), 16419–16426.
98. F. Chiti, M. Stefani, N. Taddei, G. Ramponi and C. M. Dobson, *Nature*, 2003, **424**, 805–808.
99. C. M. Dobson, *Trends Biochem. Sci.*, 1999, **24**, 329–332.
100. V. N. Uversky, J. R. Gillespie and A. L. Fink, *Proteins*, 2000, **41**, 415–427.
101. A. Esteras-Chopo, L. Serrano and M. Lopez de la Paz, *Proc. Natl. Acad. Sci. U.S.A.*, 2005, **102**, 16672–16677.
102. M. Lopez de la Paz, G. M. de Mori, L. Serrano and G. Colombo, *J. Mol. Biol.*, 2005, **349**, 583–596.
103. S. Jones, J. Manning, N. M. Kad and S. E. Radford, *J. Mol. Biol.*, 2003, **325**, 249–257.

104. K. Tenidis, M. Waldner, J. Bernhagen, W. Fischle, M. Bergmann, M. Weber, M. L. Merkle, W. Voelter, H. Brunner and A. Kapurniotu, *J. Mol. Biol.*, 2000, **295**, 1055–1071.
105. L. O. Tjernberg, J. Naslund, F. Lindqvist, J. Johansson, A. R. Karlstrom, J. Thyberg, L. Terenius and C. Nordstedt, *J. Biol. Chem.*, 1996, **271**, 8545–8548.
106. S. Ventura, J. Zurdo, S. Narayanan, M. Parreno, R. Mangues, B. Reif, F. Chiti, E. Giannoni, C. M. Dobson, F. X. Aviles and L. Serrano, *Proc. Natl. Acad. Sci. U.S.A.*, 2004, **101**, 7258–7263.
107. M. von Bergen, P. Friedhoff, J. Biernat, J. Heberle, E. M. Mandelkow and E. Mandelkow, *Proc. Natl. Acad. Sci. U.S.A.*, 2000, **97**, 5129–5134.
108. M. Monti, B. L. Garolla di Bard, G. Calloni, F. Chiti, A. Amoresano, G. Ramponi and P. Pucci, *J. Mol. Biol.*, 2004, **336**, 253–262.
109. A. Fontana, M. Zambonin, P. Polverino de Laureto, V. De Filippis, A. Clementi and E. Scaramella, *J. Mol. Biol.*, 1997, **266**, 223–230.
110. P. Polverino de Laureto, N. Taddei, E. Frare, C. Capanni, S. Costantini, J. Zurdo, F. Chiti, C. M. Dobson and A. Fontana, *J. Mol. Biol.*, 2003, **334**, 129–141.
111. V. J. McParland, A. P. Kalverda, S. W. Homans and S. E. Radford, *Nat. Struct. Biol.*, 2002, **9**, 326–331.
112. Y. Mazor, S. Gilead, I. Benhar and E. Gazit, *J. Mol. Biol.*, 2002, **322**, 1013–1024.
113. S. Ventura, E. Lacroix and L. Serrano, *J. Mol. Biol.*, 2002, **322**, 1147–1158.
114. R. Azriel and E. Gazit, *J. Biol. Chem.*, 2001, **276**, 34156–34161.
115. U. Baxa, K. L. Taylor, J. S. Wall, M. N. Simon, N. Cheng, R. B. Wickner and A. C. Steven, *J. Biol. Chem.*, 2003, **278**, 43717–43727.
116. M. Tanaka, Y. Machida, Y. Nishikawa, T. Akagi, I. Morishima, T. Hashikawa, T. Fujisawa and N. Nukina, *Biochemistry*, 2002, **41**, 10277–10286.
117. W. C. Wigley, B. D. Stidham, N. M. Smith, J. F. Hunt and P. J. Thomas, *Nat. Biotechnol.*, 2001, **19**, 131–136.
118. Y. He, H. Tang, Z. Yi, H. Zhou and Y. Luo, *FEBS Lett.*, 2005, **579**, 1503–1508.
119. Y. K. Chae, K. S. Cho and W. Chun, *Protein Pept. Lett.*, 2002, **9**, 315–321.
120. A. Musacchio, M. Noble, R. Pauptit, R. Wierenga and M. Saraste, *Nature*, 1992, **359**, 851–855.
121. M. T. Pastor, A. Esteras-Chopo and M. Lopez de la Paz, *Curr. Opin. Struct. Biol.*, 2005, **15**, 57–63.
122. M. T. Pastor, N. Kummerer, V. Schubert, A. Esteras-Chopo, C. G. Dotti, M. Lopez de la Paz and L. Serrano, *J. Mol. Biol.*, 2008, **375**, 695–707.
123. A. Esteras-Chopo, M. T. Pastor, L. Serrano and M. Lopez de la Paz, *J. Mol. Biol.*, 2008, in press.

CHAPTER 11

Scenarios for Protein Aggregation: Molecular Dynamics Simulations and Bioinformatics Analysis

RUXANDRA DIMA,[a] BOGDAN TARUS,[b] G. REDDY,[d] JOHN E. STRAUB[b] AND D. THIRUMALAI[c,d]

[a] Department of Chemistry, University of Cincinnati, Cincinnati, OH 45221, USA; [b] Department of Chemistry, Boston University, 590 Commonwealth Ave, Boston, MA 02215; [c] Biophysics Program, Institute of Physical Sciences & Technology, University of Maryland, College Park, MD 20742; [d] Department of Chemistry & Biochemistry, University of Maryland, College Park, MD 20742, USA

11.1 Introduction

Increasing numbers of diseases including Alzheimer's disease,[1] transmissible prion disorders,[2] and type II diabetes are linked to amyloid fibrils.[3] The mechanism of amyloid fibril formation starting from the monomer is still poorly understood. During the cascade of events in the transition from monomers to mature fibrils a number of key intermediates, namely soluble oligomers and protofilaments, are populated. It is suspected that the conformations of the peptides in this aggregated state differ substantially from the isolated monomer, which implies that the monomer undergoes large

---

RSC Biomolecular Sciences
Protein Folding, Misfolding and Aggregation: Classical Themes and Novel Approaches
Edited by Victor Muñoz
© Royal Society of Chemistry 2008

inter-peptide interaction-driven structural transformations.[4] The need to understand the assembly kinetics of fibril formation has become urgent because of the realization that soluble oligomers of amyloidogenic peptides may be even more neurotoxic than the end product, namely the amyloid fibrils.[5] In order to fully understand the routes to fibril formation one has to characterize the major species in the assembly pathways. The characterization of the energetics and dynamics of oligomers (dimers, trimers, *etc.*) is difficult using experiments alone because they undergo large conformational fluctuations. In this context, carefully planned molecular dynamics simulation studies,[6–9] computations using coarse-grained models,[10] and bioinformatic analysis[11,12] have given considerable insights into the early events in the route to fibril formation. Here we describe progress along this route using examples taken largely from our own work.

In this chapter, we focus on the following aspects of protein aggregation using Aβ-peptides and prion proteins as examples.

- What are the plausible scenarios in the transition from monomers to amyloid fibril formation?
- What features of the amyloidogenic peptides control the growth kinetics of fibrils? Although the assembly mechanism is complex the overall growth kinetics is determined largely by the charge states and hydrophobicity of the monomers.
- Can sequence and structural analysis be used to predict specific patterns that are likely to be aggregation prone? By exploiting the sequence profiles and structures of the cellular form of prions, PrP^C, we uncover the regions that are likely to trigger the large conformational changes in the transition from PrP^C to the scrapie form, PrP^{Sc}.
- Is there an organizational principle in oligomer and fibril formation? The formation of morphologically similar aggregates by a variety of proteins that are unrelated in sequence or structure suggests that certain general principles may govern fibrillization. However, the vastness of sequence space and the heterogeneity of environmentally dependent interactions make deciphering the principles of aggregation difficult. Nevertheless, we will argue that oligomers and higher-order structures form in such a way that the number of intra- and inter-molecular hydrophobic interactions are maximized and electrostatic repulsions are minimized. The latter implies that the motifs that minimize the number of salt bridges are preferred.

For the issues raised above we formulate tentative ideas using phenomenological arguments and atom molecular dynamics (MD) simulations. Using a number of experimental observations and results from computer simulations certain general principles of amyloid formation seem to be emerging. There are a number of unresolved issues that remain despite significant progress. A few of these are outlined at the end of the paper.

11.2 Scenarios for Peptide Association

11.2.1 General Ideas

The molecular details of the cascade of events that lead to the formation of amyloid fibrils remain unknown because the species along the aggregation pathways are highly dynamic and metastable. Indeed, AFM images of protofibrils show that they undergo shape fluctuations, which implies a heterogeneous population of species. A number of experimental studies suggest that fibril formation exhibits all the characteristics of a nucleation growth process. It is suspected that the formation of a critical nucleus is the rate-determining step in the fibril formation.[13] Once the critical nucleus, whose very nature might depend on sequence as well as external conditions, forms the fibril the elongation process is essentially downhill in free energy. The nucleation characteristics manifest themselves in the appearance of a lag phase in the fibril formation. The lag phase disappears if a seed or a preformed nucleus is present in the saturated peptide solution. The seeded growth of fibrils has also been observed in simple lattice and off-lattice models of protofibril formation. These are the general features observed in experiments, but a more detailed account can be found in Chapter 10 by Esteras-Chopo, Pastor, and Serrano. From this perspective an overall scenario for explaining aggregation kinetics is in place. However, the molecular and mechanistic details of the process, including the dependence of the growth kinetics on the specifics of the sequence, are not fully understood.

Here we present two extreme scenarios[14] that describe the needed conformational changes in monomers that lead to a population of species that can nucleate and grow. The two potential scenarios, which follow from the energy landscape perspective of aggregation (see Chapter 3 by Wolynes for a description of the energy landscape perspective in protein folding), differ greatly in the description of the dynamics of the monomers. It was advocated early on that fibrillization requires partial unfolding of the native state[15] or partial folding of the unfolded state (see Scenario I in Figure 11.1). Both events, which are likely to involve crossing free energy barriers, lead to the transient population of the assembly-competent structures N^*. The better appreciated possibility is Scenario I in which environmental fluctuations (pH shifts for example) produce spontaneously the N^* conformation. For example, extensive experiments[16] have shown that the N^* state in transthyretin (TTR), which has a higher free energy than the native state N, formed upon unraveling of the strands C and D at the edge of the structure. This process exposes an aggregation-prone strand B. One can also envision a scenario in which N^* has a lower free energy than N thus making the folded (functional) state metastable. It is likely that amyloidogenic proteins, in which nearly complete transformation of their structure takes place upon fibrillization, may follow the second scenario. In both cases fibrillization kinetics result from the ability to populate the N^* species. In either scenario (TTR aggregation that follows Scenario I or PrP^{Sc} formation that follows Scenario II) growth kinetics are initially determined by the "unfolding" barriers

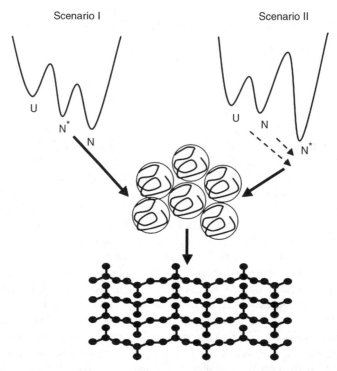

Figure 11.1 Schematic diagram of the two plausible scenarios of fibrillization based on free energy landscape perspective. According to Scenario I, the assembly competent state **N*** is metastable with respect to the monomeric native state **N** and is formed due to partial unfolding. In Scenario II **N*** is formed upon structural conversion either of the native state **N** (as in prions) or directly from the unfolded state **U** (as in Aβ-amyloid peptides). In both cases proteins (or peptides) in **N*** states must coalesce into larger oligomers capable of growth into fibrils.

separating **N*** from either **N** or **U**. The energy-landscape perspective for aggregation (Figure 11.1) suggests that the free energy of stability may not be a good indicator of fibril growth kinetics. Rather, growth kinetics should correlate with unfolding barriers.

In Scenario I, the amyloidogenic state **N*** is formed by denaturation stress or other environmental fluctuations. The production of **N*** in Scenario II can occur by two distinct routes. If **N** is metastable, as is apparently the case for PrP^c,[17] then conformational fluctuations can lead to **N***. Alternatively, formation of **N*** can also be triggered by intermolecular interactions. In the latter case **N*** can form only when the protein concentration exceeds a threshold value. As noted below, there is evidence for both scenarios in the routes to fibril states.

In order to understand the kinetics of fibrillization it is necessary to characterize the early events and pathways that lead to the formation of the critical nucleus. In terms of the energy landscape, the structures of **N***, the ensemble of transition state structures, and the conformations of the critical nuclei must be known to fully describe the assembly kinetics. Teplow and co-workers, who have followed the growth of fibrils for eighteen peptides including $A\beta_{1-40}$ and $A\beta_{1-42}$,[18] showed that the formation of amyloids is preceded by the transient population of the intermediate oligomeric state with high α-helical content. This is remarkable given that both the monomers and fibrils have little or no α-helical content. Therefore, the transient formation of α-helical structure represents an on-pathway intermediate state. Somewhat surprisingly, we found using multiple long MD simulations that in the oligomerization of $A\beta_{16-22}$ peptides[6] the oligomer assembles into an anti-parallel β-structure upon interpeptide interactions. Even in the oligomerization of these small peptides from the Aβ family the assembly was preceded by the formation of an on-pathway α-helical intermediate. Based on our findings and the work by Teplow and co-workers we postulated that the formation of oligomers rich in α-helical structure may be a universal mechanism for Aβ peptides.

Formation of the on-pathway α-helical intermediate may be rationalized using the following arguments. The initial events involve the formation of "nonspecific" oligomers driven by hydrophobic interactions that reduces the effective available volume to each Aβ peptide. In confined spaces peptides tend to adopt an α-helical structure. Further structural changes are determined by the requirement of maximizing the number of favorable hydrophobic and electrostatic interactions. Provided that Aβ oligomers contain large numbers of peptides, this can be achieved if Aβ peptides adopt extended β-like conformations.

There is some similarity between the aggregation mechanism postulated for Aβ peptides and the nucleated conformational conversion (NCC) model envisioned for the conversion of Sup35 to [PSI$^+$] in *Saccharomyces cerevisiae*.[19] By studying the assembly kinetics of Sup35, Serio *et al.*[20] proposed the NCC model, which combines parts of the templated assembly and nucleation-growth mechanisms. The hallmark of the NCC model[20] is the formation of a critical-sized mobile oligomer, in which Sup35 adopts a conformation that may be distinct from its monomeric random coil or the one it adopts in the aggregated state. The formation of a critical nucleus to which other Sup35 can assemble involves a conformational change to states that it adopts in the self-propagating [PSI$^+$]. The α-helical intermediate seen in Aβ peptides may well correspond to the mobile oligomer that has the "wrong" conformation to induce further assembly.

11.2.2 The Assembly of $A\beta_{16-22}$ Oligomers

11.2.2.1 The $A\beta_{16-22}$ Monomer is a Random Coil

The small size of $A\beta_{16-22}$ peptides, which adopt an anti-parallel β-sheet structure in the fibril state, is ideal for exploring in detail the mechanism of oligomer formation. Using fairly long and multiple trajectories,[6] the assembly pathways for $3A\beta_{16-22} \to (A\beta_{16-22})_3$ were probed using all-atom simulations in explicit water. The simulations of $A\beta_{16-22}$ and the corresponding mutants allowed us to draw a number of conclusions that may be of general validity.

The simulations of the $A\beta_{16-22}$ monomer at room temperature and at neutral pH showed that it is predominantly a random coil. The finite size of the system gives rise to large conformational fluctuations that lead to the population of strand-like structures. There is a very low ($\sim 3\%$) probability of α-helical conformations. The study of this simple system shows that the β-sheet conformation adopted by the monomer must be due to interactions with other peptides.

11.2.2.2 Oligomerization of Three $A\beta_{16-22}$ Peptides Requires a Transient Monomeric α-Helical Intermediate

Upon interaction with other peptides substantial changes in the conformations of the individual monomers occur. The size of the monomer increases by about 50%. More surprisingly, we found that as the inter-peptide interactions increase there is a dramatic increase in the percentage of α-helical content during intermediate times. At longer times the monomer undergoes an $\alpha \rightarrow \beta$ transition. Due to the small size of the oligomer ($n = 3$) there are substantial conformational fluctuations even after the three strands are roughly in anti-parallel registry. Nevertheless, the simulations showed that the size of the nucleus for $A\beta_{16-22}$ cannot be large because even with $n = 3$ there are signatures of stable oligomers. Indeed, explicit simulations for $t > 300$ ns show that one can obtain nearly perfectly aligned $A\beta_{16-22}$ trimers in which the strands are in anti-parallel registry.[67] In these simulations $\langle u_i(t) u_j(t) \rangle$ where $u_i(t)$ is the unit vector connecting the N- and C-termini of peptide i and fluctuates around values close to -1.

The dominant pathway for $3\ A\beta_{16-22} \rightarrow (A\beta_{16-22})_3$ from the simulations showed that in the intermediate stages the monomer transiently populates an α-helix (see Figure 11.8 from reference 6). It should be emphasized that in the assembly process (especially in the early stages in the oligomerization) there are multiple routes. As a result, kinetic trapping can result in structures that are not conducive to forming the most stable anti-parallel structures. Such kinetically controlled structures have been explicitly probed in dimer formation of small fragments of $A\beta$ peptides. These studies and other simulations illustrate the complexity in dissecting the assembly of even small amyloidogenic peptides into ordered structures.

11.2.2.3 Role of Side-chain Interactions

The gross features of the fibril structures of a number of proteins and peptides whose monomer sequences and structures are unrelated are similar. This observation might suggest that the interactions that stabilize the oligomers and fibrils must be "universal" involving perhaps only backbone hydrogen bonds. It might appear that side chains, and hence sequence differences, might play a secondary role. Such a conclusion is further supported by repeated observations[21] that any protein or peptide can be made to form cross-β structures under appropriate conditions. However, experiments[22] and simulations[7,23] show that side-chain interactions are crucial in directing oligomer formation. Trimers of $A\beta_{16-22}$ are stabilized primarily by favorable inter-peptide hydrophobic interactions between residues in the central hydrophobic cluster (LVFFA) and secondarily by inter-peptide salt-bridge formation between K and E.

The importance of side chains can be demonstrated by examining the effect of mutations on the trimer formation in $A\beta_{16-22}$ peptides. We showed using simulations that the mutant GLVFFAK, which eliminates the formation of intermolecular salt bridge, entirely destabilizes the trimer. Similarly replacement of L, F, F by S also destabilizes the trimer. These simulations show that the sequence plays a key role in the tendency of peptides to form amyloid fibrils. Although no general role has emerged it seems that sequences with enhanced correlation between charges[12] or a preponderance of contiguous (>3) hydrophobic residues might be amenable to amyloid formation on finite timescales.

11.2.3 Dimerization of $A\beta_{10-35}$ Peptides

11.2.3.1 Generation of Putative Dimer Structures

In contrast to $A\beta_{16-22}$ fibrils the longer peptide $A\beta_{10-35}$ adopts a parallel β-sheet conformation in the amyloid state. It is now suspected the monomer in the fibril state is stabilized by an intra-molecular salt bridge between Asp23 and Lys28. In order for this salt bridge to form there has to be a bend in the monomeric structures involving the residues VGSN. The importance of a stable turn, which was experimentally determined in the NMR structures, was emphasized in MD simulations as well.

In a recent study,[8] we used a number of computational methods to probe dimer formation. We first generated a putative set of dimer conformations that is based on shape complementarity. The work on $A\beta_{16-22}$ showed that both inter-peptide hydrophobic interactions and the creation of favorable electrostatic contacts are required to produce marginally stable oligomers. In order to dissect their relative importance we generated two homodimer decoy sets by maximizing the number of contacts between the monomer interfaces. The first 2000 dimer structures of each set were selected by minimizing the interaction energy between the monomers. In order to distinguish between desolvation and electrostatic interactions we used two distinct energy functions. The φ-dimer (Figure 11.2(a)) minimizes the desolvation energies of the dimer at the interface whereas the ε-dimer (Figure 11.2(b)) corresponds to structures that have the highest inter-peptide electrostatic interactions. The structure of the φ-dimer is dominated by contacts between hydrophobic segments of the monomers. The hydrophobic core, LVFFA(17–21), and the hydrophobic C-termini of both monomers are buried at the dimer interface. The contacts at the interface of the φ-dimer are conserved over the lowest energy dimer structures. The ε-dimer interface is characterized by electrostatic inter-monomeric interactions, among which the salt bridge Glu11(A)–Lys28(B)

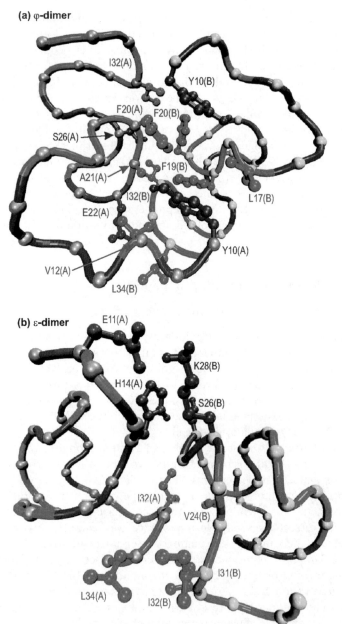

Figure 11.2 The putative dimer structures corresponding to the φ-dimer (a) and ε-dimer (b), respectively. The side chains at the dimer interface are depicted explicitly. The positively and negatively charged, polar, and hydrophobic residues are colored in blue, red, purple, and green, respectively. The C_α atoms of the monomers A (left) and B (right) are colored in cyan and yellow, respectively.

has the largest contribution. Contrary to the φ-dimer, the contacts observed at the ε-dimer interface are not conserved across the set of the low-energy dimers due to the increased specificity and strength of the electrostatic interaction.

11.2.3.2 Interior of Aβ Oligomers is Dry

Insights into the assembly mechanism of the φ-dimer and ε-dimer can be obtained from the Potential of Mean Force (PMF). The PMF for the dimerization process was obtained along the center of mass of the two monomers as Figure 11.3 indicates. For each free-energy profile, one can distinguish three distinct intervals. In the outer interval, the PMF value is nearly constant, from 6.5 Å–7.0 Å to maximum separation, which in our case is 9.0 Å. At a distance of 6.5 Å for the ε-dimer and 7.0 Å for the φ-dimer, the first solvation shells of the monomers come into contact, and for both dimers the energetics of desolvation of the associating monomers is unfavorable. In the second interval for the ε-dimer, the value of the

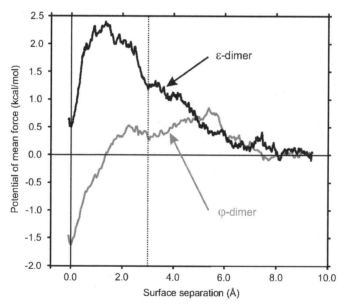

Figure 11.3 The Potential of Mean Force (PMF) is plotted for two different relative orientations of the monomeric peptide within the dimer. The PMF is computed as a function of the surface separation, $\delta = \xi - \xi_{cont}$, along the distance between the centers of mass (DCOMs) of the two monomers, where ξ and ξ_{cont} are the DCOMs of the two monomers when they are at an arbitrary separation and in contact, respectively. The black curve corresponds to the free energy surface computed using the ε-dimer as the starting structure. The gray curve is similarly computed using the φ-dimer as the starting structure. The difference between the two surfaces suggests that hydrophobic interactions may be more essential to stabilization of the dimer structure than electrostatic interactions.

PMF continues to increase up to $1.2\,\text{kcal}\,\text{mol}^{-1}$ at a $3.0\,\text{Å}$ separation; for the φ-dimer, the potential energy reaches a value of $0.8\,\text{kcal}\,\text{mol}^{-1}$ at $5.5\,\text{Å}$, and after that the desolvation is favorable, ending in an unstable local minimum at $3.0\,\text{Å}$. For the third interval, from $3.0\,\text{Å}$ to $0.0\,\text{Å}$, there is only one solvation shell between the monomers. The water molecules are most strongly ordered near the monomers through electrostatic interactions and hydrogen bonds. As a result, the PMF for the ε-dimer increases sharply between $3.0\,\text{Å}$ and $1.3\,\text{Å}$ up to $2.4\,\text{kcal}\,\text{mol}^{-1}$. At contact, the van der Waals attraction predominates, making the overall dimerization process energetically favorable. For the φ-dimer, the solvation shell between the hydrophobic regions of the monomers is only weakly bound to the solute. After a small increase in the PMF, corresponding to the van der Waals attraction, the desolvation is entirely favorable.

If the approach along the center of mass of the monomers approximately represents a minimum energy path, then the expulsion of water in the φ-dimer must be an early event in the assembly. Explicit simulations for $A\beta_{16-22}$ oligomers[6] also show that desolvation occurs early. As a result, the interior of Aβ oligomers is dry.

11.2.3.3 Hydrophobic Interactions Between Monomers Are the Driving Force in the Association of $A\beta_{10-35}$ Peptides into Dimers

Comparing the φ-dimer and ε-dimer models for monomer association, we find that the former appears to lead to more energetically favorable dimerization than the latter. It appears to be more efficient to remove the entropically unfavorable structured water between the opposing hydrophobic regions of the two monomers than to stabilize the monomer solely through electrostatic interactions. This is in good agreement with the experimental and MD simulations observation that the mutation E22Q – where a charged glutamic acid residue is replaced by a polar glutamine residue – increases the propensity for amyloid formation.[24,25] Molecular dynamics simulations of this increased amyloidogenic activity for the E22Q mutant peptide led to the conclusion that the water–peptide interaction is less favorable for the mutant peptide.[26] Following a more detailed analysis of the structure and dynamics of the WT and E22Q $A\beta_{10-35}$, it has been suggested that a change in the charge state of the peptide, due to the E22Q mutation, leads to an increase of the hydrophobicity of the peptide that could be responsible for the increased activity.[27]

The time evolution of the φ-dimer structure was analysed and it was observed that the monomers remain in contact during the simulation. It was shown that the hydrophobic interaction between the monomers of the φ-dimer acts as a stabilizing force of the dimer. The "extended core" region 15–30 of both monomers in the φ-dimer makes the principal contribution to the hydrophobic interaction energy. The φ-dimer undergoes internal structural reorganization in the terminal regions of the monomeric peptides. Our simulations indicate that there is substantial reorganization of the peptide monomers in the N- and C-terminus regions, as expected for a dimer weakly and relatively non-specifically stabilized by hydrophobic contacts at the dimer interface. Importantly, the structure of the central hydrophobic cluster LVFFA region assumes a conformation similar to that observed for the monomeric peptide in both experiment[28] and simulation.[29] Our simulations suggest that the preservation of the structure of the LVFFA central hydrophobic cluster plays an important role in the stabilization of the φ-dimer structure.

The finding that the φ-dimer may constitute the ensemble of stable $A\beta_{10-35}$ dimer has important implications for fibril formation. The initial event in the dimerization involves, in all likelihood, contacts between the central hydrophobic clusters. In this process, expulsion of water molecules in the interface might be a key event just as in the oligomerization of $A\beta_{16-22}$ fragments.[6] Since this process involves cooperative rearrangement of ordered water molecules, it is limited by an effective free-energy barrier. Based on our results, we conjecture that events prior to the nucleation process themselves might involve crossing free-energy barriers which depend on the peptide–peptide and peptide–water interactions (Figure 11.1).

11.2.4 Initial Stages in the PrP^C Conformational Transition

11.2.4.1 Experimental Observations and Theoretical Considerations

Prion proteins are extracellular globular proteins that are attached to the plasma membrane by a glycosylphosphatidylinositol (GPI) anchor. They have been linked to various transmissible spongiform encephalopathies (TSEs) including bovine spongiform encephalopathy, scrapie disease in sheep, and Creutzfeldt–Jakob disease in humans. The causative agent in these diseases is believed to be the aggregated form (PrP^{Sc}) of the cellular prion protein (PrP^C).[30] The transition to the scrapie form involves a large conformational change from the mainly α-helical PrP^C to the PrP^{Sc} state that is rich in β-sheet. According to the "protein-only" hypothesis[2] PrP^{Sc} serves as a template in inducing conformational transitions in PrP^C that can subsequently be added to PrP^{Sc}. The "protein-only" hypothesis implies that the conformational change leading to the PrP^{Sc} formation from the normal cellular form PrP^C may be spontaneous or might involve interactions with unidentified protein X.[31] Prion proteins, encoded by a single gene, consist of about 250 residues of which the first 22 form a signal sequence. This segment is followed by unstructured, but likely helical, Cu^{2+} binding octarepeats rich in glycine.[2] The NMR[30,32,33] and X-ray[34] structure of PrP^C in various species (human, mouse, syrian hamster, bovine, and sheep) shows that the ordered C-terminal part is composed of a short antiparallel β-sheet that contains 8% of the residues in the (90–231) fragment and three helices representing 48% of the secondary structure (Figure 11.4). Fourier transform infrared spectroscopy measurements[35,36] indicate that PrP^{Sc} (90–231) has 47% β-sheet and 24% α-helical content.

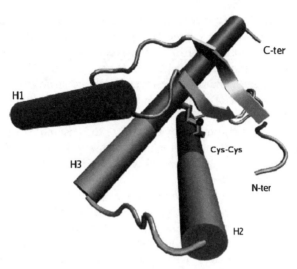

Figure 11.4 Cartoon representation of the structure of human PrP^C (PDB entry 1QLX). The three helices in the 90–231 ordered region of PrP^C are shown in dark gray, while the short β-sheet is in light gray. The two cysteine residues (179 and 214) involved in the disulfide bond that connect H2 with H3 are indicated in bond representation. The C-term end of H2 and the N-term end of H3, which we believe to be implicated in the initial stages of the α → β transition, are represented in light gray. The figure was produced with packages VMD[66] and PovRay (http://www.povray.org/).

11.2.4.2 Protein Regions Involved in the Conformational Transition

We have suggested using structural, bioinformatic, and molecular dynamics simulations that formation of PrP^{Sc} follows Scenario II (see Figure 11.1). This implies that, either spontaneously or in the presence of a seed of PrP^{Sc}, the metastable cellular form, PrP^C, undergoes a transition to the PrP^{C*} state that is capable of further aggregating or adding to an already present PrP^{Sc} particle. Experiments[37] and scenarios of protein aggregation[14] suggest the proposal that the conformational transition involving the formation of PrP^{C*} is energetically driven (*i.e.* PrP^{C*} is more stable than PrP^C). The transition from the metastable PrP^C → PrP^{C*}, which involves crossing a substantial free-energy barrier on the order of 20 kcal mol^{-1},[17,38] results in a state that can nucleate and polymerize to the protease resistant form.

We also identified the putative regions that are involved in the PrP^C → PrP^{C*} transition. Comparison of a number of structural characteristics (such as solvent accessible area, distribution of (Φ,Ψ) angles, mismatches in hydrogen bonds, nature of residues in local and non-local contacts, distribution of regular densities of amino acids, clustering of hydrophobic and hydrophilic residues in helices) between PrP^C structures and a databank of "normal" proteins shows that the most unusual features are found in helix 2 (residues 172–194) followed by helix 1 (residues 144–153).[11] In particular, the C-terminal residues in H2 are frustrated in their helical state. Application of the recently introduced notion of discordance, namely incompatibility of the predicted and observed secondary structures, also points to the frustration of H2 not only in the wild type but also in mutants of human PrP^C. This suggests that the instability of PrP^C proteins may play a role in their being susceptible to the profound conformational change.

We showed[11] that, in addition to the previously proposed role for the segment (90–120) and possibly H1, the C-terminus of H2 and possibly the N-terminus of H3 may play a role in the $\alpha \to \beta$ transition. Sequence alignments show that helices in avian prion proteins (chicken, duck, crane) are better accommodated in a helical state, which might explain the absence of PrP^{Sc} formation over finite timescales in these species. From the analysis it is clear that the conformational fluctuations in the C-terminal end of helix 2 (H2) and in parts of helix 3 (H3) are involved in the transition to PrP^{C*}. Because the stability of PrP^C arises from the structures in the C-terminal end, the transition to PrP^{C*} requires global unfolding of PrP^C,[39] which explains the origin of the high free-energy barrier separating PrP^C and PrP^{C*}.[11] NMR experiments[37,40] showed that conformational fluctuations that originate in the C-terminal part of H2 are essential in the formation of PrP^{C*}. Structural and mutational studies have also shown that the relatively short helix 1 (H1) is stable over a range of pH values and solvent conditions, and hence is unlikely to undergo conformational change in the transition to PrP^{C*}.[41-43]

The required conformational fluctuations in PrP^C needed to populate PrP^{C*} suggest that the earliest event involves extensive unfolding of the monomeric PrP^C. We used results from a database search of sequence patterns in helices of PrP^C and extensive all-atom molecular dynamics (MD) simulations of helical fragments from the mouse prion protein (mPrP^C) to shed light on the nature of instabilities that drive the PrP$^C \to$ PrP^{C*} transition.[44] Previously MD simulations have been used to probe other structural aspects of prion proteins including structures of protofibrils.[45] The 10-residue H1, with an unusual sequence pattern (highly charged and presenting the largest percentage of salt bridges in any α-helix in the PDB, see below), remains helical for the duration of the simulation ($\approx 0.09\,\mu$s). The double mutant (D147A, R151A), which eliminates one of the three salt bridges in H1, is less stable than the wild type. Multiple MD trajectories of peptides encompassing H2 and H3 (together with their connecting loop) with intact disulfide bond (Cys179–Cys214) showed that residues in the second half of H2 clustered around positions 187–188 have large conformational flexibility and non-zero preference for β-strand or coil-like structures. Instability in H2 propagates to H3 especially from position 214 onwards. Based on these results, we mapped the plausible structures of the aggregation prone PrP^{C*}. Despite the limitations (short simulation time and the expected variations of results with different force fields) of all-atom simulations, different computational approaches yield qualitatively similar results, namely the initial conformational transition must involve at least partial unfolding of parts of H2 and H3.

11.2.4.3 Structural Insights from Bioinformatic Analysis

The Pattern of Charges in H1 is Rare. The distribution of $R(+,-)$ for the 2103 helices from the DSMP shows that *no other natural sequence* has as many $(+,-)$ pairs at positions $(i, i+4)$ as H1 from PrP^C. The search of the entire PDBselect database for the H1 charge pattern shows that in only 56 (4.6%) sequences this pattern occurs at least once, with the total number of patterns being 63. If we restrict the search to be the exact pattern of H1, *i.e.* $I = -$, $i+3 = -$, $i+4 = +$, $i+7 = +$ and $i+8 = -$ the number of sequences is a mere 23 (or 1.9%). Ziegler et al.[43] arrived at a similar conclusion based on an H1 pattern search in the PDB. The 23 rare sequence fragments are either α-helical (83%) or in a random coil state (17%). Analysis of the yeast genome shows that 828 (or 9.2%) of sequences have the general pattern of H1 with only 253 (2.8%) having the exact pattern. In the *Escherichia coli* genome the numbers are 158 (3.7%) for the general charge pattern and 51 (1.2%) for the exact match. These results suggest that the sequence of H1 in PrP^C is unusual not only in its high charge content, but also in the *positioning of charges along the sequence*. More importantly, for the 23 proteins with known 3D structures, the exact charge pattern results overwhelmingly in α-helices. Even more interestingly, analysis of the 19 sequences with mostly α-helical structure reveals that the majority (88%) of $(+,-)$ pairs of residues found at positions $(i, i+4)$ form salt bridges. These results indicate that the unusual stability of the short helix H1 is possibly associated with its ability to form the highly stabilizing salt bridges involving $(i, i+4)$ residues.

Pattern of Hydrophobicity in H2 is Rare. There are very few sequences that share the pattern of hydrophobicity of H2. In PDBAstral40[46] (proteins in the PDB having at most 40% sequence similarity) there are only 12 (0.2%) such sequences. In the *E. coli* genome the number is 46 (1%), while in the yeast genome it is 122 (1.4%). Inspection of the structures of the 12 proteins from PDBAstral40 shows that the sequence is never entirely helical. For example, in only 13% of these proteins the last five residues are found in a helix. A characteristic pattern seen in H2 from mammalian prion proteins is TTTT (positions 190–193). In the PDBAstral40 this pattern occurs in only 18 proteins, including the prion sequence. In an overwhelming number of these cases (15 of the 18 proteins) the TTTT pattern is found in a strand and/or loop conformation (irrespective of the identity of the flanking amino acids). These results add further support to our proposal[11] that the second half of H2 would be better accommodated in non-helical conformations.

"Frustrated" Secondary Structural Elements May be Harbingers of a Tendency to Polymerize. The ease of aggregation and the morphology of the aggregates depend not only on the protein concentration, but also on other external conditions such as temperature, pH, and salt concentration. Although most proteins can aggregate under suitable conditions, the observation that several disease-causing proteins form amyloid fibrils under physiologically relevant conditions raises the question: Is aggregation or the need to avoid unproductive pathways encoded in the primary sequence itself? It is clear that sequences that contain a patch of hydrophobic residues are prone to form aggregates.[47] However, it is known that contiguous patches (three or more hydrophobic residues) occur with low probability in globular proteins.[48] For example, sequences with five hydrophobic residues (LVFFA in Aβ peptide) in a row are not well represented. Similarly, it is unusual to find hydrophobic residues concentrated in a specific region of helices such as is found in helix 2 in PrP^C.[11]

It is natural to wonder if secondary structure elements bear signatures that could reveal amyloidogenic tendencies. Two studies have proposed that the extent of "frustration" in the secondary structure elements (SSE) may be harbingers of amyloid fibril formation.[11,49] Because reliable secondary structure prediction requires knowing the context-dependent propensities and multiple-sequence alignments (such as used in PHD, Profile network from Heidelberg[50]), it is more likely that assessing the extent of frustration in the SSE rather than analysis of sequence patterns is a better predictor of fibril formation. Frustration in SSE is defined as the incompatibility of the predicted (from PHD, for example) secondary structure and the experimentally determined structure.[49] For example, if a secondary structure is predicted to be in a β-strand with high confidence and if that segment is found (by NMR or X-ray crystallography) to be in a helix, then the structure is frustrated (or discordant or mismatched). The α/β discordance, which can be correlated with amyloid formation, can be assessed using the score $S_{\alpha/\beta} = [1/L]\Sigma_{i=1}^{L}(R_i - 5)$, where R_i is the reliability score predicted by PHD at position i of the query sequence, 5 is the mean score, and L is the sequence length. The bounds on $S_{\alpha/\beta}$ are $0 \leq S_{\alpha/\beta} \leq 4$ with maximal frustration corresponding to $S_{\alpha/\beta} = 4$. Similarly, the measure $S_{\beta/\alpha}$ gives the extent of frustration of a stretch that is predicted to be helical and is found experimentally to be a strand. Using $S_{\alpha/\beta}$ and other structural characteristics, one can make predictions of the plausible regions that are most susceptible to large conformational fluctuations.

PrP^C and Dpl. Using the above concept of SSE frustration the 23 residue sequence (QNNFVHDCVNITIKQHTVTTTTK) in mouse PrP^C, with a score of 1.83, was assessed to be frustrated or discordant.[11] Other measures of quantifying the structure also showed that the maximal frustration is localized in the second half (C-terminal of H2).[11] The validity of this prediction finds support in the analysis of mutants of the PRNP gene associated with inherited TSEs (familial CJD and FFI). According to SWISS-PROT[51] seven disease-causing point mutations (D178N, V180I, T183A, H187R, T188R, T188K, T188A) are localized in H2. We have used the sequence numbering for the $mPrP^C$. A naive use of propensities to form helices, à la Chou-Fasman,[52] would suggest that with the exception of D178N all other point mutations should lead to better helix formation. However, the $S_{\alpha/\beta}$ scores for the mutants are 1.94, 1.80, 1.30, 1.80, 1.54, 1.94, and 1.94 for D178N, V180I, T183A, H187R, T188K, T188R, and T188A respectively. Thus, in all these mutants H2 is frustrated making it susceptible to the conformational fluctuations that have to occur prior to fibrillization. The differences in $S_{\alpha/\beta}$, which can be correlated with local stability, suggest that stability alone might not be a good indicator of the kinetics of amyloid formation.

The gene coding for the Doppel protein (Dpl), termed Prnd,[53] is a paralog of the prion protein gene, Prnp, to which it has about 25% identity. Normally, Dpl is not expressed in the central nervous system, but it is up-regulated in mice with knockout Prnp gene. In such cases, over-expression of Dpl causes ataxia with Purkinje cell degeneration,[53] which in turn can be cured by the introduction of one copy of wild-type PrP mouse gene.[54] NMR studies of the three-dimensional (3D) structure of mouse Dpl[55] showed that it is structurally similar to the structure of PrP^C (Figure 11.5). However, PrP^C and Dpl produce diseases of the central nervous system using very different mechanisms: PrP^C causes disease only after conversion to the PrP^{Sc} form, while simple over-expression of Dpl, with no necessity to form the scrapie form, causes ataxia.

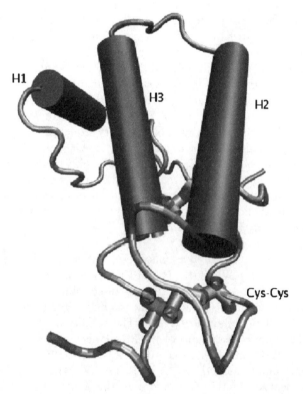

Figure 11.5 Cartoon representation of the structure of human Doppel protein (Dpl) (PDB entry 1LG4). The three helices in the 24–152 ordered region of Dpl are represented in dark gray. The four cysteine residues (94, 108, 140, and 145) involved in the two disulfide bonds that connect H2 with H3 and the loops preceding them are indicated in bond representation and colored in light gray. The figure was produced with packages VMD[66] and PovRay (http://www.povray.org/).

Scenarios for Protein Aggregation

The markedly different disease mechanisms of PrP and Dpl would suggest, in light of the findings for PrP^C, that the mouse Dpl (PDB code 1i17) would not be frustrated. Indeed, prediction of secondary structure by PHD[50] on mouse Dpl correlates well with the experimentally derived structure. The only difference between the predicted and the derived structure in Dpl is found in the first β-strand region, which is predicted to be helical by PHD. But the corresponding $S_{\beta/\alpha} = -3.0$ indicating that this α-helix prediction is unreliable as this sequence has low complexity. Also, the analysis of 1i17 with the WHAT CHECK program[56] reveals that, on average, there are only eight unsatisfied buried hydrogen-bond donors/acceptors representing 7.4% of all residues in mouse Dpl. This is comparable with the average value of 6% found in normal proteins, but it is quite a lot smaller than the 14% value seen in mPrP (PDB code 1ag2). This analysis rationalizes the lack of observation of scrapie formation in Dpl.

11.2.4.4 Conformational Fluctuations from Molecular Dynamics Simulations

Helix 1 in mPrP is Stable. In order to dissect the stability of PrP^C fragments that were identified using bioinformatic analysis, we used MD simulations of H1, H2, and H3 from the PrP^C state. With the exception of residues 150–152, the propensities of the interior residues for α-helical or β-strand conformations show that the helical structure is overwhelmingly preferred. The distribution of distances between residues at positions $(i, i + 4)$ averaged over the five trajectories shows that, with the exception of residues in the second half of H1, the helical structure is preserved. Snapshots of typical conformations at various moments along one of the trajectories show that even the C-terminal end of H1, which becomes disordered after ∼12 ns, returns to the helical conformation towards the end of the run. Small fluctuations in a short helix are unusual because it is known that isolated helices are at best marginally stable.[57,58]

In order to check if the predicted stability of H1 depends on the force field we generated two trajectories for a total of 40 ns using the CHARMM27 parameter set with the package NAMD. The backbone RMSD with respect to the PDB structure stabilizes around 2.5–3.0 Å after ∼10 ns. The RMSD for the backbone of the 144–149 fragment of the chain remains close to 0.5 Å for the duration of the run, which is in very good agreement with the previous set of simulations. The difference between these two sets of simulations is only in the fraying of the C-terminus residues. These results, which are consistent with the MOIL simulations, also show that the fraction of helix content in H1 is high.

Mutations of Residues in the Second Salt Bridge (D147–R151) Enhance Conformational Fluctuations. The pattern searches suggest that the three $(i, i + 4)$ salt bridges ((Asp144,Arg148), (Asp147,Arg151), and (Arg148,Glu152)) in H1 should stabilize the isolated H1. To probe the importance of the second salt bridge (Asp147,Arg151), we simulated the double mutant H1[D147A,R151A]. Replacing D and R by A should not compromise the local helical propensity because Ala is the best helix-former among the amino acids.[59] Consequently, any loss of stability in the structure can be attributed largely to the loss of the salt bridge. From relatively long MD simulations for H1[D147A,R151A], we find that the double mutant has increased conformational flexibility compared to the wild-type chain. Most residues, except position 145, have non-zero β-strand propensity.

The larger conformational fluctuations result in extended states with only the first turn of the helix still present. Based on these findings, we conclude that H1[D147A,R151A] populates two basins of attraction: one that is predominantly α-helical with a radius of gyration ∼6 Å, and the other being mostly RC with a radius of gyration of ∼7.7 Å. Time evolution of distances between ($i,i+4$) residues (data not shown) shows that the conformational change starts towards the C-terminus part of the sequence and proceeds in a highly cooperative manner. Our findings are in agreement with recent experiments,[43] which showed that the peptide huPrP(140–158)D147A is destabilized compared to wt-huPrP(140–158). The decreased stability of the mutant could result in the efficient conversion of PrP^C (90–231) to the protease resistant form.

By classifying the structures generated in the MD simulations as helical,[6] we find that the helical fraction, f_H, of the mutant is 0.55 while f_H for the WT is 0.64. The value of f_H is 0.63 using the CHARMM parameters. We should emphasize that the absolute values of f_H might be overestimated and could depend upon the force field. However, meaningful conclusions can be drawn using the relative values. Using the f_H values we can estimate the free energy of stability using $\Delta F = -RT \ln([(f_H)/(1-f_H)])$. For the WT $\Delta F_{WT} \sim -0.37$ kcal mol^{-1}, whereas for H1[D147A,R151A] $\Delta F_M \sim -0.13$ kcal mol^{-1}. If f_H from the CHARMM parameter set is used then $\Delta F_{WT} \sim -0.34$ kcal mol^{-1}. The relative difference $\Delta\Delta F = \Delta F_{WT} - \Delta F_{M\Delta} \sim -0.24$ kcal mol^{-1}, which arises from the salt-bridge formation in WT. Interestingly, this estimate for free-energy gain due to salt bridge formation is in the range of the values reported in the literature.[60]

Second Half of H2 is Susceptible to Conformational Fluctuations. The trajectories, obtained using the NAMD package for a total of 185 ns, showed a drastic reduction in the amount of helical structure accompanied by an increase in β-strand content. The conformational transition starts in the second half of H2 and propagates towards its N-terminal, while H3 unwinds concomitantly at its two ends. The propensities of residues for α-helical or β-strand conformations show that only positions 178 and 179 (H2) and residues 205 to 212 (H3) maintain their native α-helical structure. The extent of the conformational transition is also reflected in the behavior of the backbone RMSD from the PDB structure (1ag2), which increases monotonically from 3 Å to 6 Å in about 5 ns and reaches 11 Å in the next 70 ns.

The conformational transitions are correlated with an increase in the angle between the axes of the two halves of H2 that changes from 20° to 90° (in the first 10 ns) followed by rapid oscillations between these values for the remainder of the trajectory. The transition is initiated in the second half of H2 where the distances between $(i, i + 4)$ positions increase from 5 to 14 Å in about 10 ns. At longer timescales ($t \sim 60$ ns) the distances between $(i, i + 4)$ residues in the first half of H2 also increase from 5 to 13 Å. These motions in H2 are correlated with fluctuations in H3, where the distances between the first four $(i, i + 4)$ pairs of residues in H3 and between positions 212 and 218 (with the exception of Cys214) increase from 5 to 13 Å in about 10 ns. Almost complete loss of helical structure occurs towards the end of the trajectory (Figure 11.6). Thus, the conclusions based on bioinformatic analysis are consistent with the results of MD simulations.

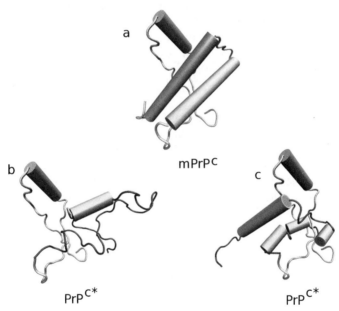

Figure 11.6 Schematic representation of $PrP^C \to PrP^{C*}$ transition, where the conformation of PrP^C is taken from the PDB file 1ag2 (light gray). The conformations of PrP^{C*} contain H1 from 1ag2 while the residues encompassing H2 + H3 are shown in a conformation (dark gray) reached towards the end of our MD simulations using the NAMD package (b) or the simulations using the MOIL package (c). The schematic PrP^{C*} structures are representatives from ensembles of fluctuating conformations. In the representative PrP^{C*} structure obtained using NAMD simulations the H1 region, together with the adjacent loops and the β-strands, and residues (205–212) from H3 retain their original conformations and are therefore depicted with the same color as in PrP^C. In the MOIL representative PrP^{C*} structure the H1 region, together with the adjacent loops and the β-strands, and residues (175–179), (184–188), (193,194) from H2 and residues (203–218) from H3 retain their original conformations and are therefore depicted with the same color as in PrP^C. The figures are rotated such that the orientation of H1 is the same in all of them. The figures were produced with packages VMD[66] and PovRay (http://www.povray.org/).

Proposed Structures for PrPC*. Our simulations[44] and recent experiments[42,43] strongly suggest that H1 is unlikely to change conformation in the $PrP^C \rightarrow PrP^{C*}$ transition. The most drastic change occurs in the second half of H2 and parts of H3. Based on the assumption that alterations in the conformation of H2 + H3 do not significantly affect the rest of the protein, we have constructed a plausible ensemble of structures for PrP^{C*} (Figure 11.6(b) and (c)). In PrP^{C*} (90–231) obtained from the NAMD trajectories (Figure 11.6(b)) the helical content is ~20% (a lower bound), and in PrP^{C*} (90–231) reached during the MOIL simulations (Figure 11.6(c)) the helical content is ~30% compared to 48% in $mPrP^C$ (90–231).

The overall characteristics of these structures are consistent with those proposed by James and co-workers.[37] It remains to be seen if formation of PrP^{C*}, with fluctuating regions in H2 + H3, is required for oligomerization of PrP^C i.e., if PrP^{C*} is an on-pathway monomeric intermediate on the route to fibrillization. We should emphasize that the conformation of the prion protein in PrP^{Sc} need not coincide with PrP^{C*}.

Comparison of PrPC* Structure with the Human Prion Protein Dimer. In an important paper Knaus et al.[61] announced a 2 Å crystal structure of the dimeric form of the human prion protein (residues 90–231). The structure (Figure 11.7) suggests that dimerization occurs by domain-swap mechanisms in which H3 from one monomer packs against H2 from another. In fact, Eisenberg and co-workers have suggested that the domain-swapping mechanism may be a general route of amyloid fibril formation.[62] The electron density map seems to suggest structural fluctuations in the residues 189–198, which coincides with the maximally frustrated region predicted theoretically. The dimer interface is stabilized by residues that are in H2 in the monomeric NMR structures. The header of the PDB file of the monomeric structure of human PrP^C indicates that H2 ends at residue 194 and H3 begins at 200. The domain-swapped dimer structure shows that residues 190–198 exist largely in a β-strand conformation. The $\alpha \rightarrow \beta$ transition minimizes frustration. An implication of the dimer structure is that oligomerization occurs by domain swapping, which in PrP^C also might implicate the disulfide bond between Cys residues at 179 and 214. The role of the S–S bond in the PrP^{Sc} formation remains controversial. In our full-atom MD simulations of reduced mouse PrP^C (data not published), we found structures that closely resemble the monomeric structure in this dimer. For example, H1 remains mostly intact, while H2 breaks into two smaller helices, one running from its normal N-term end to position 187 and the other being formed by the C-term end residues of the original H2 and residues from the loop connecting it to H3. These findings suggest that the dimer structure is a likely route to unfolding and self-assembly of monomeric PrP^C.

Sequence pattern matches and long multiple molecular dynamics simulations of helix 1 in $mPrP^C$ using two force fields show that the stability of H1 is due to the formation of stabilizing internal salt bridges. In view of the high propensity of α-helix observed in the isolated H1 in conjunction with supporting

Scenarios for Protein Aggregation 261

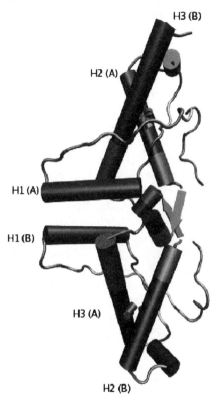

Figure 11.7 Cartoon representation of the X-ray crystal structure of the human PrP^C dimer (PDB entry 1I4M). For each chain, A and B, the three helices in the 90–231 ordered region of PrP^C are colored in dark gray, while the short β-sheet is in light gray. The two cysteine residues (179 and 214) involved in the disulfide bond that connect H2 with H3 are indicated in bond representation. The C-term end of H2 and the N-term end of H3, which we believe to be implicated in the initial stages of the α → β transition, are colored in light gray. We notice that, in contrast to the monomeric PrP^C structure from Figure 11.4, here this region is no longer entirely helical, but contains a short stretch of β-strand structure and a shorter helix as well. The figure was produced with packages VMD[66] and PovRay (http://www.povray.org/).

experimental results[41–43] it is clear that alterations in the conformation of H1 are unlikely in the PrP^C → PrP^{C*} transition.

The predicted tendency for the second half of H2 to be involved in the formation of PrP^{C*} is also consistent with the observation that a number of mutations at 187 and 188 (H187R, T188R, T188K, and T188A) are associated with various prion diseases. Based on our findings we proposed that regions 186–190 and 214–226 must play a central role *in the initial stages* that involve the PrP^C → PrP^{C*} transition. The large conformational change is likely to be accompanied by stretching and rotation of the two halves of H2 and by the unwinding of the N-terminal end of H3. The formation of the domain-swapped structure in the dimer structure of human PrP^{C61} might be facilitated by these large-scale motions.

11.3 Conclusions

The development of methods to envisage the structure of amyloid fibrils has enabled us to obtain molecular insights into the assembly process itself. Computational and experimental studies are beginning to provide detailed information, at the residue level, about the regions in a given protein that harbor amyloidogenic tendencies. We have harnessed these developments to propose tentative ideas on the molecular basis of protein aggregation. These principles (or, more precisely, rules of thumb) may be useful in the interpretation and design of new experiments.

Examination of the stable structures of oligomers and fibrils obtained using experimental restraints and simulations show that these must be stable conformations that maximize the inter-peptide interactions and minimize electrostatic repulsions. Broadly, this is the only amyloid self-organization principle (ASOP) that seems to be obeyed. From the ASOP it follows that the formation of amyloid fibrils should indeed be a generic property of almost all proteins and peptides under suitable conditions. If this were the case then it is remarkable that during normal function aggregation is avoided most of the time. The lack of preponderant protein aggregation may well be due to the efficiency of cleaning mechanisms operating in the cell. This may explain the lack of aggregation of PrP^{C*} under most circumstances. We conjecture that because of efficient degradation processes only mild sequence constraints are needed to prevent oligomer formation during the typical life cycle of newly synthesized proteins.

From a biophysical perspective there are a number of open problems. Are there common pathways involved in the self-assembly of fibrils? Because of the paucity of the structural description of the intermediates involved in an aggregation process a definitive answer cannot be currently provided. The energy landscape perspective, summarized briefly in Figure 11.1 (see also Chapter 3 by Woylnes), suggests that multiple scenarios for assembly must exist. Although the generic nucleation and growth governs fibril formation, the details can vary considerably. The microscopic basis for the formation of distinct strains in mammalian prions and in yeast prions remains a mystery. Are these merely associated with the heterogeneous seeds or are there unidentified mechanisms that lead to their growth? What factors may determine the variations in the fibrillization kinetics for the wild type and the mutants? A tentative proposal is that the kinetics of polymerization is determined by the rate of production of N^* (Figure 11.1),[63] which in turn is controlled by barriers separating N and N^*.[11,64] In this scenario the stability of N plays a secondary role. The generality of this observation has not yet been established. Finally, how can one design better therapeutic agents based on enhanced knowledge of the assembly mechanism? Even in the case of sickle cell disease viable therapies began to emerge only long after the biophysical aspects of gelation were understood.[65]

References

1. D. J. Selkoe, *Physiol. Rev.*, 2001, **81**, 741–766.
2. S. B. Prusiner, *Proc. Natl. Acad. Sci. U.S.A.*, 1998, **95**, 13363–13383.
3. F. Chiti and C. M. Dobson, *Annu. Rev. Biochem.*, 2006, **75**, 333–366.
4. E. H. Koo, P. T. Lansbury and J. W. Kelly, *Proc. Natl. Acad. Sci. U.S.A.*, 1999, **96**, 9989–9998.
5. R. Kayed, E. Head, J. L. Thompson, T. M. McIntire, S. C. Milton, C. W. Cotman and C. G. Glabe, *Science*, 2003, **330**, 486–489.
6. D. K. Klimov and D. Thirumalai, *Structure*, 2003, **11**, 295–307.
7. J. Gsponer, U. Haberthur and A. Caflisch, *Proc. Natl. Acad. Sci. U.S.A.*, 2003, **100**, 5154–5159.
8. B. Tarus, J. E. Straub and D. Thirumalai, *J. Mol. Biol.*, 2005, **345**, 1141–1156.
9. N. V. Buchete, R. Tycko and G. Hummer, *J. Mol. Biol.*, 2005, **353**, 804–821.
10. R. I. Dima and D. Thirumalai, *Protein Sci.*, 2002, **11**, 1036–1049.
11. R. I. Dima and D. Thirumalai, *Biophys. J.*, 2002, **83**, 1268–1280.
12. R. I. Dima and D. Thirumalai, *Bioinformatics*, 2004, **20**, 2345–2354.
13. J. D. Harper and P. T. Lansbury, *Annu. Rev. Biochem.*, 1997, **66**, 385–407.
14. D. Thirumalai, D. K. Klimov and R. I. Dima, *Curr. Opin. Struct. Biol.*, 2003, **13**, 146–159.
15. A. L. Fink, *Fold. Des.*, 1998, **3**, R9–R23.
16. J. W. Kelly, *Curr. Opin. Struct. Biol.*, 1998, **8**, 101–106.
17. I. V. Baskakov, G. Legname, S. B. Prusiner and F. E. Cohen, *J. Biol. Chem.*, 2001, **276**, 19687–19690.
18. M. D. Kirkitadze, M. M. Condron and D. B. Teplow, *J. Mol. Biol.*, 2001, **312**, 1103–1119.
19. P. Chien and J. S. Weissman, *Nature*, 2001, **410**, 223–227.
20. T. R. Serio, A. G. Cashikar, A. S. Kowal, G. J. Sawicki, J. J. Moslehi, L. Serpell, M. F. Arnsdorf and S. L. Lindquist, *Science*, 2000, **289**, 1317–1321.
21. F. Chiti, P. Webster, N. Taddei, A. Clark, M. Stefani, G. Ramponi and C. M. Dobson, *Proc. Natl. Acad. Sci. U.S.A.*, 1999, **96**, 3590–3594.
22. I. M. Ivanova, M. R. Sawaya, M. Gingery, A. Attinger and D. Eisenberg, *Proc. Natl. Acad. Sci. U.S.A.*, 2004, **101**, 10584–10589.
23. D. K. Klimov, J. E. Straub and D. Thirumalai, *Proc. Natl. Acad. Sci. U.S.A.*, 2004, **101**, 14760–14765.
24. L. Miravalle, T. Tokuda, R. Chiarle, G. Giaccone, O. Bugiani, F. Tagliavini, B. Frangione and J. Ghiso, *J. Biol. Chem.*, 2000, **275**, 27110–27116.
25. W. P. Esler, A. M. Felix, E. R. Stimson, M. J. Lachenmann, J. R. Ghilardi, Y. A. Lu, H. V. Vinters, P. W. Mantyh, J. P. Lee and J. E. Maggio, *J. Struct. Biol.*, 2000, **130**, 174–183.

26. F. Massi and J. E. Straub, *Biophys. J.*, 2001, **81**, 697–709.
27. F. Massi, D. K. Klimov, D. Thirumalai and J. E. Straub, *Protein Sci.*, 2002, **11**, 1639–1647.
28. S. Zhang, N. Casey and J. P. Lee, *Fold. Des.*, 1998, **3**, 414–422.
29. F. Massi, J. W. Peng, J. P. Lee and J. E. Straub, *Biophys. J.*, 2001, **80**, 31–44.
30. R. Riek, S. Hornemann, G. Wider, M. Billeter, R. Glockshuber and K. Wuthrich, *Nature, London*, 1996, **382**, 180–182.
31. G. C. Telling, M. Scott, J. Mastrianni, R. Gabizon, M. Torchia, F. E. Cohen, S. J. DeArmond and S. B. Prusiner, *Cell*, 1995, **83**, 79–90.
32. D. G. Donne, J. H. Viles, D. Groth, I. Mehlhorn, T. L. James, F. E. Cohen, S. B. Prusiner, P. E. Wright and H. J. Dyson, *Proc. Natl. Acad. Sci. U.S.A.*, 1997, **94**, 13452–13457.
33. R. Zahn, A. Liu, T. Luhrs, R. Riek, C. von Schroetter, F. L. Garcia, M. Billeter, L. Calzolai, G. Wilder and K. Wuthrich, *Proc. Natl. Acad. Sci. U.S.A.*, 2000, **97**, 145–150.
34. L. F. Haire, S. M. Whyte, N. Vasisht, A. C. Gill, C. Verma, E. J. Dodson, G. G. Dodson and P. M. Bayley, *J. Mol. Biol.*, 2004, **336**, 1175–1183.
35. B. W. Caughey, A. Dong, K. S. Bhat, D. Ernst, S. F. Hayes and W. S. Caughey, *Biochemistry*, 1991, **30**, 7672–7680.
36. M. Gasset, M. A. Baldwin, R. J. Fletterick and S. B. Prusiner, *Proc. Natl. Acad. Sci. U.S.A.*, 1993, **90**, 1–5.
37. K. Kuwata, H. Li, H. Yamada, G. Legname, S. B. Prusiner, K. Akasaka and T. L. James, *Biochemistry*, 2002, **41**, 12277–12283.
38. I. V. Baskakov, G. Legname, M. A. Baldwin, S. B. Prusiner and F. E. Cohen, *J. Biol. Chem.*, 2002, **277**, 21140–21148.
39. L. P. Hosszu, N. J. Baxter, G. S. Jackson, A. Power, A. R. Clarke, J. P. Waltho, C. J. Craven and J. Collinge, *Nature, Struct. Biol.*, 1999, **6**, 740–743.
40. K. O. Kuwata, K. Kamatari, K. Akasaka and T. L. James, *Biochemistry*, 2004, **43**, 4439–4446.
41. A. Liu, R. Riek, R. Zahn, S. Hornemann, R. Glockshuber and K. Wuthrich, *Biopolymers*, 1999, **51**, 145–152.
42. J. O. Speare, T. S. Rush III, M. E. Bloom and B. Caughey, *J. Biol. Chem.*, 2003, **278**, 12522–12529.
43. J. Ziegler, H. Sticht, U. C. Marx, W. Muller, P. Rosch and S. Schwarzinger, *J. Biol. Chem.*, 2003, **278**, 50175–50181.
44. R. I. Dima and D. Thirumalai, *Proc. Natl. Acad. Sci. U.S.A.*, 2004, **101**, 15335–15340.
45. M. DeMarco and V. Daggett, *Proc. Natl. Acad. Sci. U.S.A.*, 2004, **101**, 2293–2298.
46. J. M. Chandonia, N. S. Walker, L. L. Conte, P. Koehl, M. Levitt and S. Brenner, *Nucleic Acids Res.*, 2002, **30**, 260–263.
47. M. W. West, W. Wang, J. Patterson, J. D. Mancias, J. R. Beasley and M. H. Hecht, *Proc. Natl. Acad. Sci. U.S.A.*, 1999, **96**, 11211–11216.
48. R. Schwartz, S. Istrail and J. King, *Protein Sci.*, 2001, **10**, 1023–1031.

49. Y. Kallberg, M. Gustafsson, B. Persson, J. Thyberg and J. Johansson, *J. Biol. Chem.*, 2001, **276**, 12945–12950.
50. B. Rost and C. Sander, *J. Mol. Biol.*, 1993, **232**, 584–599.
51. A. Bairoch and R. Apweiler, *Nucleic Acids Res.*, 2000, **28**, 275–284.
52. P. Y. Chou and G. D. Fasman, *Annu. Rev. Biochem.*, 1978, **47**, 251–276.
53. R. C. Moore, I. Y. Lee, G. L. Silverman, P. M. Harrison, R. Strome, C. Heinrich, A. Karunaratne, S. H. Pasternak, M. A. Chishti, Y. Liang, P. Mastrangelo, K. A. Wang, A. F. Smit, S. Katamine, G. A. Carlson, F. E. Cohen, S. B. Prusiner, D. W. Melton, P. Tremblay, L. E. Hood and D. Westaway, *J. Mol. Biol.*, 1999, **292**, 797–817.
54. N. Nishida, P. Tremblay, T. Sugimoto, K. Shigematsu, S. Shirabe, C. Petromilli, S. P. Erpel, R. Nakaoke, R. Atarashi, T. Houtani, S. Sakaguchi, S. J. DeArmond, S. B. Prusiner and S. Katamine, *Lab. Invest.*, 1999, **79**, 689–697.
55. H. Mo, R. C. Moore, F. E. Cohen, D. Westaway, S. B. Prusiner, P. E. Wright and H. J. Dyson, *Proc. Natl. Acad. Sci. U.S.A.*, 2001, **98**, 2352–2357.
56. R. W. W. Hooft, G. Vriend and C. Sander, *Nature, London*, 1996, **381**, 272–273.
57. A. Bundi and K. Wuthrich, *Biopolymers*, 1979, **18**, 299–311.
58. H. Dyson and P. E. Wright, *Nat. Struct. Biol.*, 1998, **5**, 499–503.
59. T. E. Creighton, *Proteins: Structures and Molecular Properties*, W. H. Freeman and Company, New York, 1993.
60. G. I. Makhatadze, V. V. Loladze, D. N. Ermolenko, X. Chen and S. T. Thomas, *J. Mol. Biol.*, 2003, **327**, 1135–1148.
61. K. J. Knaus, M. Morillas, W. Swietnicki, M. Malone, W. K. Surewicz and V. C. Yee, *Nature: Struct. Biol.*, 2001, **8**, 770–774.
62. Y. Liu, G. Gotte, M. Libonati and D. Eisenberg, *Nature Struct. Biol.*, 2002, **8**, 211–214.
63. M. Ramirez–Alvarado, J. S. Merkel and L. Regan, *Proc. Natl. Acad. Sci. U.S.A.*, 2000, **97**, 8979–8984.
64. P. Hammarstrom, X. Jiang, A. R. Hurshman, E. T. Powers and J. W. Kelly, *Proc. Natl. Acad. Sci. U.S.A.*, 2002, **99**, 16427–16432.
65. W. A. Eaton and J. Hofrichter, *Science*, 1995, **268**, 1142–1143.
66. W. Humphrey, A. Dalke and K. Schulten, *J. Mol. Graphics*, 1996, **14**, 33–38.
67. P. H. Phuong, M. S. Li, G. Stock, J. E. Straub and D. Thirumalai, *Proc. Natl. Acad. Sci. U.S.A.*, 2007, **104**, 111–116.

Subject Index

Page references to *figures*, *tables* and *text boxes* are shown in *italics*.

3_{10}-helix 3

acetylation 5–6
acid denatured state, characterization 73–4
activated condensed phase reactions 111–12
activation barrier, polypeptide chains 111, *112*
AGADIR, LR-based helix–coil model 12, 39
AK peptides 4
α-helix
  formation 28–46
  history of study 28–30
  kinetics 31–9
  stability 1–2, 13–21, 38
  structure 1–2
  see also helix–coil kinetics; peptide helices, design of
$α_L$ motif 2
Alzheimer's disease (AD) 215–16
AMBER force field variants, computer simulations 42, 175, 178–9, 195
amidation 6
amide hydrogen exchange *72*
  β-helix structure *72*
  Linderstrøm–Lang model, folded proteins 72–3
  unfolded polypeptide *72*
amino acid helix propensities *14, 16–17*

amyloid formation 214–35
  amyloid induced toxicity 224–6
  cytotoxicity studies 228–9, 234–5
  experimental study techniques 226–8
  fibril structural architecture 221–4
  histopathological stains 216
  historical perspective 216
  molecular basis 217–21
  supramolecular structure 222
amyloid model systems 229–35
  protein amyloidogenic regions 230–5
  protein family variations 229–30
amyloid precursor protein (APP) 217, 218–20
amyloid protofilament core 222–4
  cross-β motif *223*
  β-helix structure *224*
  zipper spine model *224*
amyloid self-organization principle (ASOP) 262–3
amyloid stretch hypothesis 232–4
amyloidoses, human 70, *215,* 215–16
Anfinsen's rule, protein folding 70, 189
apomyoglobin (AMb)
  amide hydrogen exchange 74
  collapse times 122–4
  pulsed-amide H/D exchange 78
atomic force fields 164–6
atomic force microscopy (AFM) 140, 227
autocorrelation function, concentration fluctuation 147–8

Subject Index 267

BBA5 mini-protein 125, 175, 180–1
BBL mini-protein 122–3, 128–9
β-amyloid formation *see* amyloid formation
bovine spongiform encephalopathy (BSE) 251
Bryngelson scenario, downhill folding 127, *128*
buffed energy landscape 54

capillarity 61–4
capping box 2
capping motifs 2–3
 α-helix stability 14–15, 20
cellular prion protein *see* PrP^C conformational transition
CHARMm, computer simulations 195, 257–8
circular dichroism (CD)
 amyloid formation 227–8
 helix–coil kinetics 32
co-factors, design studies 202
computational protein design 193–8
 backbone structure 194
 directed evolution 203–4
 energy function 195
 examples *200*
 residue degrees of freedom 194–5
computed *vs.* experimental folding barriers 113–14
computer simulations and investigation tools 161–84
 AGADIR, LR-based helix–coil model 12, 39
 α-helix formation 34–5
 AMBER force field variants 42, 175, 178–9, 195
 challenges 163–4
 CHARMm 195, 257–8
 coarse-grained models 170–1
 constraints 161–3
 eCodonOpt 203
 estimating rates, two-state approximation 173–6
 future developments 182–4
 graph-based methods 171

Gromos 195
high-temperature unfolding 170
implicit solvation models 166–7, 170
kinetics simulations 179–81
low-viscosity models 170, 172
Markovian State Models (MSMs) 171–2, 176–7
MD simulation, free-energy landscape 41–3, *168, 169*
minimalist models 167–8, 170–1
model accuracy 168–9
NAMD 257–60
nucleation-elongation model, free-energy landscape 39–41
protein folding 161–84
Replica Exchange Molecular Dynamics (REMD) *168,* 169, 182
SCADS (statistical computationally assisted design strategy) 199, 201–2
SCHEMA 203–4
SIRCH 204
thermodynamics simulations 181–2
Tightly Coupled Molecular Dynamics (TCMD) 169
timescales *163*
TIP3P model 175, 178
T-jump studies, helix–coil kinetics *43*
unfolding rate predictions 177–8
validation 172–9
concentration determination 5–6
confocal single molecule experiment *142*
cooperativity, side-chain interactions 18
Coulomb's law 165
Creutzfeldt–Jakob disease 251
Csp*Bc* cold shock protein 153
Csp*Tm* cold shock protein, kinetics 143, *144,* 146, 147, 153–4
C-terminal β hairpin, protein G, folding simulations 178–9, 180
C-terminus 2
 amino acid propensities *16–17*
 LR model 10
 stability affects 14–15
Rd-apocytochrome $b_{562,}$ native-state hydrogen exchange 83

cytochrome c (cyt c)
  amide hydrogen exchange 75
  collapse time 123
  native-state hydrogen exchange 82
  pulsed-amide H/D exchange 78
  rapid-flow mixing experiments 114–15

denaturation (irreversible), free-energy barriers 100–3
differential scanning calorimetry (DSC) 85–103
  denaturation free-energy barriers 100–3
  folding free-energy barriers 96–8
  partition functions 87–93
  protein folding–unfolding thermal equilibria 85–7
  two-state equilibrium model 93–6
diffusion coefficient, α-helix formation 44–5
dimers, Aβ peptide 247–51, *248, 249*
directed evolution, protein design 192, 203–4
DNA polymerase 192
DNA shuffling 203
Doppel protein (Dpl), human, and PrP 255–7, *256*
downhill folding 127–30
  evolutionary selection 129–30
  intermediates 130
Dps (DNA binding protein from starved cell) 202–3
dye binding techniques, amyloid formation 228

eCodonOpt, computer simulations 203
electron microscopy (EM), amyloid formation 227
energy landscape 39–41
  folding kinetics 64–6
  long evolved proteins 55–61
  low barrier *vs.* high barrier free-energy surfaces *122*
  MD simulation 41–3
  minimal frustration and capillarity 61–4

nucleation-elongation model 39–41
  simulations 181–2
  statistical analysis 50–4
  theoretical basis 49–50
E22Q mutation 250–1

fast protein folding 106–31
  downhill folding 127–30
  folding timing and limits 106–8
  future developments 130–1
  intermediates 108
  microsecond protein folding 115–27
  multiple pathway scenarios *108*
  polypeptide chain dynamics 110–15
  'speed limit' 106, 108–9
  study rationale 108–10
  tailored mutations for ultrafast folding, 198–9
fibril formation *see* amyloid formation
fluctuation-dissipation theorem 146
fluorescence, helix–coil kinetics 31
fluorescence spectroscopy 120, 139
  confocal technique, freely diffusing molecules *142*
  fluorescence correlation spectroscopy (FCS) 140, 143, 146–8
folding time *vs.* temperature *58,* 58–61
formation, α-helix 28–46
  diffusion coefficient 44–5
  free-energy landscape 39–41
  mechanisms 41–3
  reaction coordinates 43–4
Förster radius, FRET 149–50
Förster resonant energy transfer *see* FRET
free-energy landscape *see* energy landscape
free-energy profile *vs.* distribution probabilities *40*
Freire–Biltonen partition function 86–7, 91–3, 99
FRET 113, 120
  distance dynamics 153–4
  Förster radius 149–50
  RNase analysis 154
  single molecule experiments 149–50

Subject Index

timescales and distance distributions 150–3
transfer efficiency 149, 151–2
'frustrated' secondary structural elements, PrP 254–5
'funnel' diagram, foldable protein *57*

generalized Born (GB) continuum solvent model 41–2
Genome Project 71
glycinamide ribonucleotide (GAR) transformylases 204
glycosylphosphatidylinositol (GPI) anchor 251
gossamer molecules 50
Gromos computer simulation 195
growth phase, amyloid fibril formation *220*, 220–1

hairpin structures 174–5
heat capacity 85–93
  heat capacity *vs.* temperature *86, 100*
helix dipole design 5
  LR model 11
helix length 5
helix propensities, amino acids *14*
helix templates 7
helix–coil kinetics
  energy landscape theory 64–6
  experimental study techniques 31–4
  theoretical studies 34–5
  theory 7–11, 29–31
  T-jump studies 35–7
  transitions 37–9
  see also α-helix
hen egg white lysozyme (HEWL), pulsed-amide H/D exchange 79–80
heteropolymers 4, 5
  random 50–6
homopolymer(s) 4, 5
  collapse models 126–7
  statistical weight 8
host–guest studies 4
human prion protein dimer 260–2, *261*
hydrogen exchange (HX) experiments 70–83

amide hydrogen exchange 72–5
native-state hydrogen exchange method 80–3, *82*
pulsed-amide H/D exchange method 75–80
hydrogen exchange processes, three-state system *81*
hydrophobicity in PrP 254

infrared (IR) spectroscopy, helix–coil kinetics 31
inverse folding problem *see* protein design, hierarchical
ionic strength, α-helix stability 20–1
islet amyloid polypeptide (IAPP) *215, 216*
isolated-pair hypothesis 10
isotropic probability density, FRET 152

kinetics
  computer simulations and investigation tools 179–81
  ensembles *vs.* single molecules 141–3
  helix–coil 31–9, 64–6
  protein kinetic stability 100–3
  rate constants and probabilities 143–6
  single molecule spectroscopy 141–6

λ-repressor 125, 129
Landau free energy functionals *95, 96–9, 100*
Lennard–Jones energy (E_{LJ}) 165
Levinthal, Cyrus 106
Lifson–Roig (LR) model 8–11
  AGADIR 12, 39
Linderstrøm–Lang model, amide hydrogen exchange in folded proteins 72–3
Lomize–Mosberg Model 13
long evolved proteins, energy landscape 55–6
Lumry–Biltonen–Brandts analysis, two state model 94–6, *95*

MABA (4-methylaminobenzoic acid), T-jump studies 35

Markovian State Models (MSMs) 171–2, 176–7
MD simulations 41–3, *168,* 169
membrane proteins, design studies 201–2
metal binding 3
microscopy
  atomic force microscopy 140, 227
  electron microscopy 227
  scanning tunnelling microscopy 140
  total internal reflection fluorescence (TIRF) microscopy 141
microsecond protein folding 115–27
  case studies 121–7
  history 115–17
  spectroscopic signatures 119–21
  sub-millisecond instrumentation and techniques 117–19
  techniques vs, timescale *117*
minimal frustration principle 55, *59,* 60, 61–4
Monte Carlo sampling methods 197–8, 199
MTT (3-[4,5,dimethylthiazol-2-yl]-2,5-diphenyltetrazolium bromide) 228–9, *229*
Muñoz scenario, downhill folding 127–8, *128*
mutations promoting amyloid formation 219
myoglobin *see* apomyoglobin (AMb)

N1, N2, and N3 preferences, LR model 11
N* state 243–4, *244,* 262–3
NAMD computer simulations 257–60
native-state hydrogen exchange method 80–3, *82*
NMR relaxation experiments, helix–coil kinetics 32
non-biological components, design studies 202
N-terminus 2
  amino acid propensities *16–17*
  LR model 10
phosphorylation 15–18
α-SH3 variants, amyloid formation 232–4, *234*
stability effects 14–15
nucleation-elongation model, free-energy landscape 39–41

Onsager's fluctuation–dissipation theorem 117–18

Parkinson's disease (PD) 215–16
PC12 pheochromocytoma cells 228
Aβ peptide *215,* 216
  in amyloidosis 217, 242, 245
  assembly 245–7
  dimerization 247–51, *248, 249*
peptide helices, design of 4–7, 191
  see also α-helix; helix–coil kinetics
peptide model systems, amyloid formation 231–2
phosphorylation, α-helix stability 15–18
phosphoserine 15
π-helix 3–4
Poisson–Boltzmann equation 166–7
poly[N^5-(4-hydroxybutyl)-L-glutamine] (PHBG) 4, 13
poly[N^5-(3-hydroxypropyl)-L-glutamine] (PHPG) 4, 13
polypeptide fast dynamics 110–15
  loop formation 111–14
  protein collapse 114–15
  secondary structure formation 115
  timescales 115
polypeptides
  acid denatured state characterization 73–4
  HX rates for unfolded 71–2, *72*
polyproline II helix 10
polyproline peptides, FRET 151–2
potential of mean force (PMF) 181
  Aβ peptide dimers *249*
prion protein dimer, human 260–2, *261*
prions 242–4, 251–62
protein
  biological functions 188–9
  industrial applications 189–90

Subject Index

protein aggregation scenarios 241–63
  peptide association 243–5
protein design 71, 188–205
  combinatorial methods 191–2
  computational approaches 193–8, 203–4
  directed evolution 192
  foldability criteria 196–7
  future 204
  hierarchical 190–1
  intrinsic limitations 192–3
  negative design 196–7
  recent successes 198–204
  sequence ensembles 197–8
  solvation 195–6
  structure and sequence 199–201
  thermodynamic hypothesis 189
protein folding
  Anfinsen's rule 70
  computer simulations 161–84
  energy landscape 49–66
  fast 106–31
  folding–unfolding ensembles 85–93
  free energy profile *62*
  implications of α-helix formation 45–6
  intermediates 70–83, 108
  'speed limit' 106, 108–9
  thermodynamic hypothesis 189
protein kinetic stability 100–3
protein macrostates as microstate ensembles *95*
protein misfolding, amyloidosis
  amyloid formation, promotion *218*, 218–20
  mechanisms *220*, 220–1
  in vitro studies 217
  in vivo 217–18, *218*
proteoglycans, amyloid pathophysiology 219
PrP^C conformational transition 251–62, *259*
  bioinformatic analysis 254–62
  and Dpl 255–7, *256*
  experimental observations 251
  structure *252*

pulsed-amide H/D exchange method 75–80
  early-folding intermediate detection *76, 77*

random energy model 50–6
random sequences 50
reaction coordinates, α-helix formation 43–4
relaxation time 32–3, *33*
Replica Exchange Molecular Dynamics (REMD) *168,* 169, 182
retroviruses 192
RNA polymerase 192
RNase H
  amide hydrogen exchange 75
  FRET analysis 154
  native-state hydrogen exchange 83
  pulsed-amide H/D exchange 79
ROP (repressor of primer) dimer 65
rotamer libraries 194

SCADS (statistical computationally assisted design strategy) 199, 201–2
scanning tunnelling microscopy 140
Schellman motif 2
SCHEMA computer simulations 203–4
scrapie 242, 251
senile systemic amyloidosis 219
side-chain interactions
  α-helix stability 18–20
  cooperativity 18
  energies *19*
  LR model 11
  Aβ peptide assembly 246–7
single molecule fast-folding studies 110, 117–23, 131
single molecule spectroscopy 139–56
  confocal technique, freely diffusing molecules *142*
  correlation analysis 146–8
  ensembles *vs.* single molecules 139–46
  FRET 149–54
  future developments 154–6
  history 140–1
  kinetics 141–6

single sequence approximation, LR model 10
SIRCH computer simulation 204
slow nucleation phase, amyloid fibril formation *220*, 220–1
solubility 6
spectroscopy
  fluorescence 120, 139
  fluorescence correlation spectroscopy (FCS) 140, 143, 146–8
  infrared (IR) spectroscopy, helix–coil kinetics 31
  signal decay, helix–coil kinetics 32–3, *33*
  single molecule spectroscopy 139–56
  spectroscopic ruler 141
  techniques, microsecond protein folding 119–21
  ultraviolet resonance Raman spectroscopy (UVRS) 32
'spherical cow model', free energy profile *62*, 62–3
stability, α-helix 1–2, 13–21, 38
  capping motifs 20
  caps 14–15
  covalent side-chain interactions *19*, 20
  helix interior 13–14
  ionic strength 20–1
  non-covalent side-chain interactions 18, *19*
  phosphorylation 15–18
  temperature 21
steady-state phase, amyloid fibril formation *220*, 220–1
stratified energy landscape 55–6, *57*
symmetric structures, design 202–3
synucleins *215*, 216

temperature
  α-helix stability 21, 35
  folding time vs. temperature *58*, 58–61
  heat capacity vs. temperature *86*, *100*
  high-temperature unfolding 170
  'U' well dynamics vs. temperature *112*
terminal motifs 2–3
thermodynamic hypothesis, protein folding 189
Tightly Coupled Molecular Dynamics (TCMD) 169
timescales
  folding *163*
  sampling methods 169–72
  viscosity dependence *173*
T-jump studies, helix–coil kinetics 35–7
  simulations *43*
total internal reflection fluorescence (TIRF) microscopy 141
transfer efficiency, FRET 149, 151–2
transmissible spongiform encephalopathies (TSEs) 251
transthyretin (TTR) 219, 241
Trp-cage protein 148, *173*, 199
two-state equilibrium model 7, 93–6
  estimating rates 173–6
  free-energy surfaces *168*
  Lumry–Biltonen–Brandts analysis 94–6, *95*
two-state irreversible model 102

'U' well dynamics vs. temperature *112*
ultrafast folding, tailored mutations for 198–9
ultraviolet resonance Raman spectroscopy (UVRS), helix–coil kinetics 32
unfolded proteins, LR model 10

van der Waals interaction
  loss of 4
  modelling 165
VHP-36 mini-protein 116–17, 125–6

Zimm chain 111–12
Zimm–Bragg (ZB) model 8, 9